T0320748

STRUCTURAL CERAMICS: FUNDAMENTALS AND CASE STUDIES

This book provides an introduction to the structural ceramics, their processing and properties. Five important groups of materials – porcelain, alumina, silicon carbide, silicon nitride and zirconia – are presented as case studies. Historical developments, the properties of constituent components, and relationships between production methods, resulting microstructures, and materials properties, are explained.

The structural ceramics have many commercial applications, ranging from high voltage insulation and fuel cells, to metal machining tools and surgical implants. These applications depend on combinations of chemical, physical and mechanical properties, which include structural stability over wide temperature ranges, strength, hardness, and resistance to wear.

Over 200 diagrams and photographs provide visual aids to learning, and end-of-chapter summaries pull together key points. With numerous review questions to test understanding of the topics covered, and extensive referencing, this book is ideal for those studying materials science and engineering, or starting research in the structural ceramics area.

FRANK RILEY was Professor of Ceramic Processing (and is now Emeritus Professor) at the University of Leeds where he researched and taught for over 30 years. He is a Fellow of the Institute of Materials, Minerals and Mining. He has directed two NATO Advanced Study Institutes, is editor of several conference proceedings, and has authored many research publications and review articles.

STRUCTURAL CERAMICS

Fundamentals and Case Studies

F. L. RILEY
University of Leeds, UK

CAMBRIDGE
UNIVERSITY PRESS

CAMBRIDGE
UNIVERSITY PRESS

University Printing House, Cambridge CB2 8BS, United Kingdom

One Liberty Plaza, 20th Floor, New York, NY 10006, USA

477 Williamstown Road, Port Melbourne, VIC 3207, Australia

314-321, 3rd Floor, Plot 3, Splendor Forum, Jasola District Centre, New Delhi - 110025, India

79 Anson Road, #06-04/06, Singapore 079906

Cambridge University Press is part of the University of Cambridge.

It furthers the University's mission by disseminating knowledge in the pursuit of education, learning and research at the highest international levels of excellence.

www.cambridge.org
Information on this title: www.cambridge.org/9780521845861

First published 2009

A catalogue record for this publication is available from the British Library

Library of Congress Cataloging in Publication data
Riley, F. L.
Structural ceramics : fundamentals and case studies / F.L. Riley.
p.   cm.
Includes bibliographical references and index.
ISBN 978-0-521-84586-1 (hardback)
1. Ceramic materials.   2. Ceramic-matrix composites.
I. Title.
TA455.C43R55 2009
620.1´4–dc22   2008055960

ISBN 978-0-521-84586-1 Hardback

For Mary

Contents

Preface

The technical ceramics can be divided into *electroceramics*, which, by and large, make use of the materials' electrical or magnetic properties, and the *structural* ceramics, with applications mainly (though not entirely) dependent on mechanical properties. The structural ceramics providing the case studies for this book have been chosen because they illustrate well the characteristic features of the class of structural ceramics as a whole. They have a wide range of properties, and they are of considerable technical importance. The five studies are intended to introduce the reader to this large class of materials, and the rôle they play in today's world. Each of the materials (more precisely, groups of materials) is examined systematically to provide an outline of its history and a simple picture of its development, how it can be fabricated, details of key physical and mechanical properties, and a summary of the principal applications based on these properties.

Because all the ceramics reviewed here have very high melting points, components are normally made by processing powders. Some appreciation of this aspect of the subject will be helpful before any examination of individual materials takes place. Chapter 1 therefore introduces the fundamental features of the powder sintering route to a ceramic, and the development of microstructure. Ceramics have a reputation for brittleness and a rather marked tendency to break if dropped, though in fact the best of the structural ceramics can have strengths comparable with those of the high tensile steels. Aspects of strength, fracture toughness and the general properties of ceramic materials important for the engineer and designer are also introduced here. The following chapters then examine each of the five materials in turn to identify distinguishing features, and those properties which are common to the structural ceramics as a class of material.

The oldest of the structural ceramics, by several thousand years, are the various types of what is usually termed "pottery", originally used for storage of grain, oil, and wine. Development of rudimentary production processes gradually led to refinements in quality, particularly aesthetic appeal, and strength, resulting in the development of the translucent, but strong, materials generally called "porcelain", or "china". Industrial porcelain, a "traditional" ceramic, is reviewed in Chapter 2, and although it is not particularly strong it provides a very useful introduction to some of the important features of the structural ceramics, and a standard by which the property values of the others can be judged. The more modern, or "technical", high-strength ceramics are then examined in the following four chapters. These studies show how limits on a material's properties can be determined by the fundamental nature of the components of the material itself, and assess the extent to which it might be possible to vary the properties or obtain improvements. The alumina, silicon carbide, silicon nitride and zirconia groups of materials have been developed as high-grade structural ceramics only during the last 40 years or so (though their history is actually very much longer). Alumina, discussed in Chapter 3, is by far the most widely used, but silicon carbide discussed in Chapter 4, and silicon nitride in Chapter 5, also have very important and expanding application areas. Zirconia in Chapter 6 is in one way the odd one out, because its markets at present are very much smaller than those of the others. It is included because the zirconia group of materials provides some of the highest strength and toughness ceramics yet seen outside of the ceramic composites area. In this respect therefore zirconia materials might be considered to be the best of the structural ceramics, though, as will be seen, they are not completely perfect. Chapter 7 summarises these case studies, and provides an overview of the development of the five groups of material, and their present areas of application.

This book could not have been written without the stimulus and willing cooperation of a very large number of people, who freely made available illustrations, photographs and technical information. I am particularly indebted to John Bailey, Jake Beatson, John Briggs, Rik Brydson, Francis Cambier, Dusan Galušek, Gren Goldstraw, Christine Hahn, Harry Hodgson, Peter Johnson, Heiner Knoch, Brian Lines, Roger Morrell, Trevor Page, Susan Payne, Günter Petzow, Vladimir Sida, Lance Snead, and Chongmin Wang, for help with background information, photographic illustrations and original diagrams.

January 2009 *Frank Riley*

1

Fundamentals

1.1 What are structural ceramics?

The word "ceramic" is usually associated with images of plates, mugs and cups and saucers, and concepts of brittleness and hardness. While ceramic materials are indeed often very hard, and certainly brittle (and can also be very fine works of art), this is a very narrow picture. Many ceramics have extremely important structural applications that depend on mechanical or thermal stability under a wide range of very demanding conditions. The aim of this book is to present a bigger and more balanced picture of these materials. This is done by taking five materials in the structural ceramics class – the case studies – and subjecting them to systematic and detailed examinations. The materials chosen are either the most widely used of their type, or show in some respect exceptional properties: they can therefore be considered to be the most important of the class. However, all the structural ceramics share their pattern of microstructures and properties to a greater or lesser extent with these five, which means that they are good representatives of the whole class. The small picture developed by these case studies should therefore be an accurate guide to the much larger, and also give the reader a full appreciation of the uses to which the structural ceramics are put.

Ceramic materials (which of course include the traditional whitewares) can be defined in very general terms as "high melting-point, inorganic, non-metallic materials" (Kingery, 1976). The word is usually assumed to be derived from the Greek *Keramos*, meaning clay, or ware (pottery) made from clay by heat treatment (Dodd and Murfin, 2006). By extension of meaning the term now includes the products of the silicate industries, thereby bringing in glass and cement. It has been widened further to include all inorganic materials made by the *powder sintering* route. Those materials generally called *structural ceramics* are a large group of ceramic materials with particularly marked properties of high strength, hardness, and resistance to wear. These properties may be retained from room up to high temperature ("white hot", ~1000 °C or more), over long periods of time, though in

1

Figure 1.1 A small selection of structural ceramic components, in various types of material. (Reprinted by kind permission of Kyocera Corporation.)

fact most of the materials reviewed here are generally used at much lower temperatures. While some find highly specialised, and restricted, applications, many of these materials are commercially produced on a very large scale. One simple common example of a structural ceramic is the shiny white insulating body of the spark plug used in all petrol engines, which is alumina, and of which millions are produced every week. Another, less obvious, example is silicon carbide (perhaps better known as carborundum), used in the increasingly important filters taking smoke particles out of the exhaust gases of diesel engines. Figure 1.1 is a small selection of the very large number of types of structural ceramic components now produced, illustrating their range of sizes and shapes. At this stage the materials shown in the photograph are not identified, nor are the applications for the components. Some of their applications would in any case be difficult to guess, because the small ceramic component is hidden within a much larger unit.

1.2 Compositions

Ceramic materials are based on compounds consisting of metal–non-metal combinations (oxides are common examples), and compounds of the semi-metallic elements (primarily boron and silicon). Simple two-element (*binary*) compounds form the basis, or major constituents, of four of the case studies (alumina, silicon carbide, silicon nitride, and zirconia). However, many of the chemical compounds (more usually referred to as *phases*) occurring in these materials are compounds of three or more elements (the aluminosilicates, for example, containing at least four elements, two metallic, one semi-metallic, and oxygen), and their crystal structures can be quite complex. Materials constructed entirely from single elements (for example silicon or carbon) are not normally regarded as ceramics, though in many respects they are barely distinguishable in terms of the pattern of their mechanical properties from materials conventionally thought of as ceramics. High-purity single-crystal silicon (in the form of "chips" – actually very thin slices) will be well known for its use in electronic devices and computer memories; carbon may be better known in one form as the transparent single-crystal diamond, and another as the black and much softer polycrystalline graphite (which incidentally provides an important illustration of the influence of the chemical bonding between simple carbon atoms on the physical and mechanical properties of the materials).

It is not possible to discuss any material without some reference to its chemical composition: atoms and ions are fundamental building units, from which all materials are constructed. A material's stiffness, its hardness, and thermal stability are determined by the strengths and arrangements of the bonds between its constituent atoms and ions. It is also useful to have some appreciation of the chemistry of the processes involving the raw materials from which the basic materials may be produced: this helps with understanding other important aspects of a material such as its purity, and the likely costs of the powders which are normally the starting points for the production of ceramics.

Nonetheless, ceramics certainly cannot be regarded simply as solid inorganic chemicals, although the chemical formula is commonly – and perhaps also misleadingly – used as a shorthand description for a material (for example, "Al_2O_3" for aluminium oxide). A simple statement of the types of constituent atom, and their proportions, in the compounds (phases) in a material is only the first step to providing a complete description of the material. The overall chemical composition, the phases and their possible relationships, and the microstructure of a material, are generally inseparable in the development of a material's properties, and the methods for controlling them.

Table 1.1 lists some of the more common basic chemical compounds making up the structural ceramics, and temperatures at which the pure materials form liquids (by melting or decomposition). While these numbers clearly give some indication of the potential of a material for high-temperature use, they form only a small part of the whole picture (and can by themselves even be misleading). Other factors will

Table 1.1 *Basic components of some common structural materials.*

Component	Chemical formula	Liquid formation temperature / °C
Aluminium oxide	Al_2O_3	2054
Calcium oxide	CaO	~2570
Magnesium oxide	MgO	~2800
Silicon dioxide	SiO_2	1726
Titanium dioxide	TiO_2	1850
Uranium dioxide	UO_2	~2880
Zirconium dioxide	ZrO_2	~2700
Boron carbide	B_4C	~2450
Silicon carbide	SiC	~2250
Aluminium nitride	AlN	2200
Silicon nitride	Si_3N_4	1900 (with decomposition)
Mullite	$3Al_2O_3.2SiO_2$	~1890 (with decomposition)
Iron	Fe	1527
Nickel	Ni	1452
Tungsten	W	~3360

be important. For example it must also (almost always) be possible to produce the material on a profitable commercial scale, in competition with established materials (often metals). Of the large range of inorganic non-metallic materials, very few meet all the requirements for an ideal material.

1.3 Microstructure

The internal structure of a material, the patterns provided by the microcrystalline grains and other phases present, their shapes, sizes, orientations, distributions, the types of boundary (or interface) between them is given the broad term *micro-structure*. This is the overall physical picture of the material, when examined on the micrometre, or (now more often) the nanometre scale. In a perfect *single crystal* the very regular pattern of the atoms forming the crystal lattice is uninterrupted throughout the piece (though real single crystals usually contain small-scale, local, disruptions to the pattern – the *lattice defects*). *Polycrystalline* materials are constructed from small crystals, microcrystallites – commonly referred to as *grains* – of μm dimension. The grains have been fused together during the production of the component, normally in the cases of the materials to be discussed here, by *sintering* (that is, heating to a high temperature) com-pacted fine powder, the particles of which are likely to be small single crystals.

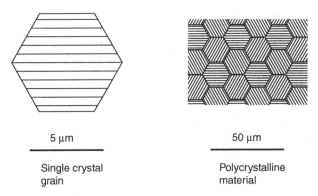

5 μm 50 μm

Single crystal   Polycrystalline
grain   material

Figure 1.2 Schematic presentations of a single crystal and a polycrystalline material.

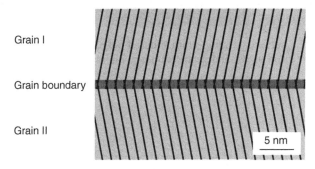

Grain I

Grain boundary

Grain II

5 nm

Figure 1.3 Schematic representation of the grain boundary region between two single-crystal grains.

The interface between two grains is the *grain boundary*, and its structure can be seen using high-resolution electron microscopy. The grains are bonded to each other either directly, or possibly through very thin (nm thickness) films which have a disordered or amorphous structure on the atomic scale. These features are illustrated schematically in Fig. 1.2 and Fig. 1.3. The sizes of the grains, and the grain boundaries, provide these materials with many of their characteristic properties, and further distinguish them from the single-crystal forms. A scanning electron micrograph of the surface of a relatively simple material, solid state sintered alumina, which has been given a heat treatment to show more clearly the boundaries between the grains, is shown in Fig. 1.4. Because the grain boundary is a region where the atoms are to some extent disordered, it also tends to act as sinks for impurity and additive atoms, so that the boundary often has its own distinct chemical compositions and physical properties. Figure 1.5 shows a transmission electron micrograph of a real grain boundary in sintered silicon

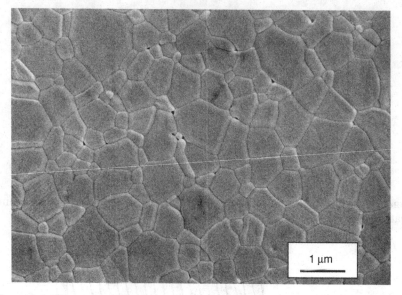

Figure 1.4 The surface of a fine-grain sintered polycrystalline alumina, showing the pattern of grain sizes, and grain boundary curvature.

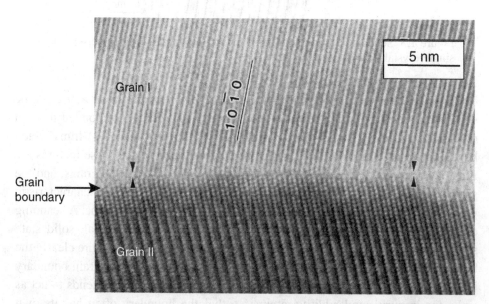

Figure 1.5 A transmission electron micrograph showing a real grain boundary in a silicon nitride ceramic. (Courtesy of Chongmin Wang.)

nitride, where the slightly disordered region, about 1 nm wide and containing in places a separate and amorphous phase, is clearly visible. Most of the materials coming under the heading "structural ceramics" have an internal structure viewed on the micrometre scale that is for the most part polycrystalline (Lee and Rainforth, 1994). The material may be built predominantly from one type of crystalline material (as in a high-purity alumina), or there may be a mixture of several phases (as in porcelain – though porcelain is not normally thought of as a polycrystalline material). Practically all ceramic materials contain more than one phase: most of the major phases will be crystalline, but varying amounts of amorphous (non-crystalline, or glass) material are almost always present. The amorphous phase is usually a silicate, or an aluminosilicate, of composition related to the main phase, or an additive used to accelerate sintering. Some of the material will be in the form of 1 to 2 nanometre thickness silicate films at grain boundaries, as in the case in the silicon nitride boundary shown above, and can be regarded as an intercrystal bonding phase. To put this dimension into perspective, the Si–O bond length is ~162 pm, so that a 1 nm thick silicate grain boundary film will be about six Si–O units across. When the amorphous material is present in large amounts it will exist as small isolated pockets, or as larger volumes of glass, dispersed between and bonded to the crystalline grains. In the porcelains it forms the major phase.

The glass, or amorphous, phases cannot be ignored: in many cases (and particularly at high temperatures) the properties of a multi-phase material may be determined more by the properties of the secondary phases and intergranular grain boundary materials, and particularly if they are amorphous, than they are by those of the major phase.

One more important microstructural feature that cannot be ignored is the *porosity* – the internal void space. This, as with the glass, has a big part in determining properties such as stiffness, strength, and thermal conductivity. The presence of porosity in sintered ceramics is almost inevitable because it is the residual void space in the compacted powder that was not quite completely removed during sintering. Most porosity is *closed*, that is inaccessible to the external environment, and is generally undesirable. Other pores are accessible (*open*), and can be essential for some applications of a material, as a gas or liquid filter, or catalyst support, for example.

To see the internal structures clearly, electron microscopy (scanning and transmission) is generally used. Optical (light) microscopy usually does not provide the necessary magnification. It is the microstructure, with all its finer details, which has a very strong influence on the mechanical and physical properties of the material. The other important influence on property and behaviour is the external environment – that is the temperature, atmosphere, loading conditions and time under load. These relationships are summarised in Fig. 1.6.

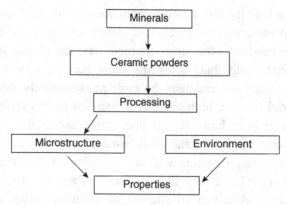

Figure 1.6 The main factors controlling the properties of a ceramic.

1.4 Powders

Most forms of the structural ceramics start their lives as fine powders, which are then compacted, shaped and sintered by one method or another. The way in which the microstructure of the final material is developed is strongly dependent on the powder from which it started. It can be said that the ceramic, in its microstructure, tends to retain a memory of its starting powder. For this reason the powder, its properties, and the processing into the ceramic, are crucial aspects of the development of a high-quality microstructure. In fact the quality of the powder processing operations can be as important for the properties, and satisfactory performance, of a ceramic component, as are the intrinsic properties of the material itself.

The main steps in the standard ceramic production process are summarised in Fig. 1.7. These are the production of a fine powder, the formation of the component shape by compaction of the powder, sintering, and any finishing operations required to refine the shape of the component or the quality of its surface. The compaction and final shaping steps can take place simultaneously (as in the *slip-casting* of a mug), or sequentially (as in the case of the spark plug insulator). Sintering is normally a separate production stage, as is indicated (Route I). However, it is also possible to combine the pressing of a powder with the sintering stage, in a process termed *hot pressing* (Route II). While this has the advantages of speed, and that a fine-grain microstructure can be obtained, it is also a more expensive process, and is therefore used only when absolutely necessary. After the sintering has been carried out, grinding and machining operations require the use of tools impregnated or coated with diamond or other ultra-hard powders. Many products, sliding parts for example, will need to be surface ground or polished as a matter of routine, because the as-sintered finish is

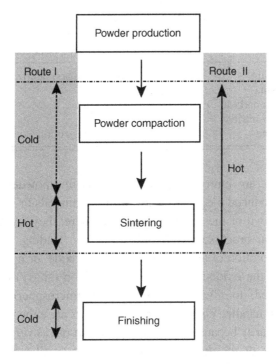

Figure 1.7 Standard ceramic processing routes: I normal pressureless sintering; II hot pressing.

usually not sufficiently smooth. In fact diamond machining of a sintered ceramic can become the most expensive stage of the whole process (although from the point of view of the manufacturer it also represents added value). It is obvious that there can be considerable advantages in trying to obtain the final required shape as closely as possible (sometimes called *near net-shaping*), while the ceramic component is still in the form of its starting powder: afterwards things become very much more difficult, time-consuming, and expensive.

1.4.1 Powder sources

The potter is fortunate in that the clay starting material, which is a major constituent of the blended powders used in the production of what is broadly termed pottery, occurs naturally as a very fine powder, the result of the weathering of igneous rocks over many millions of years. Clay particles usually have sizes of around 1 μm (though the question of size is not simple because, for reasons to do with the crystallography of the constituent minerals, the shapes of the individual clay particles tend to be thin plates). The other main ingredients of "pottery" (to which the porcelain treated in Chapter 2 comes closest), and those of the

Table 1.2 *Typical source minerals.*

Material	Typical source minerals
Porcelain	Clays, feldspars, quartz sand
Alumina	Bauxite
Silicon carbide	Quartz sand
Silicon nitride	Quartz sand
Zirconia	Baddeleyite, zircon sand

technical ceramics, are rarely found in nature as fine powders. These starting powders can be produced by grinding and milling minerals or sands. More often, as in the cases of two of the ceramics discussed here, they must be obtained by chemical extraction processes from bulk minerals, leading ultimately to pre-cipitation from a solution of an insoluble hydrated oxide, or hydroxide, followed by dehydration to the oxide. An aqueous suspension (slurry) of the oxide may then be spray-dried, to give soft, free-flowing microspherical agglomerates, which are easier to handle. Very high-purity powders are either difficult to obtain, or are expensive, partly because many materials start out as very impure minerals, and partly because of the difficulties of extracting small amounts of impurity compounds and elements from a chemically rather unreactive material. It is quite feasible to use standard aqueous phase chemistry to prepare many kinds of ultra-high purity (loosely, of chemical "analytical reagent" purity, or >99.9%) ceramic powder, but the cost of these very high-purity powders can be very high and the scale on which they are used is small.

Table 1.2 lists some of the more common minerals used as starting points for the materials treated in this book: the methods by which the prepared starting powders may be obtained from them will be outlined in each chapter. Some knowledge of the powder production process is useful in that it allows the user to be aware of the type, and amount, of impurity to be expected in a commercial powder, and hence the ceramic derived from it. One of the many difficulties faced by manufacturers is that the purity (and indeed other important characteristics) of the powder provided by the supplier tends to vary slightly from batch to batch. It should also be noted that impurities in the powder at the "parts per million" level can have significant effects on the surface chemistry of the particle and thus on the rheology of particle suspensions in water, and consequently on what can very broadly be termed the *processability* of the powder into a ceramic component. Only slightly larger quantities can have detectable and important influences on sintering characteristics, and the high-temperature strength of the finished prod-ucts, as was clearly shown during the development of sintered silicon nitride materials (Iskoe *et al.*, 1976).

1.4.2 Powder processing

Structural ceramic materials are normally (though not always) produced from compacted very fine powders, by a high-temperature heat treatment, traditionally called *firing* from the days when wood or coal fired kilns were used, and more generally and technically as *sintering* – kilns and furnaces are now gas, or electrically heated (Rahaman, 1995; Rahaman, 2008). The powder sintering process converts a powder directly into a dense polycrystalline solid: essentially the particles of the powder become the bonded crystalline grains of the microstructure, retaining to a large extent the pattern of particle size and arrangement that was present in the powder (Eisele, 1996). Although almost all ceramic materials can be melted (some decompose before they melt – see Table 1.1), the main reason that components are not usually made this way is that their melting-points are inconveniently high (inconveniently, in this context: metals are often cast in moulds made from materials which are essentially ceramic). It is certainly possible to melt and cast a material such as aluminium oxide (this is routinely carried out by some refractories industries, and for the production of large single crystals), but for the mass-production of small components there would be severe difficulties with finding suitable moulds, and with preventing the development of an extremely coarse-grained microstructure during cooling: a general requirement for high strength, and wear resistance, is a very fine grain size (on the μm scale or less).

1.4.3 Shape formation

The use of clay as the basis for pottery will be fairly well-known, though it may not be appreciated that the plastic and sticky material that clogs tyres and boots in wet weather, but becomes almost rock hard when dry, is simply a very fine (μm dimension) powder, the particles of which are bonded, or lubricated, by very thin surface films of water. For a damp clay powder to be plastic and mouldable it must contain about 30% by volume of water, though *clay* is in fact a family of materials, some members of which develop plasticity more readily than others (Worrall, 1986). The starting powders for the technical ceramics are not normally naturally plastic, or readily bonded, in this kind of damp state, but judicious use of small amounts of a range of organic polymers (a wax, or polyvinyl alcohol, for example), *binders*, often provides the necessary physical adhesion and formability. This is important because components, which are essentially just compacted powders at this stage, usually have to be handled or machined before sintering (see Fig. 3.7). Most smaller shapes (~10 cm maximum dimension) can be formed by *uniaxial pressing* in a die at pressures of up to ~200 MPa, free-flowing granulated, or spray-dried, powders, containing binders and plasticisers in amounts of ~1–3% (Read,

1995). These organic materials burn off in the early stages of the sintering process. A well-compacted, fine, powder, with the addition of a binder, can be made to have the strength and physical characteristics similar to those of a hard biscuit (with a bend strength of the order of 1–10 MPa). The process can readily be automated so that small components can be produced at the rate of several thousand per hour. Dimensional accuracy can be very good when required, though a common procedure is to produce a larger billet of compacted powder, which can then be machined to final shape and dimension in a standard lathe (with a dust-extraction facility, and the "dust" – the powder – can be recycled).

A major problem with uniaxial pressing is that the pressure gradients (caused by friction) within the compacting powder lead to density variations in the final powder component. This is particularly severe when the component aspect (axial) ratio is very high (>3:1). These density variations then result in shape distortion during sintering. One answer is to use *isostatic* or *hydrostatic* pressing, in which multi-axial pressure is applied through a fluid acting on a flexible plastic mould. The external profile of the compacted powder billet is then not so precise, and more extensive machining will usually be required. Objects with a very high aspect ratio, but of constant cross-section, such as tubes and rods, can be formed by extruding mixtures containing more plasticiser. A further extension of the principle, through the use of much larger volumes (typically 10–20%) of polymer, is to injection-mould the component, a technique common in the plastics industry (where cheap inert fillers are often incorporated into the polymer, to give colour or opacity). The polymer must now be burned out very slowly, to avoid delamination ("onion skinning"), or bloating of the powder component.

Another common starting point for the formation of complex shapes in a powder is *slip-casting*. The *slip* is a high solid concentration, but very fluid, slurry formed by deflocculating (that is, dispersing) fine powder particles in a liquid medium. The basis is the achievement of strong particle–particle repulsion, often by the development of surface charges. Under the right conditions, with good deflocculation, the solids content of a slip can approach 50% by volume. This is the standard method used for the production of teapots and figurines: it has also been used for the production of turbine blades. All of these shape production techniques, originally developed for the traditional types of clay-based ceramic, or borrowed from the plastics industries, are now used for the production of structural ceramics (Reed, 1995; Rahaman, 2007).

1.4.4 Particle size

Clay was referred to as a "very fine" powder. It is important to know what "fine" really means when we are processing ceramic powders. Granulated sugar might be considered to be a reasonably fine powder: it forms smooth-sided conical heaps and flows smoothly off the spoon into the coffee mug (though brown and Demerara sugar, which tend to contain other organic "impurities", or binders, and be moist and slightly sticky, do not). Flour is obviously finer still: the individual particles of flour are almost invisible to the unaided eye, and they tend to adhere to each other even when dry, with the result that flour does not flow quite as smoothly as white sugar. Talcum powder is a similarly fine powder, and it may be a surprise to learn that it is also an important ceramic raw material, and technically a magnesium hydroxysilicate. Bulk talc is a soft mineral known as *steatite*, or *soapstone*, and some of the earliest low-voltage electrical insulators were machined from it.

The *fineness* of a powder is a fairly arbitrary description, and in recent years standard ceramic starting powders such as those used for the materials described in this book have tended to become even "finer", with terms such as "ultra-fine", "submicron" and "nanoparticle" being applied. Generally "fine" would be taken to mean particle sizes mostly in the range 1–20 μm. A "coarse" powder would consist of particles of sizes up to several 100 μm. Above this point terms such as "grits" or "sands" tend to be used. "Ultra-fine", or "submicron", powders will have particle sizes in the approximate range of 0.5 μm (500 nm) to 1 μm, and "nanopowders" will have sizes less than about 100 nm. The common terms *colloid*, or *colloidal particle*, simply refer to particles which, when suspended in a fluid medium, are imperceptibly slow to settle under gravity: they therefore have particle sizes less than about 2 μm. These last two terms, which are commonly used in the chemical and biological areas, should be regarded as interchangeable with "ultra-fine". A clay (or other fine powder) slip can also be regarded as a colloidal suspension. An outline illustration of this range of particle sizes, spanning nine orders of magnitude, is shown in Fig. 1.8: it includes some common non-ceramic particles for comparison (the old unit still in use is the ångström (Å), 10^{-10} m, or 100 pm).

A ceramic powder hardly ever consists of particles all of one size (though they can certainly be specially made): there is a range of particle sizes. The spread of sizes within a fine powder is often very large, ranging from zero (in principle – the size of the crystal unit cell would be a more realistic theoretical limit), to several tens of μm. Particle sizes are readily measured with extremely dilute suspensions of the powder dispersed in a liquid medium (such as water), using automated methods based on the scattering of light by the particles (Stanley-Wood and Lines, 1992; Allen, 1997). Two common ways of depicting the distribution of particle sizes in a powder are shown in Fig. 1.9. A powder is often characterised in terms of its median particle size, or the d_{50} *value*, the size for which half the particles are finer than (or equal to), and half coarser. For a more

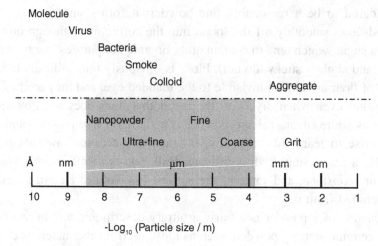

Figure 1.8 An illustration of the range of particle sizes, on a $\log_{10}$ scale, spanning nine orders of magnitude: from molecules to pebbles, with some other particles for comparison. The shaded area covers the normal range of powder sizes used for the production of the structural ceramics.

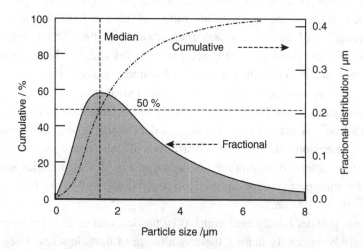

Figure 1.9 Two standard ways of showing the distribution of particle sizes in a powder.

complete picture, the particle size distribution is needed, expressed by the shape of the distribution curve and roughly represented by (for example) d_{10} and d_{90} values, the sizes corresponding to the 10% and 90% proportions "less than" (though 15.87% and 84.13% have greater statistical significance). Particle size distributions often appear to be approximated by the log-normal distribution function, and will plot to a straight line on logarithmic probability paper. This kind of treatment is, however, not usually applied to the fine powders used in the production of the technical ceramics.

The term "particle size" itself is generally used in a fairly loose way, and there are often difficulties with defining precisely (and measuring) particle "size". What is the "size" of a particle that is the shape of a jagged fragment of broken glass, a credit card, or a pencil? The terms *aspect ratio*, or *axial ratio*, can be useful in describing axially symmetric, plate-like or fibrous particles, but for others more subtle ways have to be used. This is one example of the kind of difficulties that have to be faced, and one reason why ceramic science cannot be such a precise discipline as is, for example, physics.

Another way of quantifying the size of a fine powder is by the use of its *specific surface area*. This is the summed area of the surfaces of the powder particles, expressed as area per unit mass (a_s, with units, usually, of $m^2 g^{-1}$). For a sphere of material of density ρ, and diameter d (or a cube of side d) a_s is given by the simple expression:

$$a_s = \frac{6}{\rho d}. \tag{1.1}$$

For real, irregularly shaped, particles the correlation between surface area and dimension cannot be expressed so precisely, but the specific surface area remains a useful semi-quantitative figure. The surface area of any powder can easily be measured using standard gas adsorption techniques, and because the specific surface area increases with decreasing particle size it is a particularly useful parameter for very fine powders, the particle sizes of which fall below the range of commercial instrumental methods (Stanley-Wood and Lines, 1992; Allen, 1997). For example, a powder of a material of density 3 Mg m^{-3}, with a median particle size of 100 nm, has a specific surface area of ~20 $m^2 g^{-1}$: powders of this order of dimension are widely used in the production of sintered silicon carbide and silicon nitride materials.

1.4.5 Particle packing

Any fine ceramic powder can be compacted and retains the shape when it is moistened by water or other wetting liquid: this is not a feature unique to clay particles. However, clay particles have the very useful property (and certainly from the point of view of the earliest, bronze age, potters) of behaving particularly well in this respect, partly because they can be very fine, and partly because of the morphology and surface chemistry of the individual particles. But while any one particle is certainly in contact with most, if not all, of its immediate neighbours, they are not able to fill completely the space around that particle.

This can be seen most easily with arrays of very much larger particles such as ball bearings, or tomatoes. The statement regarding space filling is, however, not quite accurate because a powder which consists entirely of special types of geometry, such as uniform sized cubes, regular hexagons, or regular octahedra with the corners removed termed tetrakaidecahedra, might in theory pack so as to fill space completely, but even then only if all the particles could be suitably arranged, and this would be pretty difficult to achieve. In theory, with the regular, ordered, *close-packing* of equal-sized spheres, the best that can be obtained is ~74.0% of space occupancy.

Statistically random packing of large mono-size spheres gives only about 60% occupancy (which is why a jar of pickled onions contains a lot of vinegar). The loose packing of typical, irregularly shaped, particles is such that the best efficiency of space filling that can usually be attained is around 50%. With very fine powders it can be much less than this, and the other 50% or more of the powder (overall or *bulk*, as opposed to *true solid*) volume must be lost during sintering. This volume of void space also corresponds to the fractional volume shrinkage ($\Delta V/V^{\circ}$) of the component after complete densification: the fractional linear shrinkage $\Delta L/L^{\circ}$ ($\Delta L/L^{\circ} \sim \Delta V / 3V^{\circ}$) would be ~15%. Components must therefore be produced over-size to allow for this shrinkage, and any irregularities in powder packing density will lead to corresponding shape distortions. (This is why it is difficult to press uniaxially components of high aspect ratio.) Because it may be difficult to achieve consistency of loss of void space and therefore shrinkage, diamond grinding and polishing of components needing to have precise ($<\pm10$ μm) dimensions will be needed.

It is often believed (erroneously) that fine particles should be able to fill space more efficiently (because, it is thought, they are smaller and therefore can somehow pack more closely together) but it can easily be shown that *theoretically*, space filling efficiency is independent of particle size. One way to appreciate this is to note that a coarse powder can be regarded as simply a fine powder viewed through a microscope, and the actual arrangement of the particles, and therefore their packing efficiency, does not change with microscope magnification. Figure 1.10 illustrates a two-dimensional view of packing (*close packing*) of circles (two-dimensional spheres) differing in size by a factor of 2. The fractional area of paper covered by the circles is ~70%, and is independent of circle radius, as can be seen by examination of the areas marked by the small triangles. Layers of spheres arranged like this and stacked exactly above each other (known as an *orthorhombic* structure) occupy space much less efficiently, and give ~60.5% filling. In practice the *finer* the powder particles the *less* efficient tends to be the particle packing – because particle adhesion caused, for example, by electrostatic attraction, hinders their redistribution into the optimum configuration.

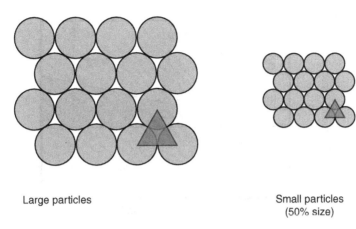

Large particles   Small particles
                                                   (50% size)

Figure 1.10 An array of close-packed discs of two sizes: the two triangles contain exactly the same proportions of disc, and void.

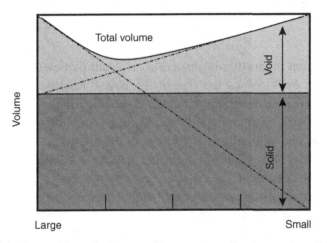

Figure 1.11 The packing of mixtures of large and small spheres: the solid volume is constant, but the total volume is determined by the changes in void volume.

The size distribution of the particles within a powder can also be important for packing efficiency. A powder consisting of mono-sized particles will in theory contain less solid per unit volume than one containing a broad distribution of sizes, and this is in practice also generally true. The reason for this is simply that finer particles are able to occupy the small spaces (the interstices) between contacting larger particles (and so on *ad infinitum*) except that again such a precise degree of particle arranging (or indeed, selection of particles of specified sizes) is not possible in practice. The improved packing as a result of blending two sets of mono-sized spheres, one large, one small, is illustrated in Fig. 1.11.

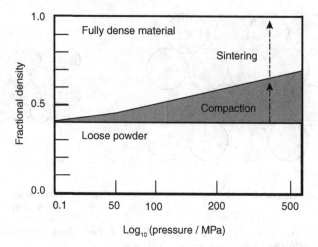

Figure 1.12 The influence of compaction pressure on powder density. The density has an approximately linear dependence on pressure on a logarithmic scale. The small change of slope is related to the crushing of soft agglomerates in the early stages of compaction.

Here the optimum proportion is about 30% of small particles (which should for optimum effect be ~15% of the size of the large). Real powders have a continuously varying size distribution, but the principle still applies: a powder with narrow particle size distribution will tend to pack (other things being equal) less efficiently than one with a broader particle size distribution.

The application of high pressures (of the order of a kilobar, or ~100 MPa, or more) to a powder might be expected to improve the efficiency of particle packing, and reduce the mean void size, but this does not happen to a significant, and really useful, extent. This is partly because ceramic particles are irregular in shape and do not easily slide over each other (though lubrication can help), and partly because they tend to be too strong and hard to be readily crushed or deformed. Increasing compaction pressure does lead to increased powder packing efficiency, but the empirical relationship found is that packing efficiency, or packing density, is approximately a function of the logarithm of the applied pressure. This means that there must be very large increases in pressure to obtain small increases in packing density, as illustrated schematically in Fig. 1.12. The slight bend in the curve at low pressures is intended; it corresponds to the crushing of small *agglomerates* of powder particles. Fine powders are very difficult to handle in bulk (pour into a die for example). For this reason they are usually spray-dried from aqueous suspensions, to form free-flowing microspheres (and they do flow like a liquid), of the order of ~100 μm in size. Each microsphere is a soft, relatively weak, agglomerate, designed to be crushed easily

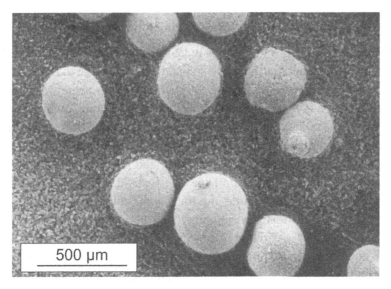

Figure 1.13 A scanning electron micrograph of spray-dried, granules: each granule is ~300 μm in diameter and consists of ~2 μm particles.

as soon as pressure is applied in a die. Figure 1.13 shows typical spray-dried microspheres, ~300 μm in diameter, and composed of ~2 μm particles (though these cannot easily be seen at this magnification).

While presses can be produced which allow the application of pressures of the order of 100 MPa for small shapes, it becomes expensive to apply such pressures to the dies needed for very large components (remember that pressure is force/area: larger areas require therefore a proportionately larger force to achieve the same pressure). Most commercial presses operate at <100 MPa. In practice, the highest particle packing densities can often be obtained, perhaps unexpectedly, by the use of a low-density powder slurry, which is cast in a porous mould – the slip-casting referred to earlier. As the liquid is absorbed by the mould, a layer of dense, compacted, powder builds up on the mould wall. This happens because the liquid-lubricated particles, pulled by surface tension and other forces, are able to move and arrange themselves into the optimum locations for packing efficiency.

The important point to note (and shown in Fig. 1.12) is that what void space cannot be removed from the powder at this stage has to be removed during the sintering process. For this reason it is better to obtain the highest powder packing density possible when a high sintered density is required. Obviously, when the sintered component is to be used as a filter or catalyst support, the controlled retention of porosity is required, and (combustible) organic particles may be introduced into the ceramic powder to help to achieve this.

1.5 Sintering

When the required shape from the consolidated powder has been formed, the next stage is to strengthen and fix the shape so that it cannot return to the powder from which it was derived (without milling, of course). This is achieved by sintering. Sintering (the word is related to *cinder*) means "heat treating", which is not very informative. In the context of the processing of ceramic powders the word defines a high-temperature heat treatment bringing about the bonding of powder particles and the development of the grain boundaries, and (usually but not always) the simultaneous elimination of much of the interparticle void space (Eisele, 1996; Rahaman, 2008). The temperatures used are high, but not high enough to melt the major phase of the material, although some liquid may be produced from impurities, or specially added ingredients. Normal clay-based pottery is produced (fired) in kilns that are at temperatures not very much higher than 1100 °C; alumina spark plug bodies are fired at around 1300 °C; silicon carbide may require 2000 °C. At these temperatures, sufficient mobility of the atoms in the microcrystals is developed, so that the (*solid*) particles of the powder are able to reshape themselves to bring about interparticle bonding with formation of grain boundaries. This reshaping process is the result of the system attempting to minimise its energy, specifically its *interfacial*, or *surface, energy*, in exactly the same way that small liquid droplets fuse to form larger drops, with spherical, minimum energy, surfaces (except that since atom mobility in the liquid is higher, rearrangement of material is much faster).

During the high-temperature sintering stage, in addition to the bonding of the powder particles, chemical reactions can occur, with the result that the sintered product may contain crystalline phases, or glasses, which were not present in the original starting powder mixture. Heating a clay to a high temperature produces a completely new phase, an aluminium silicate called *mullite* (a mineral identified in the Scottish island of Mull). The sintering operation is often therefore a very complex series of processes and steps, some physical, some chemical. These processes represent the approach of the whole system to a state of chemical and physical equilibrium, though this may not actually be achieved, because time is not allowed (and in many cases attainment of equilibrium is not actually really essential).

Discussion of sintering will start by assuming that the material is chemically pure, and that the temperature is well below its melting point. And because the material is pure there will be no secondary low-melting point phases present. This (usually idealised) type of process is generally referred to as *solid state sintering*. If all the voids in the compacted powder are lost, the material density will approach that of the pure solid material – usually referred to as the *true solid density* or *theoretical density*. From the point of view of obtaining maximum strength and stiffness in the material, this is the target. Sintering involves the physical movement or rearrangement of material, which in practical terms means the migration, by diffusion, of atoms (or ions). In a solid within which the interatomic bonding is very strong, this is not easy, and for atoms to move at a reasonable rate requires high temperatures.

In order to convert the compacted powder into a fully dense solid, the void space (all 40–50 volume% of it) remaining between the particles must be filled, and the particles bonded together; these are the essential functions of the sintering process. This is equivalent to a statement that (somehow or other) a corresponding proportion of the solid material of each particle must be moved into its adjacent void spaces. It is obvious that to fill a small individual volume of space (a small void) requires the movement of less material *per void* than filling a larger one does (see Fig. 1.10); intuitively then it would seem that it should be easier, or quicker, to fill a collection of small voids (provided they are all filled simultaneously), than a large void of the same total volume. One important factor determining void size is the size of the particles making up the powder (another factor, as seen above, is the compaction pressure used). Other things being equal, larger particles will tend to have larger gaps between them: the voids between the grains of sugar on a teaspoon are much smaller in size than those between an array of oranges on a supermarket shelf. A powder compacted to a higher *green density* (why *green*? – this term may derive from the fact that damp impure clay is often a grey/green colour) will have a smaller total pore volume, and it will also have a smaller mean pore dimension. Well-compacted particles will be in contact with a greater number of neighbours, and the opportunities for material transfer are increased. Sintering therefore tends to take place more quickly, the finer the powder, and the higher the compacted density.

In the earlier stages of the process (when the geometries of the packed particles are simpler) the time taken (t^*) to achieve some specified degree of sintering, as measured for example by the extent of particle bonding, or strength development, can be related (other things, like particle packing, being equal) to the mean particle dimension L using simple scaling laws (Herring, 1950) by expressions of the form

$$t^* = kL^{-n} \qquad (1.2)$$

where L is the dimension of the particle, and the value of n depends on the dominant material transport mechanism, as shown in Table 1.3 (Rahaman, 1995; Rahaman, 2008). This expression is very much simplified, but it draws attention to the importance of particle size for the sintering rate of a powder.

Another feature of the powder sintering process is that the quality of the sintered component, expressed in terms of its residual void (now usually

Table 1.3 *Values of the exponents, controlling the sensitivity of the initial rate of sintering to particle dimension (Rahaman, 1995).*

Material transport mechanism	n
Vapour transport, from the particle surface	2
Surface diffusion, along the particle surface	4
Grain boundary diffusion, down the boundary	4
Lattice diffusion to the neck surface	3

referred to as *pore*) content and void size, is to some extent dependent on the sizes of the voids in the starting powder. This is because in the earlier stages of sintering only a small proportion of each particle is involved in the material redistribution process: the rest remains more-or-less unchanged. Later, during the grain growth process, small particles will be subsumed into the larger particles, with reduction of overall surface energy. This is why the sintered component is able to "remember" its earlier life as a compacted powder (unlike the case of a metal which has been completely melted and solidified). For this reason the ways in which a ceramic powder (or its free-flowing microspheres) is treated and compacted during the shaping of the component are crucial for the subsequent quality of the sintered component, expressed in terms of its microstructure, and ultimately its mechanical and physical properties. The final high-temperature heat treatment, allowing void elimination and densification to occur, can also be regarded as a potential homogenisation process. There are, however, limits in practice to what is possible during sintering, and for that reason it is better not to demand too much of it. To attempt more complete elimination of voids just by sintering for longer (or at a higher temperature) usually results in excessive grain growth, and poor mechanical properties: this is not the best way to go about it.

1.5.1 Solid state sintering

The temperatures required for sintering pure compounds are far higher than those needed for the simple drying of a powder by the loss of physically adsorbed water (normally 100–200 °C), or even the loss of chemically bonded water of crystallisation from hydrated salts (probably 200 or 300 °C), or hydrated minerals such as kaolinite (about 600 °C). Aluminium oxide particles will sinter at room temperature but a great deal of patience would be required to observe the process because the times required to achieve significant interparticle bonding will be of the order of millions of years (sintering does occur, but on the geological time-scale). However, heating to 1000 °C or so causes atom mobility to become very much higher, and 1 μm particles will bond perceptibly in a few hours. At higher temperatures still (~1500 °C) sintering of a fine aluminium oxide powder becomes very fast and under the right conditions theoretical densities can be obtained in less than an hour. Atoms of other materials have different degrees of mobility, and effective sintering may then be possible at lower temperatures, or may require, like silicon carbide, even higher temperatures.

But *why* does sintering of a powder occur? And there is the immediate supplementary question – can *all* ceramic powders be sintered under reasonable conditions? By reasonable we mean temperatures that are easily obtainable in the laboratory or factory, and on a realistic time-scale (an 8 hour factory shift for example). The answer to the first question is quite easy to find. The answer to the second question is, unfortunately, no.

A powder is always thermodynamically unstable with respect to the corresponding bulk, fully dense, material. This is an immediate consequence of the excess energy present in the form of the surface, or interfacial, energy of the particles. This energy (the solid–vapour interfacial, or surface, energy γ^{sv}) corresponds to the energy (or part of it – some is lost as heat) originally required to break down the piece of solid into the very small fragments of the powder. For a solid ceramic at room temperature γ^{sv} commonly has values in the range ~0.5 to ~1 J m^{-2}. The powder will lose much of its excess surface energy when it sinters to a dense, polycrystalline solid, as shown in Fig. 1.14.

However, the polycrystalline solid still contains grain boundaries, of surface energy, γ^{ss}, with a value often assumed to be of the order of 30% of that of γ^{sv}. The system will also try to lose this, which it does by the process of *grain growth*. Theoretically the material should ultimately become one large grain, or single crystal. However, grain growth is in practice usually undesirable, because an inverse relationship is often found between grain size (G) and strength (σ), which approximates to the form:

$$\sigma = KG^{-\frac{1}{2}}. \tag{1.3}$$

Sintering is therefore normally controlled to minimise grain growth (though there are exceptions, as will be seen in Chapters 5 and 6). Because loss of γ^{sv} and γ^{ss} generally occurs simultaneously (and not sequentially as implied by Fig. 1.14), it may be necessary to compromise between attainment of full density, and minimisation of grain growth.

While Fig. 1.14 illustrates the thermodynamic driving force for sintering, the picture on the atomic scale also needs to be examined: how do individual atoms *know* what they are supposed to do in order to lower the overall energy of the system? It can easily be shown that the effect of the interfacial (surface)

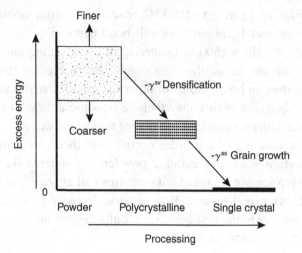

Figure 1.14 A schematic illustration of the energy changes during the sintering of a powder. The steps are usually not sequential; grain growth and particle sintering can take place simultaneously.

energy is to place the interior of a particle into a state of compressive stress (p^{i}). The effect of compressing the solid (and reducing interatomic distances) is to raise the energy level of the atoms (that is, increasing their *chemical potential*, or *molar Gibbs function*), with respect to a reference level, which is that of the atoms in a large flat slab of unstressed material (formally of infinite size). And because the atoms have a higher energy than normal they will attempt to reduce this energy, for example by moving to locations of lower mechanical stress (lower chemical potential – as will be shown). Put more precisely, the atoms will move away more quickly than they can move back again, so that there is an overall drift of material from regions of high chemical potential, to regions of low chemical potential.

The calculation of the excess energy of a powder particle, or an atom within the particle, is most easily carried out by assuming an idealised perfectly spherical particle of radius r, for which the pressure difference ($p^{i}-p^{o}$) across the interface between phases a and β can easily be shown to be:

$$(p^{i} - p^{o}) = \Delta p = \frac{2\gamma^{a\beta}}{r}. \tag{1.4}$$

a will be the solid phase of the particle, and β will normally be the vapour phase, air, or a liquid. This is illustrated by Fig. 1.15. Usually p^{o} will be atmospheric pressure, though it could also be a mechanically applied pressure. Equation (1.4) is a special case of the Young–Laplace equation for a surface with two principal

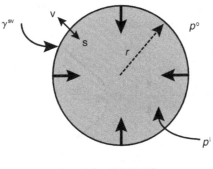

$$\Delta p = (p^i - p^o) = 2\gamma^{sv}/r$$

Figure 1.15 An illustration of the internal pressure in a solid or liquid sphere. For a soap bubble the pressure is twice this, because the bubble wall has two surfaces.

radii of curvature r_1 and r_2, for which it can be shown (not quite so easily; Adamson, 1982) that

$$\Delta p = \gamma^{\alpha\beta} \left(\frac{1}{r_1} + \frac{1}{r_2} \right). \tag{1.5}$$

Equation (1.4) is obtained from (1.5) by putting $r_1 = r_2$. Where the surface is flat ($r = \infty$), $\Delta p = 0$. For fine powders Δp becomes very large; for example with a 1 μm dimension particle of aluminium oxide, for which γ^{sv} is of the order of 1 J m^{-2}, the internal pressure will be ~4 MPa, or ~40 atm (note the sign convention being used, that a compressive stress is *positive*, tensile negative). This pressure, corresponding to a radial compressive stress acting towards the centre of the particle, is large enough to move the atoms slightly closer together, and in the case of crystalline material this corresponds to a small reduction in the size of the unit cell. X-ray diffraction measurements on very small crystals are able to detect the slight reduction in interatomic spacing between the atoms (Nicholson, 1956).

The associated change in energy of the material (that is, its atoms) in a particle can now be estimated. The incremental change in energy, the *Gibbs function* (dG), with a change in pressure (dp) or temperature (dT) in a closed system, is given by (Atkins and dePaula, 2006)

$$dG = V.dp - S.dT \tag{1.6}$$

and at constant temperature

$$dG = V.dp. \tag{1.7a}$$

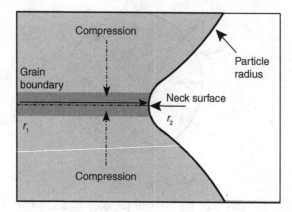

Figure 1.16 Grain boundary formation between two sintering spheres, showing the development of the neck zone by the diffusion of atoms from the spheres. The neck has two principal radii of curvature: the large r_1, which is convex, and the small r_2, which is concave.

For 1 mole of substance G can be replaced by the molar Gibbs function (symbol μ, normally with units of J mole^{-1}) and V becomes V_m, the molar volume (which for a solid is regarded as constant), and

$$\mathrm{d}\mu = V_m.\mathrm{d}p. \tag{1.7b}$$

The term μ is the *chemical potential* just introduced above. Integrating from p^o to p^i gives

$$(\mu - \mu^o) = V_m.(p^i - p^o) = 2V_m \frac{\gamma^{\alpha\beta}}{r}. \tag{1.8}$$

The extra internal energy for this 1 μm aluminium oxide powder (V_m ~25.6 cm^3 mol^{-1}) is of the order of 100 J mol^{-1}.

Where there is a solid *concave* surface, for which r has negative values, the internal stress developed is tensile. Such a *concave* surface must be developed in the zone between two fusing particles, where the particle radii of curvature change form (Fig. 1.16), and it is to this surface that material diffuses, either through the crystal lattice, or down the grain boundary (which is usually the quicker route). The neck zone can then be considered to be a thin disc, of radius r_1, with a curved re-entrant rim of radius r_2. In this case the full Equation (1.5) applies, but since r_2 is $\ll r_1$

$$\Delta p \sim -\frac{\gamma^{\alpha\beta}}{r_2} \tag{1.9}$$

and the balance of the stress within the neck zone is tensile. Equation (1.8) shows that the atoms there have reduced chemical potential, compared to atoms further

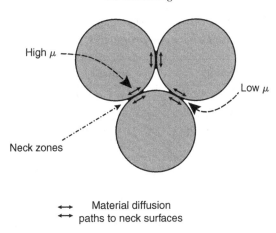

High μ

Low μ

Neck zones

↔ Material diffusion
↔ paths to neck surfaces

Figure 1.17 The diffusion of material from regions of high chemical potential (the sphere interiors), to regions of low chemical potential (the surfaces with concave curvature in the interparticle neck).

away, towards the centre of the particle: that is, $(\mu - \mu^\circ)$ is negative. Atoms will then tend to diffuse into this zone from the other regions of the particle, and the transfer of material to the neck concave surface takes the surface outwards, increasing r_1. Using the same assumption of $\gamma^{\alpha\beta} \sim 1$ J m^{-2}, and setting $r_2 = 100$ nm, shows that there is an "attraction" between the two particles corresponding to a compressive load of ~10 MPa. (This is why two perfectly flat moistened surfaces, placed face to face, can be separated only with great difficulty.)

The overall particle fusion (that is, sintering) process is shown schematically for three spherical particles in Fig. 1.17. The mechanically induced increases, or decreases, in energy (~100 J mol^{-1}) are not large when compared with other energies, such as atomic bond strengths, or the lattice energies of crystals (Table 1.4). However, they are a lot larger than nothing at all, and quite sufficient to develop a flux of atoms to regions where their energy is lower – the neck regions – and to develop interparticle bonds, which extend outwards to form an embryonic grain boundary.

These stresses, and their associated energies, simply represent thermodynamic driving forces: for actual atom migration to occur, energy barriers (represented by an *activation energy*, ΔE) must also be crossed (see Fig. 6.9). The activation energy is the energy an atom must have to make a jump from one lattice site to a suitable adjacent site (and includes any energy which might be required to create the site). This term expresses the difficulty of atomic movement, and therefore the rate at which atom diffusion can take place. The higher the temperature, the greater also is the energy of an atom, and thus the probability that an attempt at crossing the barrier will be successful. Diffusion rate, like most chemical

Table 1.4 *Forms of energy responsible for physical and chemical properties.*

Type of energy	Value/kJ mol^{-1}
Surface	~0.1 to ~1
Van der Waals	~1 to 5
Hydrogen bonding	~10 to ~20
Weak ionic	~500+
Strong ionic/covalent	~1000+

reactions, is for this reason sensitive to temperature. The rate of material transport (the *flux*) can be expressed by *Fick's first law* (the counterpart of Ohm's and Fourier's laws for electricity and heat), in which the flux (j_x) of material along the x-axis is proportional to the stress gradient (dp/dx)

$$j_x = -\frac{D_x}{RT}\frac{\mathrm{d}p}{\mathrm{d}x} \tag{1.10}$$

where D_x is the *diffusion coefficient* on the x-axis. The flux is measured in terms of moles of material crossing a plane of unit area normal to the stress gradient in unit time. Using (1.7b), at constant temperature

$$j_x = -\frac{D_x}{(V_\mathrm{m}RT)}\frac{\mathrm{d}\mu}{\mathrm{d}x} \tag{1.11}$$

where dμ/dx is the chemical potential gradient. It can easily be shown (by dimensional analysis) that a chemical potential gradient is a virtual force with units of N mol^{-1}. The diffusion coefficient (D_x) is related to thermodynamic temperature (T) by

$$D_x = D_\mathrm{o}\exp\left(\frac{-\Delta E}{RT}\right) \tag{1.12}$$

where D_o is a constant for the material. ΔE is the activation energy expressing the difficulty of atomic movement (the total energy needed for an atom to make a jump from one lattice site to a vacant adjacent site, including the energy required to create the vacancy) and is usually of the order of several 100 kJ mol^{-1}. R is the gas constant (8.314 J K^{-1} mol^{-1}). The flux, j_x, increases rapidly with temperature, because the exponential function in (1.12) is the dominant temperature term.

For shrinkage of a powder compact to occur, material must be moved from the particle–particle contact points (which is the material holding the particle centres apart) so that the particle–particle centres can approach each other. For this to happen, it is required that atoms diffuse outwards either along the grain boundary, or through the crystal lattice (which is often the slightly longer, and slower, route). On the other hand, if the material is volatile, and readily transported as the vapour, or if surface diffusion predominates, then the interparticle distance does not change. Material will accumulate in the lower chemical potential neck zone, and reduce slightly the excess energy of the particles, but there is no reduction in particle–particle centre distance, and no overall shrinkage, so that the porosity remains. This is the problem with sintering highly covalent materials such as silicon carbide and silicon nitride (discussed below), in which lattice diffusion is very slow.

A very approximate idea of how far an atom (or ion) can move in a given time can be obtained from the equation for the one-dimensional random movement of a particle, where the root mean square distance (X_{rms}) travelled in time t is given by

$$X_{rms} = (2Dt)^{\frac{1}{2}} \tag{1.13}$$

where D is the diffusion coefficient (Atkins and dePaula, 2006). Data for diffusion coefficients for most ionic species in their common crystal structures are tabulated in the literature. For a reasonable value of D (10^{-14} m^2 s^{-1}), at a typical sintering temperature of 1400 °C, after 1 hour (3.6 ks) X_{rms} is about 1 μm. Particles of typical dimension $1-10$ μm should then sinter quite well at 1400–1500 °C.

All aspects of the process of solid state sintering of ceramics have been examined in considerable detail over many years, and refined theoretical treatments have been developed (Rahaman, 2008). Solid state sintering is generally assumed to be responsible for densification in those high-purity systems either where densification aids are not needed (zirconia), or, if they are, do not form liquids at high temperature. Materials in this last category are sintered silicon carbide, using mainly boron and carbon additives, and aluminium oxide, when magnesium oxide is used to allow the development of a translucent form of the material. In practice true solid state sintering is probably rare, because of the contamination of most powders by small amounts of liquid-forming impurities.

1.5.2 Grain growth

Grain growth occurs in order to lower the total system energy, by reducing the total amount of grain boundary energy, as was shown in Fig. 1.14. This is mechanistically possible, because the same pressure (that is, chemical potential) gradients are developed within the system. Grain boundaries tend to become slightly curved (as an examination of Fig. 1.4 shows). In a simple two-dimensional

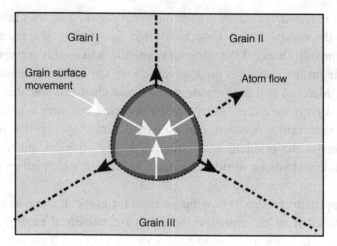

Figure 1.18 The shrinkage of a small grain. The atoms move outwards across the grain boundaries, the boundaries themselves move inwards; the small grain shrinks, and the three larger grains grow at its expense.

picture, the condition for equilibrium between the grain boundary forces at the common three-grain edge is easily shown to be

$$\gamma^{ss} = 2\gamma^{sv} \cos\left(\frac{\theta}{2}\right). \tag{1.14}$$

Then $\cos(\theta/2) = 0.5$, and θ is $120°$, the natural internal angle of a hexagon. In order to attain this internal angle in a grain with fewer than six sides (using the two-dimensional model), the three-grain edge must move slightly towards the centre of the smaller grain, and convex boundary curvature develops. This is illustrated in Fig. 1.18. For grains with more than six sides the opposite is the case. A curved grain boundary (Fig. 1.4) will have compressive stresses on the inside (convex) surface of the grain, and tensile stresses on the outside (concave) surface. Atoms will tend to migrate across the boundary from the region of compression (high chemical potential) to regions of low chemical potential, and the boundary itself will tend to move towards its centre of curvature: the atoms can be thought of as being squeezed out of the grain, through a permeable membrane, the slightly disordered grain boundary region. The grain defined by a set of convex surfaces will therefore eventually disappear, while the surrounding grains grow at its expense.

1.5.3 Void elimination

A powder has been treated so far as a collection of particles within which void space is trapped. Equally, and at times more usefully, a powder can be regarded

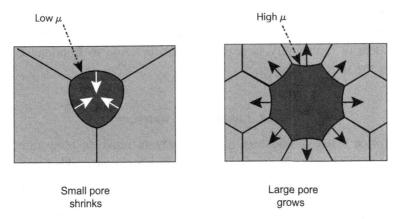

Figure 1.19 The movement of atoms across pore surfaces. Material flows inwards across the concave surface to fill the small pore; material flows outwards from the convex surfaces of the large pore, which is then enlarged.

as an assembly of voids, each created, and its *coordination number* defined, by the number of surrounding and contacting solid particles. The spatial distribution of voids, and to a certain extent void sizes, is of course determined by the homogeneity of packing of the particles (which is equivalent to uniformity of packing density). Because it is these voids that must be eliminated during sintering if full density is to be obtained, the size, and size distribution, of the voids is important. Defects in particle distribution, or poor packing homogeneity, tend to become defects in the sintered component.

Normally, as pointed out above, a small void created by small particles can fill more rapidly than a large void. But even in a fine powder (or perhaps especially in a fine powder), small voids can cluster to become, in effect, a large void. A large void of size much larger than the average (and therefore bounded by a larger number of particles – with a high coordination number) will become a larger than average size pore in the sintering material. These large pores can, theoretically, be impossible to remove during sintering. In fact they should tend to grow, to have an adverse effect on final density and strength. Figure 1.19 shows two two-dimensional voids: the one with the three convex (seen from the void side) surfaces can shrink and disappear; the one with the eight concave surfaces should grow. The reason for this is basically the same as that for the shrinkage of small grains shown in Fig. 1.18: the concave solid surface is under tension, and provides a low chemical potential region, to which atoms will tend to move (down the grain boundaries, or by lattice diffusion). This results in the grain–pore interfaces moving inwards, and pore filling. Conversely, again, a convex solid surface is under compression, the chemical potential of the surface atoms is raised, and they will try to escape, back down the grain boundaries (and in theory to an external surface). The homogeneity (that is, the quality) of powder particle packing, at the powder processing stage, and the avoidance of regions of low packing density (potentially large voids), are therefore of importance for the quality of the sintered product.

1.5.4 *Liquid phase sintering*

Oxides (often selected silicates), able to generate liquid at temperatures well below the melting-point of the main powder, are commonly added to a ceramic powder to accelerate sintering. To be an effective sintering aid, the liquid must wet the particles, and simultaneously act as a solvent for the solid phase. Silicates are very useful additives because they meet these criteria. For example, liquid is formed in the $Na_2O–Al_2O_3–SiO_2$ system at temperatures as low as 732 °C, and in the $CaO–Al_2O_3–SiO_2$ system at ~1150 °C (Levin *et al.*, 1964). The densification process is then termed liquid phase sintering, in contrast to solid state sintering, without liquid (Kwon, 1996). The practical advantages of using a liquid are that sintering is possible at lower temperatures, so that the formation of the ceramic is both faster and cheaper. The liquid aids densification in a number of ways. Initially the surface tension of the liquid–vapour interface pulls the powder particles together, creating compression at particle contact points. The liquid also functions as a lubricant to aid particle sliding, and improve packing efficiency. The liquid itself can flow to fill voids, in what is essentially a bulk *material* rearrangement process. It can easily be shown that the loss of void space will be equal to the volume of liquid which was formed from some of the (solid) powder particles, and more if improved *particle* packing results. These effects are illustrated in Fig. 1.20. But the main advantage is that the diffusion of atoms is generally much faster (at the same temperature) in a liquid than in a solid, and the redistribution of dissolved material is easier. The chemical potential of the solid is highest at particle contact points where the compressive stress resulting from the pull of the interfacial (liquid–vapour) energy is greatest. In this case the solubility of the material under stress (S) can be related to its chemical potential by relationships of the form

$$\frac{S}{S_o} = \exp\left(\frac{\mu - \mu_o}{RT}\right) \tag{1.15}$$

where S_o is the solubility of the unstressed material (of chemical potential μ_o), and μ is the (raised) chemical potential at the particle contact points (Rahaman, 1995). Solution of material takes place at regions of higher chemical potential (to provide regions of relatively high concentration in the solvent), and atoms (or ions) diffuse from there to recrystallise at surfaces of lowest chemical

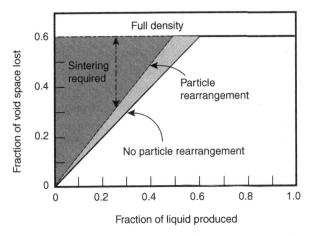

Figure 1.20 Void filling by liquid formation: the simultaneous rearrangement of particles can contribute to additional loss of void space. The balance of the void space then has to be eliminated during sintering.

potential (and where the equilibrium concentration is relatively low). Using the earlier data for the excess chemical potential of a small aluminium oxide crystal (100 J mol^{-1}) at 1400 °C, S/S_o is ~1.007. This is not an enormous increase in solubility, but quite enough to drive material migration and permit rapid sintering. Normally the movement of material will cause extension outwards of the neck zone, as it does in solid state sintering. It is of further interest that Equation (1.15) also shows that small particles will have a higher solubility than large particles, because of the inverse relationship (1.8) between chemical potential and particle dimension. This in fact provides another mechanism for sintering and grain growth in the presence of a liquid solvent, known as *Ostwald ripening*: a small particle will dissolve and the atoms (or ions) then recrystallise on a larger particle. This process operates most effectively when the powder has a very large range of particle sizes, which is probably not the case with the very fine powders used for the production of the structural ceramics.

It should be emphasised that the liquid is not merely a (more-or-less inert) solvent for the solid, but can often be a very reactive component of the whole system. This means that the crystal dissolving, and the material recrystallising, are not necessarily the same crystalline phase, or even the same material. The presence of a liquid during sintering can be helpful in assisting attainment of chemical equilibrium, but the disadvantage of liquid phase sintering, as will be seen, is that the finished component then has a corresponding volume of intergranular phase material (often a glass, because liquid silicates can be slow to crystallise) derived from the liquid-forming sintering additives. The location of the glass is influenced by interfacial energies, but it tends to exist in pockets at grain corners, and along three-grain edges (usually referred to as *triple-points*). There is often a very thin film of glass on the grain faces, separating slightly the crystal faces. This glass reduces the mechanical stiffness and strength of the material, particularly at high temperature, when under tensile and shear stresses the grains can slide away from each other. Indeed, for many structural ceramics the mechanical properties (particularly at high temperature) will be determined by the presence of the intergranular glass. If secondary crystalline phases are formed, internal stresses may develop on cooling from sintering temperature as a result of differences in thermal expansion coefficients between the secondary and the major phase. However, liquid phase sintering is a widely used process. Practically all commercial alumina and silicon nitride ceramics are made in this way: indeed, silicon nitride is almost impossible to densify without the aid of liquids, and silicate-forming systems are routinely used.

Another reason for the complexity of the microstructures of many commercially produced ceramic materials, and the spread of their mechanical properties at high temperature, is that, for reasons of cost, the starting materials used in their production are generally not absolutely chemically pure. The powders can contain a range of impurities that remain in the material after the sintering stage, and may then become further components of the material's microstructure. However, the current trend is to use higher-purity materials, because of the need for reproducibility of sintering rates, and consistency and quality of the product.

1.5.5 Vitrification

Solid state, and liquid phase, sintering both involve quite a considerable redistribution of material (~40% of the volume). If sufficient liquid could be generated (by melting some of the constituent particles) to fill all the void space, particle shape change would not, in theory, be necessary (Cambier and Leriche, 1996). This extension of liquid phase sintering is a process often termed *vitrification*, and it is responsible to a large extent for the densification of materials such as porcelain. In these materials large volumes (50% or more) of silicate liquid can be produced, which is enough to completely fill the initial void space. Even so, particle shape change does occur, because of the natural solubility of the particles (and their chemical reactions) in what are, at high temperature, very reactive liquids. In fact there will not be just shape change, but wholesale changes of chemical composition, with the production of new crystalline phases. The starting clay minerals decompose and dissolve, and do not reappear: instead, a new aluminium silicate phase, mullite, crystallises from the liquid.

These three ways of achieving full density in a sintering powder are summarised schematically in Fig. 1.21. The use of large volumes of liquid-forming

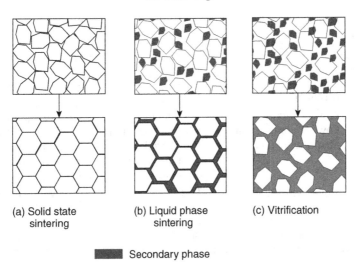

(a) Solid state
sintering

(b) Liquid phase
sintering

(c) Vitrification

■ Secondary phase

Figure 1.21 Schematic representations of (a) solid state sintering; (b) liquid phase sintering; (c) vitrification.

components, giving vitrification, is in practice confined to the production of the porcelains, and related domestic products such as bone china.

1.5.6 Reaction sintering

An important variation on the standard route to a ceramic – powder production, shape forming, and sintering – is a process termed *reaction sintering* (also often called *reaction bonding*). This consists (as the names suggest) of an *in situ* production of the required ceramic phase, by chemical reaction from its components, or raw materials, with simultaneous particle bonding (sintering). Because one of the species involved is in practice supplied from the outside, it must be mobile enough to travel easily through the pores of a compacted powder, and liquids or gases are required. To take a hypothetical example, instead of sintering silicon carbide powder, carbon or graphite particles could be compacted and then reacted at high temperature (~1500 °C) with silicon in liquid (or vapour) form supplied from an external reservoir (it would not be practicable to try to react a mixture of silicon and carbon powders). Reaction of the carbon with silicon, with simultaneous crystallisation and bonding of the silicon carbide into a strong material, would then take place. This process, illustrated schematically in Fig. 1.22, forms the basis for the production of one type of silicon carbide ceramic (though it does not quite happen as suggested; in practice a large volume of silicon carbide powder is mixed in with the graphite), and it is also used for the production of silicon nitride from compacted silicon powder, and nitrogen gas.

Table 1.5 *A summary of the processes involved in the conversion of the starting powder into a ceramic.*

Temperature	Low	High	Cooling
Processes	Mechanical	Physical and chemical	Physical
Stages	Powder mixing Powder compaction Shape refinement	Surface energy reduction Chemical reactions Pore shrinkage and void elimination Approach to equilibrium	Re-equilibration Phase crystallisation Glass formation from liquid
Results	Homogenisation and compaction of powder Compact microstructure development	Partial shape change and shrinkage Ceramic microstructure development	Microstructure modification Property development

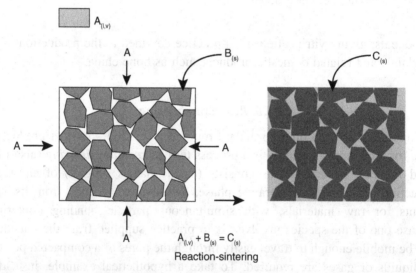

Figure 1.22 The reaction-sintering (reaction-bonding) route to a dense ceramic. Phase A infiltrates phase B, with simultaneous reaction to form the new phase C.

While in principle practically any ceramic (including alumina) could be made in this way, the conventional powder processing route is generally cheaper and easier. One major problem with the reaction-bonding route is that the chemical reactions themselves are usually highly exothermic, and the temperature, and reaction rates, can therefore be rather difficult to control.

Table 1.5 summarizes the main types of process taking place during the production of a standard structural ceramic from its starting powders, and their consequences. This shows the importance of each stage of this process, which must be closely controlled if a high-quality product is to be obtained.

1.6 Mechanical and physical properties

1.6.1 Strength

The distinguishing property of ceramics (as everybody knows) is that they are brittle: if a plate is dropped on a hard surface it usually breaks. Put more technically, ceramics are linearly elastic up to the stress required to cause failure, which is then catastrophic with unstable crack propagation. The problem with ceramic materials is that, at low temperatures at least, the stress-relieving plastic deformation processes often available to metals, such as dislocation movement, or plastic flow, are insignificant in the components of the ceramic. For this reason ductility under an applied stress is absent, and at a critical stress fast crack propagation is initiated and failure is complete. Only at very high temperature does significant plasticity become apparent, and then it is often related to the presence of intergranular glass.

The Young modulus of elasticity (E) is the slope of a plot of stress as a function of strain:

$$E = \frac{\sigma_i}{\varepsilon_i} \tag{1.16}$$

where σ_i is the stress, and ε_i the strain in the same direction i, without restraint in the orthogonal directions. When σ is increased beyond a critical value, typically $0.01E$ to $0.001E$, fracture occurs, that is, the strain to failure is small, ~0.01 to ~0.001, and for the ceramics discussed here, relatively insensitive to the material. This behaviour is illustrated in Fig. 1.23. The term *strength* (the stress required to cause fracture) is normally taken (if not specified) to mean *bend* strength. Bend strength, in three- or four-point loading, is easy to measure, once test bars have been machined to the requisite size and surface finish. The three-point bend strength (σ_{max}) for a rectangular cross-section bar is obtained from the load (F) required to cause failure using the standard expression

$$\sigma_{max} = \frac{3Fl}{2bd^2} \tag{1.17}$$

where l is the distance between the two outer knife-edges, and b is the breadth, and d the depth, of the bar. The value of σ_{max} is the maximum stress experienced by the bar, along a line on the bar face, opposite the central knife-edge, and which is where failure should occur. Tensile testing (which might allow more direct comparisons to be made with the strengths of metals and plastics) is rarely carried

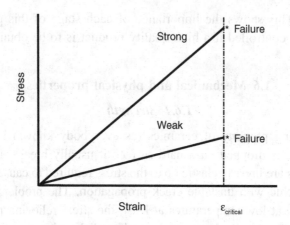

Figure 1.23 Stress as a function of strain in a brittle material; the slope is the Young modulus.

out, because of the problems with the gripping, and alignment of a brittle material (and a tendency for shear to occur), but as a rough guide a value for tensile strength is often taken to be ~60% of the measured bend strength. Values for strength measured in a compression test are also quoted, but while very high values (2–3 GPa) can be obtained these are less useful (unless the potential load-bearing capacity of kiln furniture is needed), and in any case actual failure may be in shear or internal tension. Simple uniaxial three- or four-point bend strength values reported also usually refer to short-term tests, with fast loading rates. Long-term data for time to failure under a given load, as represented in strength/probability/time diagrams, are clearly of greater relevance for service life, but for obvious reasons are harder to obtain. All these aspects are well-covered in the literature (Davidge, 1980; Morrell, 1989).

1.6.2 Flaws

The theoretical strength of a pure material can be calculated on the basis of the stiffness of the atomic bonds, and the energy of the surface created when fracture occurs. The value is of the order of 40 GPa for aluminium oxide, but this bears very little relation to the actual strength of aluminium oxide crystal, and even less to the strength of a complex, multi-phase sintered alumina. This is because the practical strength of a brittle material is very strongly controlled by its micro-structure, and the presence of defects, including scratches or damage in the surface of the test bar. The reason for this is the lack of ductility, and ability to relieve stress by plastic flow, in the ceramic (at least at low temperatures). Strength is usually treated in terms of the local interactions between the applied stress, and features of the microstructure that allow this stress to become magnified, to the extent that bond breaking becomes possible. An important relationship (Griffith, 1920; Griffith, 1924) relating the critical stress for failure, σ_{crit}, Young modulus, defect size (c), and fracture surface energy (γ_i) is

$$\sigma_{crit} = \left(\frac{2E\gamma_i}{\pi c}\right)^{\frac{1}{2}}. \tag{1.18}$$

As a measure of the *actual* stress at the tip of a sharp crack or other defect, the *stress intensity factor* (K) is a useful material parameter. For a material containing a sharp crack, uniaxially loaded normal to the crack, in crack opening mode (*mode I*), it is K_I, and related to the stress (σ) applied to the material by

$$K_I = \sigma Y c^{\frac{1}{2}} \tag{1.19}$$

where Y is a numerical parameter related to loading geometry and crack shape, and c is the length of the crack (or half-length of an internal ellipsoid crack). In practice c is often taken to be simply a flaw dimension, and which may not be easy to identify in a microstructure.

The load (σ_{crit}) at which failure occurs by fast crack propagation is then a function of the physical properties of the material, as expressed by K, and the dimension of the largest stress-concentrating feature. This relationship is commonly written as a form of the Griffith equation

$$\sigma_{crit} = \frac{1}{Y}\left(\frac{2E\gamma}{c}\right)^{\frac{1}{2}} \tag{1.20}$$

which is often expressed more simply as

$$\sigma_{crit} = \frac{1}{Y}\frac{K_{Ic}}{c^{\frac{1}{2}}} \tag{1.21}$$

where K_{Ic} for mode I stressing, now the *critical stress intensity factor*, is $(2E\gamma_i)^{1/2}$. These equations highlight the importance of the dimension (c), of the (*critical*) defect causing fracture.

The fracture energy (γ_i) can be calculated from the load required to initiate the fast fracture of a specimen containing a crack of known length. Work of fracture (γ_f) and fracture surface energy (γ_i) values can be calculated from the area under the stress–strain curve and the planar area of the fracture face. The two fracture energy values, while similar, will usually not be the same, because of the influence of crack velocity on fracture energy: values of γ_f are usually less than those of γ_i, because of the different conditions of crack growth.

In practice, fracture toughness (K_{Ic}) values are considered to be a material property, and are used as a convenient measure of the ability of a material to tolerate flaws, and other damage. The larger is K_{Ic}, the higher the stress required to cause failure for a particular defect size. Most polycrystalline ceramics have room temperature K_{Ic} values of the order of 2–5 MPa m$^{1/2}$. An exceptionally "tough" polycrystalline ceramic (such as a transformation toughened zirconia) might have values of 12 MPa m$^{1/2}$ or more. However, even these higher values of fracture toughness are still very small when compared with those for metals, such as the steels and titanium alloys, for which room-temperature fracture toughness values are in the range 40–60 MPa m$^{1/2}$. Using reasonable values for K_{Ic} (5 MPa m$^{1/2}$) and strength (300 MPa) shows that the dimension of the critical flaw size in this material would be of the order of 50–100 μm. In order to develop a strength of 1 GPa with the same material the flaw size would have to be reduced to <10 μm, or for a 50 μm flaw the toughness raised to ~12 MPa m$^{1/2}$.

Although ceramics are often regarded as weak materials, the strengths at room temperature of 500 MPa to 1 GPa and above, which can readily be obtained with the higher grade structural ceramics such as the silicon nitride and zirconia being examined here, are really very respectable. While direct comparisons of strengths measured in bend and tension are not straightforward, these values are not far short of the ultimate tensile strengths of standard structural steels, with ultimate tensile strengths in the range 400–750 MPa. 4.5% carbon cast iron has an ultimate tensile strength of ~200 MPa. The problem is in the low fracture toughness, or notch sensitivity, of the ceramic.

For many materials fracture toughness is also a function of the flaw size, and may increase with increasing crack length: this is the *R-curve effect*. In these cases K_{Ic} can be related to the initiation of unstable crack propagation.

1.6.3 Porosity

Because of the way they are made, by sintering compacted powders, ceramics tend to contain internal pores. These are the remnants of interparticle voids that were not filled during sintering. In a porous material the second important factor controlling strength is the pore fraction. It is obvious that the more internal void space a material contains, the lower must be its strength because the overall load experienced by the solid phase (expressed as N m^{-2}) must be higher, because the actual cross-sectional area of solid material available to support the load is smaller. Pores also act as a second phase, of zero modulus, and in a multi-phase material, to a first approximation, each phase contributes to the overall modulus in proportion to its volume fraction:

$$E = \sum \frac{E_i}{V_i}. \tag{1.22}$$

A reduction of E in (1.20) should lead to a decrease in strength (and porosity also decreases fracture energy, γ_i). A commonly used, purely empirical, equation expresses the relationship between strength and pore fraction (p):

$$\sigma = \sigma_o \exp(-bp) \tag{1.23}$$

where σ_o is the strength of the fully dense, pore free, material, and b a numerical constant with values for many materials again in the region of 4. This type of relationship is important because it indicates the (perhaps unexpected) importance of small amounts of porosity for the strength of a ceramic. For a pore fraction of 0.05 (which would be a good value for a porcelain), and the commonly found b value of 4, the ratio σ/σ_o is ~0.82, which represents a significant fall in strength.

It should be noted that the total *volume* of the residual porosity is a quite distinct parameter from that of the *dimensions* of the individual pores, the largest of which may turn out to be the critical defect in the Griffith equation. One large pore theoretically therefore may be able to reduce the strength to the same value given by a large volume fraction of very small pores. Total pore fraction, and the distribution of pore dimensions, while in practice likely to be related, are two separate parameters both needing to be controlled, and if possible minimised, during the powder processing and sintering stages. This is important for the reaction bonded form of silicon nitride, discussed in Chapter 5.

1.6.4 Strength distribution

Because of the importance of flaws (including internal porosity) for strength, and the normally random nature of the spatial distribution and size of the flaws, the strengths of a batch of material always have a spread of values. For a small batch of specimens, a variation of ±25% is quite normal. The spread of data can be expressed by an equation of the form:

$$P_f = 1 - \exp\left(\frac{-V(\sigma - \sigma_u)}{\sigma_o}\right)^m \tag{1.24}$$

where P_f is the probability of failure at a uniform applied stress (σ), for a sample of volume (V), σ_u is the stress below which fracture is assumed to have zero probability (often taken to be zero), σ_o is a normalising parameter, and m represents the spread of data, or consistency of the batch, the *Weibull parameter* (or *modulus*) (Weibull, 1951; Davidge, 1980a). The larger is m, the less variable the strength. The Weibull modulus for a reasonably homogeneous, consistent, batch of material will be >10; for less consistent, more variable materials, much lower

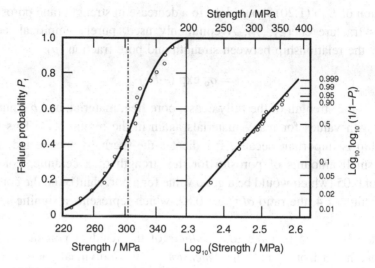

Figure 1.24 Data for a batch of materials, showing the strength distribution, and the Weibull distribution graph. (After R. Morrell, 1989. © Crown copyright material is reproduced with the permission of the Controller of HMSO and Queen's Printer for Scotland, under the terms of the Click-Use Licence.)

values are obtained. For convenience of plotting data, (1.24) can be rearranged, by taking logarithms twice, into

$$\ln \ln \left(\frac{1}{1 - P_f} \right) = m \ln(\sigma - \sigma_o) - m \ln\sigma_o + \ln V. \tag{1.25}$$

For a batch of N samples, ranked in order of ascending strength, P_f can be obtained from the relationship

$$P_f = \frac{i}{N + 1} \tag{1.26}$$

where i is the sample's rank position (Gumbel, 1959).

In this type of treatment, $\log_{10}$ can equally well be used instead of $\log_e$, and the survival probability, P_s (which is $1 - P_f$), instead of P_f. An example of this type of relationship, the strength distribution within a batch of 21 samples of material (95% alumina), is shown in Fig. 1.24. For two samples of volumes V_1 and V_2 (and setting σ_u to zero), this expression resolves into

$$\frac{\sigma_{v1}}{\sigma_{v2}} = \left(\frac{V_2}{V_1} \right)^{\frac{1}{m}} \tag{1.27}$$

where σ_{v1} and σ_{v2} are the stresses giving the same probability of failure. For a material of reasonable consistency ($m \sim 10$), increasing the stressed volume by a factor of 10 will decrease the mean strength by around 20%, which has important implications if predictions for large components are going to be made on the basis of strength measurements obtained from small-dimension test bars. This relationship also assumes (unrealistically) that flaws on the surface of the material have no greater effect than internal flaws. Because of the importance (for several reasons) of the surface for fracture initiation the volume term is often appropriately replaced by an area term. Tabulated test sample values should be regarded simply as guides to strength in a batch of material, or as means of ranking materials.

1.6.5 Poisson ratio

When an elastically isotropic body is subjected to a strain (ε) in one direction, it experiences a strain ($-v\varepsilon$) orthogonal to that direction. The term v is the Poisson ratio, and has a value of 0.5 for a perfectly plastic body. The Poisson ratio is related to the Young (E) and shear (μ) moduli by the expression:

$$v = \frac{E}{2\mu - 1}. \tag{1.28}$$

The ratio is of importance because it appears in many expressions concerned with the development of strain (and possible failure), when stresses are applied to the material. Values of v for the structural ceramics are generally in the range ~0.2 to ~0.3, with 0.25 being a common value (Morrell, 1989).

1.6.6 Thermal stresses

When the temperature of a material is changed rapidly, internal stresses develop. Expressed in simple terms, a material attempts to change its dimensions with change in temperature, because the mean interatomic distances increase with increase in the atomic vibrational energy. These changes can be expressed by the *linear coefficient of thermal expansion* (*thermal expansion coefficient*, or *thermal expansivity* for short), a_T:

$$a_T = \frac{1}{l}\frac{dl}{dT} \tag{1.29}$$

the fractional increase in length per degree rise in temperature, T. Strictly speaking a_T is the expansivity only at the particular temperature T. It is often assumed that a is independent of temperature, but this is not usually the case, and a values are therefore better expressed as the mean (a_m) over a specified temperature range. The mean thermal expansion coefficient (a_m) is expressed by the fractional change in length $(\Delta L/L_o)$ over the temperature range $(T-T_o)$ using the relationship

$$\frac{(L - L_o)}{L_o} = \frac{\Delta L}{L_o} = (T - T_o)\, a_m \qquad (1.30)$$

where L_o and T_o are the initial values for length and temperature, and L and T the final values. Thermal expansivity values are often recorded over the temperature range 25–1000 °C, and for most ceramics the mean values are between 4 and 12 MK^{-1}.

An unconstrained bar can expand or contract freely along its length. But if the bar is constrained rigidly at both ends by clamping, the potential dimensional change (a strain, ε) leads to the development of a stress, which will be compressive by heating, or tensile by cooling, through the relationship

$$\sigma = E\varepsilon = E(T - T_o)a_m. \qquad (1.31)$$

When a bar is quenched from a temperature T to a lower temperature T_o, because of the heat transfer from the bar to the surrounding medium, a tensile stress develops in the surface of the bar, when it attempts to contract but cannot because of the constraint of the hot interior. The maximum stress is developed in the case of an infinitely fast quench, for which it is assumed that there is insignificant heat flow from the centre to the surface. The surface stress (σ_{ts}) is then given by:

$$\sigma_{ts} = \frac{E\Delta T a}{1 - v} \qquad (1.32)$$

with the extra term involving the Poisson ratio (v) occurring because of the biaxial nature of the stress.

1.6.7 Thermal shock

It immediately becomes clear by putting data into (1.32) that these thermally induced surface stresses can be very large. For example, for a siliceous porcelain cooled rapidly from 500 °C, to room temperature with $E \sim 70$ GPa, $a \sim 6$ MK^{-1}, and $v \sim 0.18$, σ_{ts} is ~245 MPa. This is about 3 times the normal bend strength, and the quench is able to induce crack initiation and unstable propagation. In a very strong material, such as sintered alumina, because of the large amount of stored energy at fracture, failure can be explosive. In practice, a ceramic component will tend to be severely cracked, or completely fractured, if it is heated or cooled too quickly. The ability to resist these sudden temperature changes can be assessed by measuring the residual strength after rapid heating or following exposure of a heated test bar to a jet of cold air, or plunging into water or oil. The general

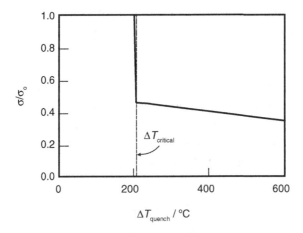

Figure 1.25 Schematic representation of the behaviour of a strong material after thermal shock treatment.

behaviour of a strong, high-density, ceramic as a consequence of thermal shock is illustrated in Fig. 1.25, where the residual strength, expressed as a fraction of the original strength (σ/σ_o) is plotted as a function of quench temperature. Typically the residual strength does not fall to zero after one shock, but for a quench temperature around the critical value (ΔT_{crit}) drops to about 20% of the value for undamaged material. With weaker materials, containing pre-existing flaws, failure is less spectacular. In this case there is a progressive loss of strength with increasing severity of thermal shock, and there may be no clearly identifiable critical quench temperature.

The critical quench temperature, ΔT_{crit}, is commonly used to rank the likely behaviour of a strong brittle material when subjected to rapid temperature changes. This is the quench temperature for which σ_{ts} equals the strength of the material, and is given by:

$$\Delta T_{crit} = \frac{\sigma_{ts}(1-v)}{Ea}.$$

(1.33)

In calculations the bend strength is commonly used. In more realistic situations the dimension of the sample, the heat transfer coefficient between the material and the quenching medium, and the thermal conductivity of the material, need to be taken into account. However, for the present purposes of simply ranking materials in terms of the value of the critical quench temperature, (1.33) is often considered to be adequate.

Because v does not vary greatly between ceramics of the types reviewed here, and because the ratio σ/E, in practice, tends not to vary greatly between materials of different types, the thermal expansivity term a is critical in determining the behaviour of a strong ceramic under conditions of severe thermal shock. This in turn determines the usefulness of the material for applications involving exposure to high temperatures, and thus the capacity to withstand fairly rapid heating and cooling.

Equation (1.33) embodies one form of what are termed thermal shock (R) parameters (Hasselman, 1969). These provide indications of the critical conditions for crack initiation and fast propagation. For the case of a strong material suddenly heated or cooled, the parameter (R) used in ranking an ability to survive without significant loss of strength would be given by

$$R = \frac{\sigma(1 - v)}{Ea} \qquad (1.34)$$

where again bend strength is usually used for σ. For the case of a weaker material already containing a crack (or serious flaw), a second form of parameter (R''') is used to express the critical condition required to propagate the crack

$$R''' = \frac{E}{\sigma^2 (1 - v)}. \qquad (1.35)$$

The assumption again is that thermal conduction is not able to reduce the thermal gradient, so that the quench is effectively instantaneous. This expression indicates that an initially weakened material should be better able to resist further thermal shock damage: this is not likely to be the case with the high-strength materials to be examined here. Equation (1.34) suggests that the ideal material for resisting thermal shock would be one for which σ was zero – but this would in other respects not be a very useful material.

1.7 The structural ceramic portfolio

Ceramics were defined earlier as inorganic, non-metallic materials of high melting-point. There are about 100 metallic and 15 semi-metallic and non-metallic elements. These immediately provide in the region of 1500 two-element ("binary") inorganic compounds formed from a metallic and a non- or a semi-metallic element, assuming that the elements have single oxidation states. However, many elements, including most of the metallic transition, lanthanide and actinide elements, have more than one stable oxidation state, which increases the scope for increasing the number of compounds. The number is enlarged further when compounds formed with more than two elements are considered, such as two metallic or semi-metallic, and one non-metallic (the magnesium silicate, enstatite, $MgSiO_3$, for example), and *vice versa* in what are termed *ternary* compounds. *Quaternary* (as in a silicon aluminium oxynitride, or sialon), and even more complex systems have to be taken into account. While all of these inorganic compounds have been examined (many in very great detail) from the points of view of their pure chemistry and their mechanical and physical properties, only a handful have yet found practical applications as structural materials, for a variety of reasons.

1.7.1 General criteria for qualification

Only a small proportion of the several thousand stable inorganic compounds are materials that potentially can be classed, or used, as ceramics. The value of a "high melting-point" is often undefined, but can be taken to be >1000 °C. For a start, not all inorganic compounds have high melting-points, and others are chemically too reactive to be useful. Many oxides and halides, particularly when the metal is in a high oxidation state, are very reactive, or have low melting-points and high volatility. For example tungsten$^{(VI)}$ oxide, WO_3, melts at about 600 °C and is then very volatile. Titanium tetrachloride, $Ti^{(IV)}Cl_4$, is a volatile liquid at room temperature, and violently reactive towards water (though it is used in the production of titanium dioxide white pigment, with a melting point of ~1850 °C); uranium hexafluoride, $U^{(VI)}F_6$, is a gas at room temperature (and is used in the enrichment of the fissile isotope of uranium, U_{235}). However, uranium dioxide, $U^{(IV)}O_2$, has a melting-point of ~2880 °C, and is used as a nuclear fuel material. Silicon forms two oxides: silicon dioxide ($Si^{(IV)}O_2$), commonly referred to as *silica*, is the basis of the stable mineral quartz, but the second, silicon monoxide $Si^{(II)}O$, is a vapour found only at high temperature, and is unstable at room temperature with disproportionation into a mixture of elemental silicon and silicon dioxide. Melting (or liquid formation) temperatures, and volatility, provide fairly clear initial rule-in or rule-out criteria. This is why a table of melting-points of materials is a common starting point for what might, or might not, be useful, although clearly it does not tell us everything we need to know. There are other criteria for the selection of ceramic materials for applications involving mechanical loading, possibly at very high temperature.

1.7.2 Chemical stability

The use of oxides as the bases for structural materials has been natural because of their ready availability. However, to be useful as structural materials, they must also have chemical stability in the operating conditions of the Earth's atmosphere, and resist deterioration under the influence of damp air, and reaction with water. Not all oxides do. Sodium oxide (Na_2O) and potassium oxide (K_2O), although they have high melting-points, react too easily with water to form the (water soluble) hydroxides to be of any use structurally, though reaction rate has to be considered. High surface area magnesium oxide (MgO) powder is also readily hydrolysed by water to magnesium hydroxide, but the reaction is acceptably slow with bulk, dense, forms of the material and magnesium oxide and the similar calcium oxide (*basic* – as opposed to *acidic*) refractory materials are important for lining steel-making furnaces (but they must be stored under dry conditions). The tendency for aluminium oxide to be hydrolysed to aluminium hydroxide, and the associated surface chemistry, is of great importance for the dispersion and behaviour of very fine aluminium oxide powder particles in water, but for many practical purposes the dense ceramic form of polycrystalline aluminium oxide (in the single form known as corundum, or sapphire) is completely inert at room temperature. However, aluminium oxide, and many other oxides, are readily attacked and dissolved by aqueous acids and alkalis at temperatures above ~250 °C (under pressure), as well as by pure water itself. (Water is a very curious substance in many ways – because of its very strong hydrogen bonding – and at low enough temperatures (<0 °C) it also becomes useful as a low-strength structural material.)

Compounds of the non-oxide elements such the borides, carbides and nitrides also have extremely useful sets of physical and mechanical properties. Many members of this group are commercially exploited, silicon carbide, for example, on the multi-thousand tonne per year scale as refractories and abrasives, and boron carbide on a much smaller scale for sand-blasting nozzles, and armour plating. Some of the highest melting-points of all are found in the nitrides, carbides, and borides of the transition metals. They can also be very much harder than tool steels, and be very good electrical conductors. These compounds are often termed *hard metals* (though they are not metals, but some do have metallic electrical conductivity). Perhaps the best-known examples of this class are the tungsten carbides, of compositions WC, and W_2C, melting at ~2850 °C, and hardness ~18–21 GPa (Storm, 1967). Hafnium carbide (HfC) has a melting point of >3800 °C. A major problem, however, is the tendency of the non-oxides to oxidise, and an important factor controlling their applications is their ability (or otherwise) to develop oxide surface films that are able to provide protection against continuing oxidation. Because the main oxide of tungsten (WO_3) is volatile, tungsten carbide cannot be exposed to high temperatures in air for long periods of time. On the other hand, silicon carbide and silicon nitride do produce protective surface film of silicon dioxide, and both materials can survive for many thousands of hours at high temperatures under oxidising conditions. It must be remembered that all non-oxides degrade (more or less quickly) in normal environments as the temperature is raised, and what is an acceptable life under specified conditions becomes a key factor in determining the choice of material.

Many of the transition metals (and particularly the third-row elements) have multiple oxidation states which provide a range of compounds, the stability of which depends on the oxygen pressure. Iron gives the oxides $Fe^{(II)}O$ and $Fe_2^{(III)}O_3$, which interconvert readily according to the system oxygen pressure and temperature. $Fe_2^{(III)}O_3$ can be sintered to full density at high temperature, but both oxides have the drawback of relatively low melting-points – $Fe^{(II)}O$ ~1370 °C, $Fe_2^{(III)}O_3$ ~ 1566 °C – and attractive though it might seem given the large quantities of iron oxide available, these have not yet found use as structural ceramics.

1.7.3 Toxicity

One material's property of importance often overlooked is toxicity. The toxic nature of "heavy" metals such as lead is now well-known: lead is no longer a common constituent of pottery glazes, nor is uranium oxide ($U^{(IV)}O_2$) used to give decorative glazes a bright yellow colour. Beryllium (in Group IIA of the Periodic Table, and therefore related to magnesium) is also very toxic and for this reason has been avoided as a component of structural materials, although its oxide, BeO, melts at ~2530 °C (beryllium is also not a readily available metal). However, beryllium oxide has been examined for nuclear reactor use because of its high thermal conductivity (Ryshkewitch and Richardson, 1985). Both uranium dioxide and plutonium dioxide ($Pu^{(IV)}O_2$) are important nuclear fuel materials, with the fuel consisting of small sintered pellets. The melting-points of the oxides ($Pu^{(IV)}O_2$, 2400 °C) are much higher than the metals, allowing the possibility of a higher-temperature reactor system. Both oxides are used as fuel in the mixed oxide (MOX) reactor. These oxides are processed in much the same way as other oxide ceramics, though of course with special care to contain the very long half-life radioactive powders and waste products.

1.7.4 Physical stability

Atoms with low atomic number are small. They may form short (~150–170 pm) strong covalent bonds, which in turn tend to give compounds (particularly the binary compounds) high thermodynamic stability, and stiffness. The more ionic compounds will tend to have very high lattice energies, and high melting-points. However, the nature of the bond is also important, and many heavier (and larger) transition metals form extremely hard, and high melting-point, binary compounds, such as tungsten carbide, with complex electron distributions. But melting-point is not the only aspect of thermal stability of importance. We are dealing with crystalline compounds, and there is the possibility that a compound may have more than one crystal structure, each of which may have different physical (and chemical) properties. Phase changes between crystal structures (which occur at specific temperatures) need to be considered: these almost always involve changes of density, and therefore volume. For example, many silicates have multiple crystalline phases: dicalcium silicate Ca_2SiO_4 melts at ~2100 °C, but has a phase change at ~520 °C (β-Ca_2SiO_4 to γ-Ca_2SiO_4) with a 13% increase in volume and a 14% shear in the crystal lattice, which causes spontaneous and spectacular disintegration of a hard sintered pellet into a fine powder (*dusting* – and responsible for the breakdown of refractories containing dicalcium silicate grain boundary phase) on cooling. The complex transformations occurring between the common crystalline forms of silicon dioxide, SiO_2 (quartz, tridymite and cristobalite, as shown in Fig. 2.18), control the structural stability of refractory materials formed by silicon dioxide, and determine their allowable operating conditions. The tetragonal-to-monoclinic transformation of zirconium dioxide (ZrO_2) at ~950 °C can, like the dicalcium silicate transformation, also be physically destructive, but when controlled properly it provides excellent mechanical strength and toughness in polycrystalline materials based on zirconia, which is why it is included in this set of structural ceramics.

1.7.5 Availability

In spite of all the restrictions outlined above, the total number of inorganic compounds potentially of interest as the basis for structural materials is still very large. The selection of ceramic materials for practical applications has been made on a number of additional, more practical, bases. The property of "availability" is a very important one, and linked closely to cost. The Earth's crust contains a high proportion of low atomic number (what are often called "light") elements, silicon, aluminium, sodium, potassium, magnesium, calcium and oxygen, as shown in Table 1.6 (Lide, 2005).

Because of the oxidising nature of the Earth's atmosphere, and their chemical reactivity, these metallic elements are mostly present in the crust as oxides. These are the compounds, the silicates and aluminosilicates, forming the rocks and natural minerals, and materials based on them are thus generally readily available in very large quantities, and at relatively low cost. Nitrogen is of course also a major constituent of the Earth's atmosphere, and carbon is available in the form of coal and oil, or as charcoal from the pyrolysis of wood (cellulose). Many of the large group of traditional ceramic materials have been produced for many thousands of years from a wide range of readily available (and low cost) raw materials such as the clay minerals, together with soft or readily crushable minerals, such as the feldspars and flint. As noted above, talc, magnesium hydroxysilicate, was at one time widely used for ceramic electrical insulators.

Table 1.6 *Abundance of the more common elements in the Earth's crust (Lide, 2005).*

Element	Symbol	Abundance in the Earth's crust / % by weight
Oxygen	O	46.1
Silicon	Si	28.2
Aluminium	Al	8.2
Iron	Fe	5.6
Calcium	Ca	4.2
Sodium	Na	2.4
Magnesium	Mg	2.3
Potassium	K	2.1
Titanium	Ti	0.6
Phosphorus	P	0.1
Strontium	Sr	0.04
Zirconium	Zr	0.02

The silicate and aluminosilicate raw materials can have complex compositions, and the minerals themselves are often a mixture of phases. While the constituent simple primary oxides may have high melting-points and thermal stabilities, the melting-points, and related mechanical property values, of the more complex derivative silicates and aluminosilicates, many of which make up the rocks of the Earth's crust, can be very low (melting temperatures <900 °C).

It is for this reason that interest has been developed in other ceramic systems, although they may not be available in an immediately usable, pure, form. In many cases, extraction from the raw material is required, or the production of the ceramic powder by chemical reactions between simpler ingredients. Some of these materials could be termed "man-made", though in reality they often occur in nature, silicon carbide and silicon nitride for example, albeit in very small quantities. The processing stages make the pure powders more expensive, and their price can form a significant proportion of the cost of the final article. It has to be remembered that economic factors are just as important for the selection of a material, as its mechanical or physical properties.

1.7.6 Processability

The powder sintering route is the standard procedure for forming components of ceramic materials. The rate at which full density can be obtained, whether or not special atmospheres are required, and whether grain growth can be contained, become important considerations. Most oxide systems are relatively easy to densify, either because they can be sintered with the aid of liquid-forming additives, or because atomic mobility in the solid is high at reasonable temperatures. Non-oxides tend to be reluctant to sinter, even at high temperature, and at one time they were considered to be "unsinterable". This difficulty is related to their strong directional covalent bonding, leading to the low mobility of their constituent atoms. Very high temperatures, and very fine powders, are needed therefore. The normal volatility of the material at these very high temperatures may also make densification difficult (though not necessarily precluding a degree of interparticle bonding, and the development of some strength). The transition metal borides are potential electrodes for the electrolysis of aluminium oxide in the production of aluminium. Zirconium diboride (ZrB_2) is a good electrical conductor and melts at ~3050 °C. However, to obtain full density from even a μm dimension powder requires hot-pressing at ~1900 °C, which is not at present commercially attractive, when the much cheaper graphite electrodes in many ways perform equally well (though they do generate carbon dioxide by oxidation).

1.8 Materials selection

When these many aspects are taken into consideration, the materials of interest for large-scale structural applications now become reduced to a relatively small number – about a dozen. The selection is determined by a combination of chemical, mechanical, and physical properties, together with the availability and cost of the raw materials, and the ease with which they can be sintered to a suitable density. Some of the basic compounds were listed in Table 1.1. However, the choice of materials for actual commercial development can be made only partially by looking at a material's properties. There are the questions of needs, and the competition from existing (often) metallic materials. For a well-established component to be replaced by one made from a new ceramic, possibly with high development and production costs, requires the ceramic to have clear advantages. The question of design has also to be addressed: a ductile metallic or plastic component cannot be exchanged for one made of a brittle ceramic, without thinking about the need to minimise stress concentrations. There must also be the potential for the mass production of components of high and consistent quality and reliability: there can be a big difference between what can be obtained in the laboratory on a small scale, and the average product of a large-scale production run. This is partly because, as noted above, the strength of a brittle material is a function of the size of the component: the larger the component the larger is the probability that a large strength-determining defect will be present.

The five materials chosen as case studies have between them a wide range of properties, representative of those of the structural ceramics as a class. They are

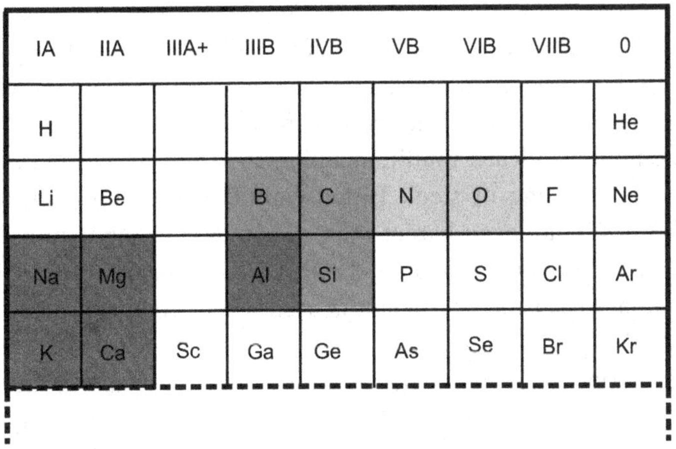

IA	IIA	IIIA+	IIIB	IVB	VB	VIB	VIIB	0
H								He
Li	Be		B	C	N	O	F	Ne
Na	Mg		Al	Si	P	S	Cl	Ar
K	Ca	Sc	Ga	Ge	As	Se	Br	Kr

Figure 1.26 The low atomic number section of the Periodic Table.

also among the most important commercially, and considerable experience in their manufacture has been developed. These are ceramics that have properties that other structural materials cannot provide. Four (two oxide and two non-oxide) are basically simple binary compounds (though there may be small amounts of secondary phases). The fifth, porcelain, chosen because it was one of the first structural ceramics and therefore provides a baseline for the newer materials, is a much more complex multi-component system (and more than half-way to being a glass). In looking for possible links between these five, an immediately clear one is that the chemical elements forming them are the lighter or more common elements. All but one contained in all the compounds and phases referred to in these chapters are in the small, highlighted, section of the Periodic Table shown in Fig. 1.26. The one element that is not, is the heavier (transition metal) zirconium, which is in Group IVA of the Table, in the fourth row, between titanium and hafnium. From this it might be deduced, correctly, that availability and cost of the raw materials is an important factor determining choice. On the other hand, that two of the five are non-oxides might suggest that absence of the instant availability of a basic constituent does not necessarily rule out a material, implying that there can be other properties of overriding importance. Looking a little more closely it is seen that the primary crystalline components of these materials have very high melting-points, and hardness. This is not surprising considering the needs for materials with mechanical stability under load, particularly at high temperature. One problem of great importance in the engineering world is wear. Intuitively, wear resistance in a material should require hardness and toughness: this is generally true, but other factors also play a part, as shown by the fact that even one type of hard ceramic can have large differences in wear rate.

There are other ceramics which have an important share of the structural ceramics market. For example, magnesium aluminosilicate, cordierite, is widely used as a porous catalyst support and filter material because of its very good resistance to thermal shock, though its melting point is relatively low (~1540 °C). The more refractory aluminium silicate, mullite, melting at ~1890 °C, has widespread uses in the continuous casting of steel. Boron carbide (2450 °C), and cubic boron nitride (> 3000 °C), are even harder than silicon carbide, and are used for their wear resistance. Aluminium titanate (~1860 °C) (Dodd and Murfin, 2006) with a very low thermal expansion coefficient is used where good resistance to thermal shock is needed.

The five materials selected to illustrate the structural ceramics are now introduced briefly: each is then examined fully in the following five chapters.

Porcelain has been known for almost 2000 years in China, but it was developed in Europe in the eighteenth century, not so much for its (relatively) high mechanical strength, but for its aesthetic qualities deriving from its translucency in thin section (though an article of thin section would also have required high mechanical strength, to have a reasonable life). It could not have been foreseen that a major practical application for porcelain would have been in high-voltage electrical insulation (in the eighteenth century the electrified railway had still to be invented! – Michael Faraday discovered electromagnetic induction, the principle behind the electric motor and generator, in 1831); but the material happened to be available at the time a need arose.

Alumina was well-known at the beginning of the last century as a high-temperature refractory material, and exploration of its uses as sintered fine-grain materials started at the same time. The incentive for its intensive development as a fine-grain structural ceramic also derived in part from the need for electrical insulators, particularly for spark-plug insulators in the new, higher compression and higher temperature internal combustion engines. Commercial exploitation required the development in the 1930s of very high-temperature furnaces to allow the sintering of alumina components to full density. The properties, and potential usefulness, of alumina then became very quickly recognised. One big advantage of alumina was that the methods for producing the precursor calcined alumina powder had already been developed, and used on a very large scale, for the production of aluminium metal, so that high-purity powders were readily available in large quantities.

Silicon carbide was also first used as a refractory material, and as a hard abrasive (carborundum). Its potential as a fine-grain structural ceramic similarly only became open to exploitation with improved understanding, when the problem of sintering fine powders to full density was solved. A major driving force for further work on silicon carbide in the 1960s was the ambition to develop new higher efficiency internal combustion engines, running at very much higher temperatures (~1370 °C) than could be sustained using normal metallic components.

Silicon nitride was also selected for detailed investigation quite deliberately, as part of a systematic search for new materials with very good thermal shock resistance, primarily on the basis of its low thermal expansivity. These were the same reasons for the choice of silicon carbide, and silicon nitride had also been developed as a refractory material several years earlier.

Zirconia had been exploited as a coarse-grained high-temperature refractory for over half a century. Its full potential as a high-strength and high-toughness material was only realised when the compositions needed to control an otherwise destructive phase transformation were identified as a result of the improved understanding of the zirconium dioxide phase changes. Zirconia provides some of the strongest, and toughest, polycrystalline ceramics yet developed.

These materials evolved as structural ceramics for the most part independently of each other, but against a background of much earlier work in related areas. In the cases of alumina, silicon carbide, silicon nitride and zirconia, there were important applications as bulk refractories. There was therefore a great deal of experience and information to support their initial developments as fine-grain, high-strength, structural, ceramics. Silicon carbide and silicon nitride, while having different early histories, were initially developed in tandem with considerable financial support from government and industrially funded development programmes in the internal combustion engine area. Other applications for these two non-oxides were then developed using different properties, and their paths tended to separate.

1.9 Exclusions

As was pointed out at the beginning of this chapter, the subject of ceramics now includes a very wide range of inorganic materials. The focus of this book is on the class known as the structural ceramics, using a very small number to illustrate the properties of the class. Other types of inorganic material coming under the ceramic heading, which clearly do have structural functions, and which in many ways seem to be closely related, have not been included in the structural ceramics category. A few words of explanation are needed because the five subject areas, structural ceramics, glasses, refractories, heavy clay and hydraulic cements, overlap to varying degrees, and treatments of these materials cannot altogether be separated. It then becomes partly a matter of almost arbitrary definition, but there are more objective reasons for deciding which materials to include in the structural ceramics category, and which not.

1.9.1 Glasses

A glass is a non-crystalline, amorphous, form of material, in which the arrangement of constituent atoms and ions outside the immediate coordination sphere
forms no regular patterns – at least, none that are very obvious on the micrometre
scale of examination. The subject of glass has traditionally been regarded as
separate from that of ceramics, with their predominant crystallinity, microstructures, and production from compacted powders. There are many volumes dealing
exclusively with the science and technology of glass (Shelby, 2005). Glasses, such
as domestic and storage ware, and sheet or plate window glass, support a range of
industries, which concentrate on the melting of raw materials, and then casting and
cooling the liquid to shape, very much as metals are. This is a distinguishing
feature of the material, and which from the scientific and technological points of
view separates the glasses from conventional ceramics. The transparency and the
absence of an obvious microstructure are features a glass has in common with a
single crystal: these can lead to common misnomers as in the terms "lead crystal"
and "quartz crystal". Though these terms are often applied to special types of clear
domestic glassware, which can be cut to provide sharp facets, which sparkle in
reflected light, they are not crystalline materials at all. On the other hand very few
ceramic materials have microstructures that are completely free from amorphous
material (that is, glass), particularly in the grain boundaries. And the first case
study – porcelain – can contain up to almost 60% by volume of glass, essentially as
the matrix within which are embedded the crystalline phases. This glass is a
dominant feature determining the physical and mechanical properties of porcelain,
and can be important for the properties of the others. It is therefore not possible to
completely ignore the glasses and their properties in considering the properties of
the structural ceramics, and some aspects of this subject will be introduced.

1.9.2 Refractories

As with bulk glass, materials conventionally regarded as refractories are produced
on a very large scale by specialist industries, and, as with the subject of glass, there
is a large volume of literature focusing on these materials. These materials are used
on the million tonne scale as structural materials in high-temperature furnaces.
There is a quite justifiable point of view that structural ceramics, since they
developed from them, are really just a special type of refractory. In fact in the
1960s, the structural ceramics (and the electroceramics) were often referred to as
special ceramics, though this was more to distinguish them from the traditional
whiteware ceramics, or pottery (Popper, 1960). The two groups do have much in
common in terms of compositions and underlying principles governing behaviour,

but exclusion of the refractories can be justified on three counts: the applications of the two groups of materials are fundamentally different, in that "refractories" have been developed almost exclusively for use at very high temperatures (such as containment of molten metals in the iron and steel industries, or liquid glass), which means a focus on selected types of property, including corrosion; conversely the fields of application (and certainly the application temperatures) of the structural ceramics are much wider, and therefore require attention to a wider range of microstructures and associated properties; the microstructures of refractories tend to be very much coarser than those of the structural ceramics, in that the grain sizes are often in the μm to mm range, rather than nm to μm, so that mechanical property values tend to be much lower than those of the structural ceramics. However, the close family relationships between the refractories, and the structural ceramics, should not be forgotten. Or perhaps they might be said to be included, but not under this label; members of the group have in part metamorphosed into the structural ceramics, and they could be regarded as present, but disguised.

1.9.3 Heavy clay

An extremely important area of industrial activity producing structural materials using clays is usually referred to as the brick and heavy clay industry (Singer and Singer, 1963). *Heavy clay* products include building bricks, drain and sewer pipes and conduits, tiles, and low-density structural insulation materials. These are generally made from impure semi-plastic or dried clays, with minimal processing or pre-treatment. Components are shaped and then fired at moderate temperatures. The aluminosilicate microstructures are usually coarsely structured, and not fully dense. Costs, and mechanical strengths, are relatively low and applications are generally at "room" temperature. The bend strength of a roof tile might be as low as 10 MPa; a high-grade engineering brick can have compressive strength of 100 MPa (this is not large by structural ceramic standards, but it would still support a brick column 4 km high). While these products can be regarded, reasonably, as structural materials, they tend to be treated in a separate category from the higher strength structural ceramics and have therefore not been considered here.

1.9.4 Hydraulic cements

Traditionally, as with the glasses and refractories, the cement industry has remained separated from the ceramics industries, although at most major ceramics conferences all four areas will be dealt with in different sections of the pro-gramme. The hydraulic cements forming the basis of well-known materials such as concrete have long been a subject in their own right (Boyd and Mindess, 1992).

While they are certainly inorganic solids (often aluminates and aluminosilicates), and the starting cement powders are produced by high-temperature heat treatments of minerals such as mixtures of clay and limestone (followed by milling to a fine powder), the final strong solid is formed through room-temperature chemical reactions between the powder and water. The chemistry, and associated physical processes, occurring in the development of the set cement microstructure are complex and their treatment has become highly specialised. The arguments for excluding these materials are similar to those for glasses and refractories: this topic is highly specialised, involves chemical and physical processes different in some way from those involved in the production of a dense polycrystalline ceramic, and the subject of cement chemistry is already well covered in its own literature. Nonetheless it must be recognised that processes occurring during the production of cement powder, its subsequent hydration reactions, and associated surface chemistry, are in principle very similar to some of those met with in the structural ceramic systems.

Questions

1.1. Define *specific surface* area. For a crystalline material of density 3.2 Mg m^{-3}, calculate the specific surface areas of (a) a rod of length 10 μm and diameter 1 μm, and (b) a cube of the same volume. What conclusion can be drawn about the relative thermodynamic stabilities of these two crystals?

1.2. What is the fractional area of paper covered by the packed discs in Fig. 1.10?

1.3. Calculate the size of the disc, relative to that of the large disc, which would just fit into the space between three discs shown in Fig 1.10.

1.4. Explain the general form of the graph shown in Fig. 1.11. Why does one sloping line extrapolate to zero, while the other does not?

1.5. A piece of single-crystal aluminium oxide is ground to a powder of mean dimension 2 μm. Assuming an interfacial energy of 1 J m^{-2}, estimate the surface energy (in J mole^{-1}) now introduced into the material. In what other ways might the powder particles contain excess energy?

1.6. Two soap bubbles, one large, one small, are separately blown (by means of a three-way valve) at opposite ends of a short length of glass tubing. Explain what will happen when the valve is opened to link the two bubbles.

1.7. Which should be more effective at increasing the initial rate of solid state sintering rate of a fine powder: (a) decreasing the particle size by a factor of 2; (b) increasing the temperature from 1300 to 1400 °C? Assume that the sintering rate exponent n in (1.2) is 4 and the activation energy for the sintering rate is 200 kJ mol^{-1}.

1.8. Show why the equilibrium internal angle at a three-grain boundary (of the type shown in Fig. 1.4) would be expected to be 120°. A crystalline inclusion has an interfacial energy (γ^{ab}) which is 1.3 times that of the main phase grain boundary energy (γ^{aa}). What now will be the equilibrium angle at a three-grain boundary formed between an inclusion, and two matrix grains?

1.9. What should be the strength of a material initially of fracture toughness (K_{Ic}) 3.0 MPa m$^{1/2}$, containing a critically large defect 50 μm? Assume Y is 1.8. Using either fracture toughness improvement, or defect size control, how could the strength be increased by 50%?

1.10. A 10 cm bar of material of Young modulus 200 GPa, and of mean thermal expansion coefficient 9 MK^{-1}, is heated to 500 °C. By how much will it expand? If the bar were constrained between two blocks, what internal compressive stress would develop?

Selected reading

Davidge, R. W. (1980). *Mechanical Properties of Ceramics*. Cambridge: Cambridge University Press.

Lee, W. E. and Rainforth, W. M. (1994). *Ceramic Microstructures: Property Control by Processing*. London: Chapman and Hall.

Rahaman, M. N. (2008). *Sintering of Ceramics*. Boca Raton, FL: CRC Press and Taylor and Francis.

Reed, J. S. (1995). *Principles of Ceramics Processing*, 2nd edition. New York: Wiley Interscience.

References

Adamson, A. W. (1982). *Physical Chemistry of Surfaces*, 4th edition. New York: John Wiley.

Allen, T. (1997). *Particle Size Measurement*, 5th edition. London: Chapman and Hall.

Atkins, P. W. and de Paula, J. (2006). *Physical Chemistry*, 8th edition. Oxford: Oxford University Press.

Boyd, A. J. and Mindess, S., Eds. (1992). *Cement and Concrete; Trends and Challenges*. Columbus, OH: American Ceramic Society.

Brook, R. J., Ed. (1996). *Materials Science and Technology: Vol. 17A, Processing of Ceramics*, Part I. Weinheim: VCH.

Cambier, F. and Leriche, A. (1996). Vitrification. In *Materials Science and Technology: Vol. 17B, Processing of Ceramics*, ed. R. J. Brook. Part II, pp. 124–44. Weinheim: VCH.

Carniglia, S. C. and Barna, G. H. (1992). *Handbook of Industrial Refractories Technology*. Park Ridge, NJ: Noyes Publications.

Chesters, J. H. (1973). *Refractories: Production and Properties*. London: The Iron and Steel Institute.

Chiang, Y.-T., Birnie, D. III and Kingery, W. D. (1997). *Physical Ceramics: Principles for Ceramic Science and Engineering*. New York: John Wiley.

Davidge, R. W. (1980a). *Mechanical Properties of Ceramics*. Cambridge: Cambridge University Press, pp. 135–9.

Dodd, A. E. and Murfin, D. (2006). *Dictionary of Ceramics [electronic resource]*, 3rd edition. Norwich, NY: Knovel Library.

Eisele, U. (1996). Sintering and hot-pressing. In *Materials Science and Technology: Vol. 17B, Processing of Ceramics*, ed. R. J. Brook. Part II, pp. 84–97. Weinheim: VCH.

Griffith, A. A. (1920). The phenomena of rupture and flow in solids. *Phil. Trans. R. Soc. Lond.*, A **221**, 163–98.

Griffith, A. A. (1924). The theory of rupture. In *Proceedings of the First International Congress on Applied Mechanics*, eds. C. B. Bienzeno and J. M. Burgers. Delft: Waltman BV, pp. 55–89.

Gumbel, E. J. (1959). *Statistics of Extremes*. New York: Columbia University Press.

Hasselman, D. P. H. (1969). Unified theory of thermal shock fracture initiation and crack propagation in brittle ceramics. *J. Am. Ceram. Soc.*, **52**, 600–4.

Herring, C. (1950). Effect of changes of scale on sintering phenomena. *J. Appl. Phys.*, **21**, 301–3.

Iskoe, J. L., Lange, F. F. and Dias, E. S. (1976). Effect of selected impurities on high-temperature strength of mechanical properties of hot-pressed silicon nitride. *J. Mater. Sci.*, **11**, 908–12.

Kingery, D. W., Bowen, H. K. and Uhlmann, D. R. (1976). *Introduction to Ceramics*, 2nd edition. New York: Wiley Interscience.

Kwon, O.-H. (1996). Liquid-phase sintering. In *Materials Science and Technology: Vol. 17B, Processing of Ceramics*, ed. R. J. Brook. Part II, pp. 101–21. Weinheim: VCH.

Lee, W. E. and Rainforth, W. M. (1994). *Ceramic Microstructures: Property Control by Processing*. London: Chapman and Hall.

Levin, E. M., Robbins, C. R. and Mcmurdie, H. F. (1964). *Phase Diagrams for Ceramists*. Columbus, OH: The American Ceramic Society.

Lide, D. R., Ed. (2005). *CRC Handbook of Chemistry and Physics*, 85th edition. Boca Raton, FL: CRC Press.

Morrell, R. (1989). *Handbook of Properties of Technical & Engineering Ceramics: Part 1: An Introduction for the Engineer and Designer*. National Physical Laboratory, London: HMSO, pp. 105–6.

Nicholson, M. M. (1956). Surface tension in ionic crystals. *Proc. Royal Soc. A, Math. Phys. Sci.*, **228**:1175, 490–510.

Popper, P. (1960). *Special Ceramics*. London: Heywood.

Rahaman, M. N. (1995). *Ceramic Processing and Sintering*. New York: Marcel Dekker.

Rahaman, M. N. (2007). *Ceramic Processing*. Boca Raton, FL: CRC Press and Taylor and Francis.

Rahaman, M. N. (2008). *Sintering of Ceramics*. Boca Raton, FL: CRC Press and Taylor and Francis.

Reed, J. S. (1995). *Principles of Ceramics Processing*, 2nd edition. New York: Wiley Interscience.

Ryshkewitch, E. and Richerson, D. W. (1985). *Oxide Ceramics: Physical Chemistry and Technology*, 2nd edition. Orlando, FL: Academic Press, pp. 550–67.

Shelby, J. E. (2005). *The Science and Technology of Glass*. Cambridge: Royal Society of Chemistry.

Singer, F. and Singer, S. S. (1963). *Industrial Ceramics*. London: Chapman and Hall, pp. 396–400.

Stanley-Wood, R. G. and Lines, R. W. (1992). *Particle Size Analysis*. Special Publication No. 102. Cambridge: Royal Society of Chemistry.

Storms, E. K. (1967). *The Refractory Carbides*. New York: Academic Press.

Weibull, W. (1951). A statistical distribution function of wide applicability. *J. Appl. Mech. – Trans. ASME*, **18**, 293–7.

Worrall, W. E. (1986). *Clays and Ceramic Raw Materials*, 2nd edition. London: Elsevier Applied Science.

2

Porcelain

2.1 Description and history

Porcelain is the term, with origins in sixteenth-century France and Italy (*porcella* or *porcellana*), used to describe a type of thin-walled and translucent (shell-like) ceramic produced using clay as a major ingredient. The first porcelains were made in China, probably around AD 100–200. Figure 2.1 shows an early eighteenth-century Chinese enamelled porcelain vase. It was only at about this time that satisfactory copies were produced in Europe (Rado, 1964; Kingery, 1986; Kingery, 1996). Porcelain would also have been considered a "strong" material, but this of course is a relative term, and the basis of comparison here is the other traditional ceramic materials collectively known as *earthenware*, or *pottery* (Dodd and Murfin, 2006). These materials are also produced using clay as one of the main raw materials, but tend to be weaker, and opaque. In the fully finished state porcelain articles normally have a transparent glaze, and coloured decoration provides additional important aesthetic appeal. Porcelain is one of the traditional ceramics with important structural applications both in the domestic and in the industrial areas. Figure 2.2 is a familiar typical example of modern domestic translucent porcelain.

The standard production method for earthenware articles usually involves a two-stage process. The shaped article (the *green body*) is first heated at a temperature in the region of 1100–1300 °C, to complete the processing, and to achieve the required microstructure and strength (Carty and Senapati, 1998). A coating of powdered glass is then applied to the surface of the article, which in a subsequent lower-temperature stage is melted to form a thin impervious glass film (the *glaze*). Although this two-stage heating process is commonly used for smaller articles of domestic and technical porcelain, the maturing of the body (that is, the development of its microstructure) and the glazing can also be carried out simultaneously in a single-stage process (*once-firing*). This procedure is the standard one used for the production of very large electrical insulators, which are

50 mm

Figure 2.1 An example of *Doucai* Chinese porcelain (coloured enamel on white porcelain) from the reign of Emperor Yongzheng (AD 1723–35). (Reprinted by kind permission of the British Museum.)

shaped, coated with powdered glaze and then completely reacted at temperatures up to ~1300 °C. Simultaneously, melting and spreading of the glaze occurs to form a bonded uniform glass film of thickness of the order of 100 μm. Special high-temperature glazes must be used for porcelain, in contrast to normal pottery glazes, which melt and mature at much lower temperatures (<900 °C). The strength in bend of standard glazed siliceous porcelain is in the region of 125–150 MPa (Bloor, 1970a, 1970b; Brown, 1991) and, as will be seen, the glaze does not simply add aesthetic appeal but also makes a significant contribution to the strength of the material.

Strictly, "porcelain" is not one material, but a family of materials (Rado, 1971; Rado, 1975). Other members of this general class of translucent material are *fine china*, and *vitreous china* (Dodd and Murfin, 2006). In this chapter the single term *porcelain* will be used to represent the general class of translucent, clay-based, ceramic material. The distinctions between the different types of porcelain (such as *hard*, *soft*, *dental*) are subtle, and for the most part will be ignored. The term *siliceous* is applied to porcelains prepared with silicon dioxide (silica), as a

Figure 2.2 A familiar example of modern domestic porcelain ware. (Reproduced by kind permission of Royal Worcester Co. Ltd.)

major ingredient, usually in the form of powdered quartz. This term encompasses most of the traditional types of porcelain. It is also common practice to replace part or all of the quartz by aluminium oxide (Austin *et al.*, 1946); these materials are usually designated *aluminous porcelains*, and merit special attention because of their much higher strength. Porcelain used for electrical insulation is often of this type.

The whiteness of the fired basic porcelain body depends on the purity of the raw materials. Particularly detrimental is the presence of small amounts of transition metal oxides such as iron and manganese. Modern domestic porcelain bodies tend to be very slightly blue/grey, in contrast to bone china, which is often pale cream. Many porcelains used as high-voltage electrical insulators appear coloured; this colouring is in fact usually that of the glaze (often dark brown), and the result of incorporating metal oxides such as manganese oxide used for the control of insulation properties.

2.2 General features of porcelain

2.2.1 Translucency

An immediate question, because of its importance for the early aesthetic appeal of porcelain, is why is porcelain translucent, when ordinary whiteware is not?

The microstructure of porcelain normally consists of a small number (usually two or three) of distinct crystalline phases derived from the starting materials, and a "glass" (there may in fact be several glasses of different compositions – but this will be explained later). The simple basis for the translucency is the very high proportion of glass (which can be 60% or more), the small differences in refractive index between the glass and the crystalline phases, and the low volume of internal void space (that is, *porosity*). The glass phase is naturally transparent, and the extent of scattering of light at glass–crystal and glass–pore interfaces on its passage through the body is relatively small; this results in a high proportion of in-line transmission of light, as opposed to internal reflection and general scattering (Dinsdale, 1976). Porcelain materials are often termed *fully vitrified* (that is, "glassified") indicating the formation of a high proportion (though not actually 100%) of glass during fabrication. It should be noted that "translucency" requires articles to be thin in section, as they normally are in domestic ware and figurines. Large industrial porcelain components are normally opaque, and white; the basic translucency only becomes apparent when thin sections have been prepared.

The next questions are: why does porcelain contain such a large proportion of glass (in contrast to other types of ceramic materials based on clay); what crystalline phases are present; and what factors are responsible for its relatively high mechanical strength? Because fully vitrified porcelain has a much higher intrinsic mechanical strength than ordinary earthenware, domestic articles can be produced with a much smaller wall thickness which enhances their aesthetic appeal. The reasons for the improved strengths of porcelain will be discussed below, but one factor is certainly the lower level of porosity (that is, higher *bulk*, or overall, density). This in turn requires a careful choice of starting materials, and (for a traditional type of ceramic) a very high firing temperature, in order to generate the large volume of liquid, to aid the elimination of the void space, and on cooling to convert to the glass.

2.2.2 Technical applications

The choice of porcelain, with its roots in the world of the aristocracy, and decorative vases, figurines, and tableware, as a case study in a volume dealing with modern structural ceramics developed specifically for technical applications, needs a few words of explanation. There are three immediate reasons for using porcelain. The first is that it does have technical applications. Porcelain, although a later development in the traditional ceramic materials used for structural applications (wine and grain storage, and cooking vessels), and historically used primarily for its aesthetic qualities (particularly when decorated with coloured enamels and glazes), has very important modern industrial applications (Johnson

Figure 2.3 Large porcelain insulators in use in the National Grid. (Reproduced by kind permission of Allied Insulators Group Ltd.)

and Robinson, 1975). Early technical porcelain was used as laboratory-ware, because of its resistance to chemical attack. The first porcelain insulators were used by Werner von Siemens to insulate the telegraph line from Frankfurt am Main to Berlin in 1849 (Liebermann, 2001). Today porcelain is very widely used as an electrical insulator in electricity distribution and transmission systems, including the 275 kV and 400 kV grid lines. A familiar example of the use of porcelain insulators, in high-voltage electrical supply lines, is shown in Fig. 2.3. On a much smaller (though in one sense perhaps wider) scale, porcelain is used in restorative dental work as a replacement tooth material (Kelly *et al.*, 1996; Kelly, 1997). The second reason is that porcelain, a *fully vitrified* material, contains a lot of glass – over 50% in fact. Most structural ceramics also contain glasses of varying compositions, though in very much smaller quantities. It will be instructive therefore to use porcelain to try to identify the specific effects of glass on the properties of the technical type of ceramic, particularly at higher temperatures. The third reason is that porcelain, as an example of the higher-strength, traditional, structural ceramics, provides a useful baseline material with which all the more recent developments during the last 40 years or so of the four "technical" structural ceramics included here can be compared. Porcelain is selected from the range of the traditional (clay and mineral based) ceramics partly because it has the highest strengths in this group of materials, and might therefore be considered, from the standpoint of the mechanical and physical properties important for a structural material, to be the "best" of the group.

The typical, traditional, porcelain can be described in simple terms as a composite material consisting of glass and various crystalline phases in, very approximately, the proportions 3:2 by weight. One major crystalline phase is an aluminium silicate known as *mullite*, of slightly variable composition, but approximately $3Al_2O_3.2SiO_2$. The presence of the glass is an important factor determining the physical and mechanical properties of porcelain, especially so at high temperatures.

2.3 Compositions and production methods

Early evidence for the production of ceramic articles – *pottery* – goes back at least as far as 10 000 BC in Japan (Rado, 1969). Raw materials were readily available: the early pottery was produced simply by digging naturally plastic damp clay from the ground, shaping and drying it, and then heating the articles in simple wood-burning kilns. The temperatures in these kilns might have reached "red heat", or around 900 °C. By modern ceramic production standards this is a very low temperature, but it would have caused the partial decomposition of the mineral constituents of the clay, and more importantly some melting of impurity minerals (silicates and aluminosilicates) which contained sodium and potassium ions. On cooling, at the *glass transition temperature*, T_g (of ~800 to ~900 °C), the liquid silicate becomes a glass. The microstructures would then consist of crystalline particles or grains bonded by a small proportion of glass: the materials would be porous, and mechanically weak, though very much stronger than the dried, but unfired, clay. Strengths (using the standard three- or four-point bend methods) of porous fired whiteware can easily reach 50 MPa, some 10 times greater than those of the well-compacted, unfired, fine powders from which it has been produced (which typically have bend strengths in the region of 1–10 MPa – a hard biscuit). With improvements in kiln technology, and understanding of the function of low melting aluminosilicate fluxes, it became possible to fire materials at much higher temperatures, and to increase the proportion of glass, and hence translucency, in an article. This led to the development, in China in the second and third centuries AD, of the materials that were later to be classified as "porcelains".

Modern technical porcelains, produced by heating at temperatures up to 1400 °C, normally meet internationally agreed standards (International

Electrotechnical Commission (IEC) Specification 60–672). Porcelains for bulk structural applications are subdivided into siliceous porcelains and aluminous porcelains. Porcelains in which feldspar minerals provide the liquid (the *flux*), and containing silicon dioxide in the forms of quartz or cristobalite (the traditional composition), are classed as C110; the aluminous porcelains, containing feldspar and where quartz is partially replaced by aluminium oxide, are classed C120; where aluminium oxide is the principal filler (up to 50% of added aluminium oxide), the class is C130.

2.3.1 Starting materials

Porcelain has traditionally been made with a complex blend of raw materials, using quite a broad range of proportions. Even for one type of material no two manufacturers will use exactly the same formulation. The types of raw material widely used today in the production of standard siliceous porcelain remain essentially unchanged from those which have been used over the last 400 years, though many efforts have been made to modify the starting materials (and to introduce new types of higher-purity material). In all cases the object is to produce a glass-rich and dense material, with a minimum of total porosity, and for technical applications <0.5% of accessible porosity (accessible, that is, to the external environment). The starting compositions usually consist of three types of material (and for this reason the designation *triaxial porcelain* has been applied): each has a separate and distinct function (Iqbal and Lee, 2000). These are usually a clay, a liquid-generating *flux* (which is often a low melting-point feldspathic mineral, an aluminosilicate containing sodium and potassium oxides) and a relatively inert *filler*, which is silicon dioxide (quartz) in the siliceous porcelains, and aluminium oxide (or a mixture of aluminium oxide and silicon dioxide) in the aluminous porcelains (Singer and Singer, 1963).

The proportions of the three components of traditional triaxial porcelain are determined by a number of factors, related to raw material quality, and the final properties needed. Typically, however, the clay, flux, and filler are present in the very approximate proportions 2:1:1 by weight, and this is the assumption which will be used in the following treatment of siliceous porcelain and its properties. Primary considerations in choice of composition are the purity of the raw materials and their particle sizes, the particle and void size ranges in the compacted powder, and the heating-time–temperature schedule to be used. For this reason it is not possible to set out a single precise recipe.

In the discussions of siliceous porcelain that follow, the system will for convenience be simplified to that of the ternary Al_2O_3–SiO_2–K_2O system, based on a flux consisting of an idealised *potash feldspar* (of composition $K_2O.Al_2O_3$.

$6SiO_2$). This system has the additional advantage, in that the phase equilibrium diagram can be drawn in two dimensions, rather than requiring three. The lowest liquid-forming temperature in this ternary system (which corresponds to the use of potash feldspar – mineralogically microcline or orthoclase) is 695 °C (Levin *et al.*, 1979, Fig. 407). In practice, present-day commercial flux systems are more complicated, and contain sodium and other oxides. Some of the earliest fluxes were based on compounds of calcium oxide, such as chalk.

2.3.2 Clay minerals

This review of the production of porcelain and its microstructure starts by examining the raw materials from which it is formed, of which clay is a major component. Other additives will be discussed in subsequent sections. An important feature of a more technical nature is that the term *clay* really describes a class of complex crystalline hydrated aluminosilicate minerals, rather than one specific phase (Worrall, 1986). These minerals have been produced by the weathering of primary igneous rocks such as granites and feldspars, carried by water and deposited in firm beds of material. They are therefore usually in a very fine particulate form, with an average particle dimension typically of the order of a micrometre or so (Grimshaw, 1971). Illustrations of the complexity of these materials are provided by the compositions of three common groups of clay mineral in Table 2.1.

A second distinguishing feature of most clay minerals is the property, in the wet state, of "plasticity". The physical description of a plastic "clay" mineral would therefore be a very fine (approximately 100 nm to 5 µm particle size) powder, the particles of which are weakly bonded by surface adsorbed water, to form an easily shaped mass. The water required to give satisfactory plasticity to a clay depends on the minerals present in the clay, and their particle sizes, but is of the order of 30% by volume. Some moist clays are naturally more plastic than others: pure, white, *china clay* is not very plastic, a *ball clay* (originally cut into large balls) is much more so, and for this reason is often blended with a china clay to improve handling characteristics (Powell, 1996). When the physically adsorbed water is removed from a shaped plastic clay article by low-temperature drying, the fine powder particles become strongly bonded to each other by van der Waals and electrostatic forces, and the result is a strong mass of dried hard clay. This material can then be further shaped using conventional metal machining tools (with dust extraction), before being heated (fired) for several hours in a high-temperature furnace.

All the clay minerals are built up from silicon, aluminium and oxygen atoms, bonded together to form thin sheets: technically they are *layer silicates* (Singer and Singer, 1963a; Grimshaw, 1971). Hydroxide ions, and cations such as those

Table 2.1 *Examples of common clay mineral groups, and their idealised compositions, and compositional ranges.*

Group name	Nominal (idealised) composition
Kaolinite	$Al_2O_3.2SiO_2.2H_2O$ or $Al_2Si_2O_5.(OH)_4$
Montmorillonite	$M_nO_3.4SiO_2.H_2O$ or $Al_{1.67}Mg_{0.5}Si_4O_{10}.(OH)_2$
	(n is 2 for M = Al, and n is 3 for M = Mg)
Illite	$(OH)_4.K_y(Si_{8-y}.Al_y)(Al,Fe)_4O_{20}$ (y = 1 to 2)

Table 2.2 *A typical china clay analysis, expressed as proportions of component oxides, and the loss in weight on heating. The weight loss in any clay is due primarily to the loss of adsorbed and combined water, and the combustion of carbon (with which most clays are contaminated).*

Component	Weight%
Silicon dioxide	47.7
Aluminium oxide	37.2
Potassium oxide	1.1
Sodium oxide	0.1
Calcium oxide	0.1
Magnesium oxide	0.3
Iron^{III}oxide	0.4
Others	0.3
Weight loss at 1025 °C	12.8
Total	100.0

of the Group I (commonly sodium, Na^+, and potassium, K^+) and Group II (commonly magnesium, Mg^{2+}, and calcium, Ca^{2+}) elements may then be incorporated into, or between, the layers. Normally, a clay as mined or extracted is a complex mixture of phases, the nature and proportions of which depend on its origins. Examples of the clay minerals are kaolinite (from the Chinese place-name *Kao-Lin*) and montmorillonite where the layers are hydrogen bonded using the OH^- ions, and the illites (*Illinois*, USA), the layers of which are held together by cations such as potassium. Common impurity minerals are mica particles, iron oxide and titanium dioxide. Silicon dioxide as quartz (the crystalline phase stable at room temperature) is commonly present in significant amounts as a particulate impurity, and must be taken account of in calculations of the phase composition of a porcelain system. For these reasons both the rheology and the high-temperature chemistry of clay are extremely complex subjects. Again, discussion of

Table 2.3 *The compositions of typical siliceous porcelains, expressed as the range of (a) mineral components, (b) the constituent oxide equivalents.*

(a)

Constituent	Nominal weight / %	Range / weight%
Clay	50	42–66
Feldspar	25	17–37
Quartz	25	12–30

(b)

Constituent	Range / weight%
Silicon dioxide	58–73
Aluminium oxide	18–36
Potassium and sodium oxides	1–8
Calcium and magnesium oxides	0–4

the production and properties of porcelain will be simplified, and the mineral kaolinite (with idealised formal composition $Al_2O_3.2SiO_2. 2H_2O$ – which can alternatively be expressed as $Al_2Si_2O_5.(OH)_4$) will be used to represent the class. The chemical analysis of a standard kaolinite, expressed as proportions of component oxides, and the loss in weight on heating (mainly the loss of water, and combustion of carbon), are shown in Table 2.2. Typical compositional ranges for the three ingredients of this triaxial system, and the corresponding chemical analysis expressed as the component oxides, are given in Table 2.3. These tables show that, despite there seeming to be a large quantity of feldspathic flux (~25% by weight), the actual content of the Group IA and Group IIA metal oxides (Na, K, Mg, and Ca) is quite low (~5% by weight). Nonetheless, the amounts are significant for liquid generation at high temperature.

2.3.3 Kaolinite

The basic constituent of all porcelains is a clay (and normally a high proportion of this would be kaolinite). This is essentially a low-cost powder, with the very important property of being readily moulded or compacted to shape in the moist, or even dry, state. Expression of the "size" of a particle with the form of a thin hexagonal plate presents difficulties, but the maximum dimension of a kaolinite crystal is normally in the range of 300 nm to 4 μm, and the thickness 50 nm to 2 μm. The specific surface area (defined in Chapter 1) is of the order of 18–30 m^2 g^{-1}, indicating a very fine powder (an equiaxed kaolinite particle of dimension

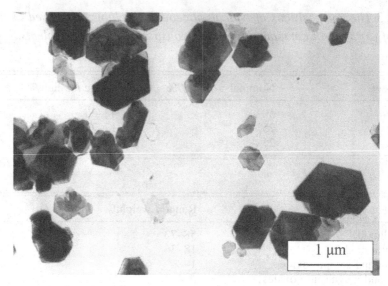

Figure 2.4 Transmission electron micrograph of kaolinite: showing the characteristic, plate-like morphology of the μm dimension crystals.

500 nm would have a specific surface area around 5 m^2 g^{-1}). A transmission electron micrograph (TEM) of kaolinite crystals illustrating the characteristic morphology is shown in Fig. 2.4. Most kaolinite is extracted, or purified, using wet techniques, to provide a damp, plastic mass, the *cake*. The free water can be removed completely by heating at ~120 °C, to leave the clay as a dry powder (now considered for the purposes of this discussion to be kaolinite). At this stage, the addition of water to the dried clay regenerates its plasticity; the loss is therefore fully reversible. Up to ~450 °C dry kaolinite is thermally stable, but further increase in temperature causes breakdown of the crystal structure and the irreversible release of structural hydroxyl (OH) groups as water vapour:

$$Al_2Si_2O_5(OH)_4 = Al_2O_3.2SiO_2 + 2H_2O_{(g)}. \qquad (2.1)$$

The product, of composition $Al_2O_3.2SiO_2$, and often of uncertain crystal structure, is termed *metakaolin*. Heating at higher temperatures still (> ~550 °C) brings about complex chemical reactions. Formation of metakaolin is completed, and at ~950–1000 °C, the decomposition takes place of metakaolin to the aluminium silicate *mullite*, and free silicon dioxide in a very finely divided (and very reactive) state:

$$3(Al_2O_3.2SiO_2) = 3Al_2O_3.2SiO_2 + 4SiO_2. \qquad (2.2)$$

Mullite has slightly variable composition but corresponds approximately to $3Al_2O_3.2SiO_2$ (Schüller, 1964; Klug *et al.*, 1987; McConville *et al.*, 1998). To differentiate this mullite, formed by the decomposition of the metakaolin,

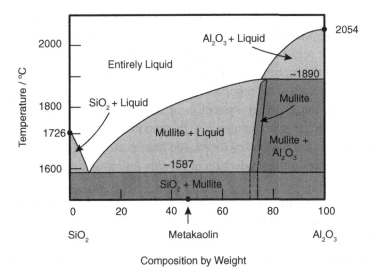

Figure 2.5 The Al_2O_3–SiO_2 binary phase equilibrium system: the liquid forming regions. (After Klug *et al.*, 1987. Reprinted with permission, Blackwell.)

from that developed later in the heating process from other sources, the term *primary* mullite is applied. At these lower temperatures, the silicon dioxide is amorphous and cannot be detected by X-ray diffraction examination, but at higher temperatures crystallisation of the silicon dioxide may occur, normally to cristobalite. Cristobalite formation is more extensive when the reaction temperature exceeds 1250 °C.

Mullite with its high incongruent (with decomposition) melting-point of ~1890 °C is the common constituent of a class of high-temperature refractory materials. Heating a very pure clay mineral (such as kaolinite) in the region of 1000 °C, leads to the production of a weak solid, with no appreciable reduction in porosity compared to the initial compacted powder. The reason for this can be understood by reference to the binary Al_2O_3–SiO_2 phase equilibrium system in Fig. 2.5, showing for specific overall compositions and temperature which phases (and their amounts) are present at equilibrium (Klug *et al.*, 1987). This is strictly a *pseudo-binary* system, because there are really three components, Al, Si, and O, but because the Al-O and Si-O ratios are fixed, there are effectively only two components, Al_2O_3 and SiO_2, and it is treated as a binary system. For simplicity the *pseudo* will be dropped for descriptions of this, and similar, oxide phase equilibrium systems. Kaolinite itself – $Al_2O_3.2SiO_2.2H_2O$ – is of course not shown in this binary diagram which does not allow for the existence of water, but the composition corresponding to metakaolin is located at ~45.9 weight% Al_2O_3. The lowest eutectic composition in this binary system is at ~94 weight% SiO_2, with the eutectic temperature ~1587 °C. This is (at the temperatures used for porcelain production) a dry system. Heat treatment of kaolinite powder at lower temperatures might therefore be regarded, in essence, as a solid state sintering process (Section 1.4), and because the mobility of the atomic species in this aluminosilicate system is low, it is also a very slow process. The refractoriness (that is high-temperature stability) of this simple binary system is of importance in the refractories field. High-purity "china" clay containing a high proportion of kaolinite is used as the basis of a range of "fireclay" refractories, and heat-treated kaolinite powder, known as *molochite*, is used in the production of moulds for casting molten metals (Dodd and Murfin, 2006).

The natural availability of starting materials in such a fine particulate form is unusual. In the cases of the other additives (quartz and feldspars) used in porcelain, and the powders for the structural ceramics to be discussed in later sections, the raw materials are not found as fine powders. It is therefore necessary to use milling, or more expensive chemical extraction routes, to produce starting powders of the micrometre dimension needed. Small amounts of binding agents (such as water, or organopolymeric materials) are usually incorporated to provide the coherence and strength necessary to allow compacted powders to be handled or machined. These organic additives, like the water and carbon in the clay, are burned off during the early heating stages.

2.3.4 Firing

The high-temperature heat treatment of a shaped powder component has traditionally been referred to as *firing*, naturally so because it would have used wood or coal. Electric and gas-fired furnaces are now used exclusively (apart from the occasional production of art pottery and archaeological simulations) in the manufacture of porcelain, but it is difficult to replace the term by "sintering", because much of the high-temperature process depends on chemical reactions, and the flow of liquids. *Reaction sintering* (Section 1.5) is therefore rather more appropriate – for the reason that the final constituents are formed simultaneously with densification – but it is not often used in the context of porcelain. The term *firing*, meaning high-temperature processing, will therefore be retained in this chapter. In industrial processing the concept of *heat-work* is often applied. This rather imprecise term equates to the total thermal energy available to the porcelain system during the firing cycle. It is useful in expressing the fact that it is the integrated temperature–time cycle which is important for determining the extent to which high-temperature chemical (and physical) processes are able to approach equilibrium. There may be the same outcome (the degree of vitrification for example) at a lower temperature with prolonged heating, as in a shorter time at a higher temperature (or not, depending on the system and conditions applied). The chemical processes taking place can be understood by examining the appropriate phase equilibrium diagrams (Bergeron and Risbud, 2006). These show which phases ought to be present at equilibrium – for kaolinite, mullite and silicon dioxide – and provide a guide for the journey of the starting powder blends towards their equilibrium state.

2.3.5 Heating schedules

In order to understand the types, and proportions, of the phases present in a fired porcelain body, it is necessary to examine what happens when the primary particulate constituents are heated. This can most easily be done by considering one simple clay mineral, kaolinite, and the phase equilibrium system formed by an idealised flux, potash feldspar (though no ingredients are chemically or phase pure, and commercial systems will be very much more complex). For the moment the porcelain will be assumed to be the traditional, siliceous, type.

Porcelain firing schedules differ in detail from manufacturer to manufacturer (depending for example on body composition, and the type of component – and it is often necessary to fire different compositions at the same time, so compromises have to be made), but have important features in common. Modern porcelain is normally produced by heating at temperatures mainly in the range 1150–1300 °C, but the exact conditions depend on the body composition, and the type of component being produced. The maximum temperature would be experienced for times of the order of 1 hour or more, depending on the need for heat to penetrate to the centre of the component, followed by slow cooling to room temperature. This is therefore not an isothermal process (Funk, 1982). Because of the need to form liquids the time spent at temperatures above 600 °C is the most important. Other critical features are the rates of heating and cooling, and the times at maximum temperature and any intermediate holding stage. The temperature can be a linear function of time, or the rises can be a series of steps, by control of the furnace temperature zones. A complete cycle of heating to, and cooling, from ~1250 to 1300 °C is likely to require between 5 and 100 hours, depending on the type of porcelain, and the size of the furnace and the type and size of component. It is in this period that thermally activated processes occur, and the system moves in the direction of chemical and physical equilibrium, though these will not be fully attained.

The important point here is that the final product, described in terms of the nature and quantities of the components (crystalline or glass), is not that of the starting powder mixture: those components had to a very large extent (though not entirely) lost their individual identity (physically and chemically) during the firing process. This is generally not the case with the other structural ceramics to be discussed, and where the major phase is usually largely unchanged during the processing of the powders into the ceramic. Because chemical changes take place during firing and vitrification, the phase composition of the product is very (in some types totally) different from that of the starting powders. Porcelain provides an extreme example, but the chemistry of all ceramic production processes is generally complicated, not least because it involves very high temperatures, and it can be difficult to model in simple terms.

2.3.6 *Fluxes*

The first forms of pottery (and the modern clay-based materials known collectively as earthenware) were not translucent because translucency requires the presence of a large volume of glass, and a low level of internal porosity. This implies extensive melting of the mineral ingredients, so that sufficient liquid is formed to fill most of the interparticle voids in the compacted powder, leaving only a small volume of internal porosity. In turn this requires much higher firing temperatures (~1100 °C and higher) than would have been available in the early kilns. These temperatures are considerably higher than the eutectic and liquidus temperatures in the ternary system K_2O–Al_2O_3–SiO_2, which can be regarded as forming the basis of common porcelain systems, and useful in explaining the high-temperature chemistry involved in the production of porcelain (Levin *et al.*, 1979; Figs. 407, 501, and 786). Under these conditions considerable volumes of liquid are generated, producing vitrification, and provide the translucency. On cooling the liquid silicate does not fully crystallise (though mullite with characteristic needle morphology usually crystallises from it), and it remains as a glass.

Examination of the simple ternary phase equilibrium diagram for the K_2O–Al_2O_3–SiO_2 system (Levin *et al.*, 1979, Fig. 407) shows that there are a number of low-melting ternary eutectic compositions, one at a temperature as low as 695 °C. The extent of the compositional region containing a high proportion of silicon dioxide, which is entirely liquid at 1000 °C, is shown in a much-simplified version of the ternary diagram, in Fig. 2.6. As the temperature is raised this liquid region progressively creeps further to the right, towards the Al_2O_3–SiO_2 axis. In reality therefore sufficient fluid and wetting aluminosilicate liquid (which will become a glass on cooling) will be generated in an impure clay at temperatures in the region of 700–1000 °C to give some bonding and the development of strength. These materials can therefore be regarded as coming with their own "built-in" liquid phase sintering aids. In contrast, the structural ceramics produced from very high-purity raw materials require the deliberate addition of

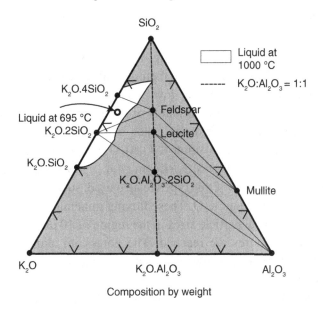

Figure 2.6 Simplified phase equilibrium diagram for the $K_2O–Al_2O_3–SiO_2$ ternary system: silica-rich compositions entirely liquid at 1000 °C, and the position of the lowest melting eutectic of 695 °C. (After Levin *et al.*, 1979. Reprinted with permission of The American Ceramic Society, 1964. All rights reserved.)

oxides or silicates if adequate liquid phase sintering is to be possible. The amounts of flux reach 20–30% by weight for electrical porcelains, and more, up to ~80%, in the case of dental porcelains (Singer and Singer, 1963).

It can easily be shown that the reduction in porosity in a compacted powder, simply as a result of liquid formation, is approximately equal to the volume of liquid formed (illustrated in Fig. 1.20 in Chapter 1). Further loss of porosity will occur as the particles rearrange. For complete elimination of the 40–50 volume% of void space in a compacted clay powder, the volume of liquid formed must therefore also be 40–50%, and this cannot be realised solely through reliance on the natural impurities in a clay, particularly with purer and whiter clays. A second ingredient is therefore needed, with the specific function of generating more liquid: this is the flux – for historical reasons often termed *stone*, though now more likely to be a high-purity feldspar.

Most compacted clays develop strength after relatively low-temperature firing. This is because either the clay mineral itself may contain cations such as sodium, potassium, magnesium and calcium interposed in, or between, the silicate layers, or it is impure, and contains small amounts of other minerals. The phase equilibrium diagrams describing the high-temperature behaviour of these materials are therefore very much more complex. A crucial feature is the presence of the components Na_2O and K_2O, which give very low melting-point phases and eutectic temperatures.

A wide range of aluminosilicate fluxing minerals is available, but all have a common feature, in that they contain in their crystal structures the oxides closely associated with the formation of very low eutectic temperatures, primarily K_2O and Na_2O, and sometimes smaller amounts of CaO and MgO. Examples of very commonly used fluxes are the feldspars, such as potash feldspar (*orthoclase* or *microcline*), of theoretical composition $K_2O.Al_2O_3.6SiO_2$, and the sodium oxide equivalent (*albite*), $Na_2O.Al_2O_3.6SiO_2$ (Levin *et al.*, 1979, Fig. 501). Other more complex feldspathic minerals of this type are also used, such as *nepheline syanite*, which contains both Na_2O and K_2O. These fluxing minerals are normally ground to fine powders, with mean particle sizes in the region of 10 μm, and react quickly as the eutectic temperatures are reached. The potassium aluminosilicate (*feldspathic*) liquid is of relatively low viscosity at temperatures above ~1000 °C. Very early European porcelains were prepared with fluxes based on the higher-melting anorthite – $CaO–Al_2O_3–SiO_2$ – system (see Chapter 3, Fig. 3.7), obtained through adding calcium carbonate (*chalk*).

Although the pure feldspars have fairly high melting-points, much lower melting eutectics are formed in the presence of silicon dioxide, which is readily available to the developing porcelain system both from decomposition of the kaolinite (and similar minerals of the clay group), and from natural impurity quartz in the clay. Potash feldspar in the pure state melts *incongruently* (that is, with decomposition), with the formation of a liquid containing about 40% of silicon dioxide, and leucite (of composition $K_2O.Al_2O_3.4SiO_2$) at ~1150 °C (Kingery *et al.*, 1975). However, in contact with silicon dioxide, liquid is formed at a considerably lower temperature, ~990 °C. This feature is illustrated by Fig. 2.7, which is a very small section of this system, the leucite–feldspar–silicon dioxide subsystem. It is important because it shows how liquid can be produced at a very low temperature (~990 °C) from apparently quite refractory materials. This can happen at the contact points between feldspar particles and the very fine decomposed kaolinite (metakaolin) crystals which contain a high proportion (~36% by weight, ~47% by volume) of silicon dioxide.

To illustrate more fully the important chemical processes taking place during firing this simple mixture (and ultimately the complete porcelain system), it will be useful to start by considering a simplified model system, consisting only of feldspar and kaolinite. To do this, a section through the full $K_2O–Al_2O_3–SiO_2$ ternary diagram with temperature as the vertical axis (not actually shown) is used (Fig. 2.8). This depicts the situation at 1200 °C, and the phases present. At this temperature the potassium feldspar has disappeared, to be replaced by a silicon dioxide-rich aluminosilicate (feldspathic) liquid, with extreme compositions

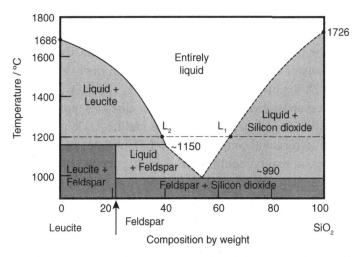

Figure 2.7 The leucite–feldspar–silicon dioxide join in the K_2O–Al_2O_3–SiO_2 system, which passes through potash feldspar. Leucite has the composition K_2O. $Al_2O_3.4SiO_2$; feldspar $K_2O.Al_2O_3.6SiO_2$. The liquid compositions L_1 and L_2 correspond to those shown in Fig. 2.22. (After Levin *et al.*, 1979. Reprinted with permission of The American Ceramic Society, 1964. All rights reserved.)

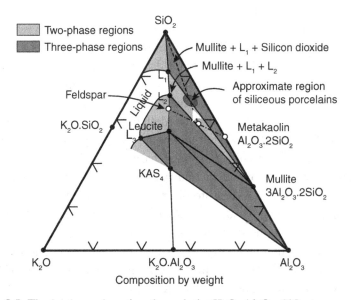

Figure 2.8 The isothermal section through the K_2O–Al_2O_3–SiO_2 ternary system at 1200 °C: the location of the siliceous porcelains. (After Kingery *et al.*, 1975. Reprinted with permission of John Wiley & Sons, Inc.)

denoted by the symbols L_1 and L_2. The liquid of composition L_1 can coexist with silicon dioxide (or mullite); liquid L_2 can coexist with leucite (or mullite). The composition of the liquid in equilibrium with mullite alone, the two-phase region, and the proportions of liquid and mullite, depend on the overall composition of the system (that is, the proportions of K_2O, Al_2O_3 and SiO_2), and vary over the range between compositions L_1 and L_2. The siliceous porcelain system lies mostly in the mullite–L_1–L_2 triangle, near the mullite–L_1 join. This means that it consists of a liquid (of composition close to L_1), mullite, and (probably) silicon dioxide. However, in reality, equilibrium is in practice never attained during the production of porcelain (not enough time is allowed).

The chemical reactivity of the flux is typical of ceramic systems; the liquid is not simply an inert solvent for one of the major phases. It is often an active participant in a set of complex chemical reactions, either supplying, or absorbing, ionic species as sintering proceeds. The phase crystallising from the flux is not always the phase which dissolved in it, as the next section describing some of these processes for the porcelain system shows.

2.3.7 Ion diffusion

The section across the full K_2O–Al_2O_3–SiO_2 ternary diagram at 1200 °C (Fig. 2.8) explains the chemical changes taking place when a mixture of two of the idealised ingredients of porcelain (kaolinite and feldspar in the proportions 2:1 by weight) is heated to 1150 °C and above. The diagram shows that for this composition at 1200 °C, at equilibrium, neither metakaolin nor feldspar exists; the feldspar has been replaced by a (feldspathic) liquid with a composition somewhere between L_1 and L_2, but nearer to L_1, and the metakaolin has turned into a mixture of silicon dioxide and mullite. The composition of this mixture (2:1, kaolinite to feldspar) lies on the dotted line, about half-way between metakaolin and feldspar. To arrive at this point the feldspar has had to lose some potassium oxide (as the mobile K^+, and charge balancing O^{2-} ions), and the kaolinite gains potassium oxide. In feldspar the Al_2O_3/SiO_2 molar ratio is 1:6, and in metakaolin it is 1:2 (Al/Si atomic ratios 1:3 and 1:1). Because the glass network-forming species, Al and Si, are relatively immobile, the Al_2O_3/SiO_2 ratios in these two regions remain approximately constant. These ratios then provide the system with a "memory" of its previous existence, and the Si and Al atoms (and their oxygen ions) can be thought of as forming a kind of lattice, or framework, within which transport of the K^+ and O^{2-} ions occurs. This process is shown schematically in Fig. 2.9.

The two regions are remembered as the *feldspar relicts*, and the *kaolinite relicts*. In fact they are physically, as well as chemically, easily recognisable in

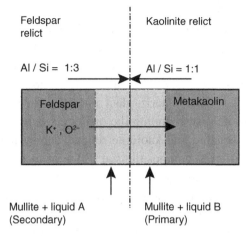

Figure 2.9 A schematic diagram of the reaction between potash feldspar and metakaolin, indicating the diffusion of potassium and oxide ions from feldspar to the metakaolin.

the final fired material. On cooling from production temperatures, further changes take place. These include the conversion of the liquids to glasses at the glass transition temperature, and the crystallisation from the feldspathic liquid of more mullite (the *secondary mullite*, to distinguish it from the *primary mullite* formed by metakaolin decomposition).

2.3.8 Fillers

The full triaxial formulations contain a third ingredient, the filler, which in the siliceous porcelains is usually quartz. This is obtained by crushing and milling flint, which is a natural microcrystalline form of quartz. Filler particles can be relatively coarse, with sizes of up to ~50 μm, though they are usually much finer than this. The incorporation of silicon dioxide can be thought of as pulling the composition of the two-component mixture upwards, towards the SiO_2 corner of the ternary diagram (Fig. 2.8), and along the second dotted line shown. This diagram shows that the area expressing the porcelain compositions, in terms of the three component oxides, lies very close to the boundary line separating the silicon dioxide–L_1–mullite, and mullite–L_1–L_2 compositional triangles. This means that crystalline silicon dioxide may or may not be stable in this triaxial system at 1200 °C, depending on the exact proportions of the three major ingredients used. In practice extensive reaction (30% or more) of the quartz takes place during firing, and in particular that of the finer, higher surface area, particles. This can be thought of as the solution of quartz in the feldspathic liquid. However, because silicon dioxide is not very mobile in the liquid at temperatures of ~1200 °C, the disappearance of the quartz is more likely to result from the diffusion of potassium ions (and charge balancing oxide ions) into the quartz crystal lattice, with simultaneous formation of a liquid potassium silicate. Chemical analyses of residual quartz–glass interfaces in the porcelain confirm that the glass immediately surrounding the quartz contains more silicon dioxide then those in the feldspathic and kaolinite relicts. The system overall therefore contains glasses of different compositions at different locations.

The viscosity of the silicate liquid is very sensitive to composition, as shown in the next section. When the finer quartz particles react with the feldspathic liquid, there is a slight increase in the total volume of liquid present, but because the liquid has a high SiO_2 content, it is of much higher viscosity than the basic feldspathic liquid. The viscosity of the feldspathic liquid is also increased, because it has been depleted in K_2O. An important additional function of the quartz therefore is to increase the viscosity of the liquids generated by the fluxes. If the quartz were not present, there is a danger that the large volume of low-viscosity liquid formed (although needed to bring about densification of the porcelain) would lead to unacceptable dimensional instability of the component, and unwanted distortion during firing of a large component through sagging under its own weight. In effect, the quartz provides a form of negative feedback, to make temperature control less critical for mechanical stability during firing.

Under the conditions of temperature and time allowed in practice for the formation of porcelain, full system equilibration, that is, complete rearrangement of the constituent Al, Si, K, and O atoms (or ions), is not possible (if it were the relicts would disappear). The microstructure strongly retains a memory both of the chemical compositions, and of the original morphologies, of the crystalline particles in the starting powder mixture. Porcelain provides an extreme case, but in principle this applies to all ceramic systems. The process of firing has been referred to as the "chemistry of arrested reactions". It can equally well be said to consist of arrested physical processes, which in this case would be the homogenisation of mullite crystal morphology and size, and the redistribution of the liquid.

2.3.9 *Liquid viscosity*

A distinction can be made between the viscosity of a liquid silicate, and the effective viscosity of the body (which can be considered to be its creep resistance). Dispersion of a powder in a fluid increases overall (that is, the system) viscosity. There are semi-empirical equations for the viscosity of very dilute suspensions of particles, for example the Einstein equation (Einstein, 1906) for the viscosity (η) of a suspension of small uncharged particles of fractional volume concentration c in a liquid of viscosity η_o gives

$$\eta = \eta_o(1 + kc) \tag{2.3}$$

where k is ~2.5 for spheres, when c is <0.02. However, the developing porcelain system, with a high concentration of solid particles (c will be of the order of 0.4) and with extremely viscous liquids, is much more difficult to model because particle–particle and particle–fluid interactions become very important. There are no satisfactory expressions for the viscosity of highly concentrated suspensions in high-viscosity fluids, though empirical equations such as (2.4) have been used for concentrated aqueous suspensions, where a is a constant and when c can be as high as 0.5 (Bergstrøm, 1994):

$$\eta = \eta_o \exp\left(\frac{ac}{1 - kc}\right) \tag{2.4}$$

(for c values of ~0.5 all the particles will be practically in contact with each other). Because the kaolinite relicts contain a high concentration of primary mullite crystals, they have very high viscosities, and effectively are solid (though at high temperature slightly plastic) particles. The deformation of a porcelain at high temperature is therefore really controlled by the (much lower) viscosity of the feldspathic liquid of the feldspar relicts, in which are dispersed both the secondary mullite needles, and the more equiaxed primary mullite crystals in the kaolinite relicts.

2.4 Physical aspects of porcelain production

So far attention has been focused on the chemistry of the firing process, the loss of starting phases with the development of important liquids, and completely new crystalline phases. The physical aspects of firing now need examination. The densification process can be considered to be a form of liquid phase sintering, because of the rôle of the liquid phase in assisting particle movement, allowing some particle shape adjustment, and the filling of the porosity between starting powder particles (or their relicts). As illustrated in Chapter 1, *vitrification* represents a special case of liquid phase sintering, where the volume of liquid generated in the porcelain system is quite sufficient to fill all the voids of the compacted powder and therefore allow the system to attain practically full density (Fig. 1.20). There is no necessity for the particle shape change characteristic of normal liquid phase sintering, though in reality, as has been seen, considerable chemical changes involving all the starting powders are taking place. The phases present in the fully sintered material are for the most part not those present in the starting powder mixture.

2.4.1 Shrinkage

The first processes observed during the early stages of heating, at temperatures between 100 and 450 °C, are the evolution of gases (primarily water vapour and carbon dioxide), and a loss of weight. This is the result of dehydration of the clay minerals, and the combustion of organic binders used in the body and the natural carbonaceous material in the clay, which can be present in amounts of 10% or more, all recorded as the *loss on ignition*. The first signs of sintering shrinkage (and the loss of specific surface area of the powders) occur at a temperature of around 750 °C. This is much lower, as pointed out above, than the eutectic temperatures expected of the simple feldspar–metakaolin binary system, and is the consequence of the many oxides, present in small amounts in the raw materials. Marked shrinkage commences at ~900 °C, when the bulk of the liquid forms rapidly through reactions involving the added feldspathic fluxing agents, and mineral impurities in the clay such as mica. Maximum shrinkage, and bulk density, is then reached within a few minutes. The firing schedule is designed to ensure the complete closure of porosity. There should be 0.5% or less of accessible porosity (to prevent water penetration into the fired body), and a minimum of total porosity in order to maximise strength. A target figure for total porosity would be around 5%, corresponding to a bulk density in the region of 2.4 Mg m^{-3}. Measurements of the volumes of accessible, and total, porosity show that both decrease together, with the sealing of accessible pores taking place just before the maximum density is attained. Because the initial packing density of the starting powder mixture will not be much higher than 50% of theoretical, attainment of ~95% density (~5% residual porosity) requires a volume shrinkage of ~45%, which corresponds to a linear shrinkage of ~15% (as can easily be shown by considering a shrinking cube).

2.4.2 Bloating

A continued rise in temperature beyond the point of maximum density commonly leads to a reversal of shrinkage, and an expansion of the body, termed *bloating*. This is caused by an increase in the volume of internal closed (*inaccessible*) porosity from which gas cannot quickly escape. The reasons for bloating have been debated, but a generally accepted explanation is that it is the result of gas release into the closed pores by the decomposition of small amounts of

1	Dehydration of kaolinite to metakaolin
2	Decomposition of metakaolin to primary mullite and silicon dioxide
3	Initial liquid formation by the feldspar–silicon dioxide reaction
4	Initial rearrangement of unreacted particles, melting of particles, loss of void space and overall shrinkage
5	Diffusion of K^+ and O^{2-} ions from feldspar relicts into the kaolinite relicts and quartz crystals
6	Continuing formation of liquid, with loss of void space; approach to local equilibrium
7	Crystallisation of secondary mullite in the feldspar relicts
8	Slow solution of aluminium oxide particles
9	Conversion of liquids to glasses during cooling.

Figure 2.10 A summary of the main physical and chemical steps in the formation of porcelain.

impurities, such as the unstable iron oxide Fe_2O_3 (oxygen), or carbonates (carbon dioxide) and sulphates (sulphur trioxide):

$$Fe_2O_{3(s)} = 2FeO_{(s)} + \tfrac{1}{2}O_{2(g)} \qquad (2.5)$$

$$CaSO_{4(s)} = CaO_{(s)} + SO_{3(g)}. \qquad (2.6)$$

These reactions are facilitated by the subsequent reactions of the oxides to form silicates. When it is remembered that 1 mole of gas at STP (*standard temperature and pressure*) occupies 22 414 cm^3, or about 104 litres at 1000 °C, it is clear that even ppm (parts per million) quantities of gas-generating impurity can cause large expansions in closed pore volume. The joining of small closed pores to form a lesser number of large pores, a purely physical process, will also lead to a small increase in closed pore volume. This is because the equilibrium pressure of gas in a pore is determined by the pore radius (Equation (1.4) in Section 1). One larger pore forming by the fusion of n gas-filled pores, with the total gas volume considered to be maintained constant, will initially contain gas at too high a pressure for the size of the new pore. The pore must then expand to its new equilibrium radius with a corresponding equilibrium gas pressure, and there is an increase in the volume of the pore. On the assumption that r is sufficiently small (about 1 μm for a liquid aluminosilicate with $\gamma \sim 300$ mN m^{-1}), so that $p_i \gg p_o$, the application of Boyle's law (for a fixed amount of gas at constant temperature, $p_1 V_1 = p_2 V_2$) gives the fractional volume increase as $(n^{1/2} - 1)$, or about 40% for $n = 2$.

A simplified summary of the main chemical and physical processes occurring during the firing stages is shown in Fig. 2.10.

2.5 Microstructure

The important feature of porcelain is that the composition, expressed in terms of the types and quantities of the major phases present, has been almost totally transformed during firing (Iqbal and Lee, 1999, 2000). These changes are the consequence of the attempt of the porcelain system to reach equilibrium during heating. However, because of the relatively low temperatures and restricted times allowed, attainment of equilibrium is incomplete. Phase changes are common during the production of the structural ceramics, but porcelain presents an extreme example. X-ray diffraction analyses of porcelain made at room temperature show that the main crystalline phases are mullite and α-quartz. The proportions vary with composition and firing schedule, but a typical X-ray diffraction analysis of a porcelain fired at ~1250 °C shows ~20% each of mullite and quartz, confirming that glass is the main phase. With higher firing temperatures there can be much smaller amounts of α-cristobalite, and some of the secondary mullite may disappear.

The porcelain microstructure is complex and generally on a very fine scale. Standard low power light microscopy is of limited usefulness, because of the small sizes of the mullite crystals, particularly the primary mullite in the kaolinite relicts, and the variations in composition that occur within the glass phase. Useful examinations of a fired porcelain microstructure can be made using scanning electron microscopy with a facility for chemical analysis. A typical example, Fig. 2.11, shows clear features corresponding (in terms of size and morphology) to the original feldspar and kaolinite particles. These have the expected Al/Si atomic ratios (Fig. 2.9), and both contain K^+. Transmission electron microscopy is required if the finer scale detail of crystal growth, and variations in composition, are to be resolved. Both types of relict contain mullite crystals, of distinctive morphologies: in the kaolinite relicts the crystals of primary mullite are small and roughly equiaxed; in the glassy feldspar relicts there is a smaller proportion (~10% by volume) of larger, and needle-like secondary mullite crystals. Overall, most of the glass is in the feldspar relicts: the larger proportion of mullite is in the kaolinite relicts. However, because of the more fibrous (or whiskery) nature of the secondary mullite, this type has been considered to be more desirable from the point of achieving maximum strength. Close examination of the two types of relict shows that the compositions of the glass phases present in each also differ slightly: the feldspathic glass has a slightly higher K^+ content.

2.6 Physical and mechanical properties

The following discussion outlines the physical and mechanical properties of importance for porcelain as a structural ceramic (Bloor, 1970a: Bloor,

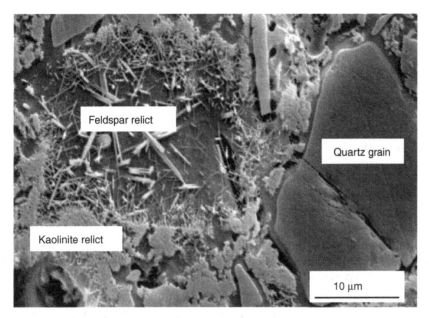

Figure 2.11 A scanning electron micrograph of a typical siliceous porcelain showing the major features: feldspar and kaolinite relicts, the remnant of a large quartz crystal, and the primary (smaller) and secondary (larger) mullite crystals.

1970b). Basic links between microstructure, the properties of the constituents, and the overall, resulting, properties of porcelain are examined. The principles introduced here are generally applicable to all ceramic materials, and this section will need to be referred to when the properties of the other structural ceramics are discussed later.

2.6.1 Constituents of a composite

Siliceous porcelain can be regarded as a particulate–glass matrix composite. The primary crystalline constituents of the microstructure of a siliceous porcelain, mullite and silicon dioxide (commonly in the form of quartz, but sometimes cristobalite), both have their own sets of key properties, a selection of which is shown in Table 2.4. In particular, they are of low density, high stiffness (high modulus), and refractory. Mullite and silicon dioxide (as cristobalite) have high melting-points (mullite ~1890 °C, and silicon dioxide 1726 °C). It might be expected that these properties would have a strong influence of those of porcelain. Put another way, it might be expected that the properties of porcelain would be some form of mean of those of the constituents. However, this view is too

Table 2.4 *Selected properties of the main constituents of porcelain.*

Component	Young modulus at 25 °C / GPa	Melting-point / °C	Density / Mg m^{-3}
Aluminosilicate glass	60–70	–	~2.3
Mullite	220	~1890	~3.2
Quartz	97 (*c*-axis)	1726*	2.6
	76 (*a*-axis)		
Aluminium oxide	390	2054	4.0
Product porcelain	~70	–	~2.5

* as cristobalite.

simplistic. These two crystalline phases form only around 40% of the material making up the porcelain; an aluminosilicate glass is the major phase. The properties of the glass strongly influence high-temperature behaviour. Glass, either as a major and interconnecting phase (matrix), or merely as a thin grain-boundary film, is also important for the high-temperature properties of poly-crystalline ceramics. Its influence is seen strongly in porcelain because there is so much of it.

2.6.2 Glass properties

Glass is a subject in its own right, but a brief discussion of the essential properties of the glass phases making up the microstructure is needed, if the bulk properties of porcelain are to be understood. A glass, an amorphous material, always has a tendency to convert to a crystalline material, in order to reduce its energy. However, silicate glass can be very slow to crystallise (window glass can last for centuries). A crystalline phase, with its regular atomic repeat patterns of the crystal lattice, converts to a liquid at its melting-point; the liquid normally has a disordered arrangement of atomic species. In contrast, a glass does not melt; there is no abrupt change in property with changing temperature, only a gradual change in viscosity. An aluminosilicate glass changes from a relatively soft, extremely viscous fluid, above the glass transition temperature (T_g), and becomes a brittle, but potentially very strong, solid below it. Another essential difference between the liquid and the glass is that the liquid is in a state of thermal equilibrium, the glass is not, because atomic movement at low temperatures is too sluggish for the structure to be able to follow a change in temperature. Over a short temperature range below the theoretical melting-point the atomic species of the silicate liquid continue to be able to adjust to the decreasing temperature (though without

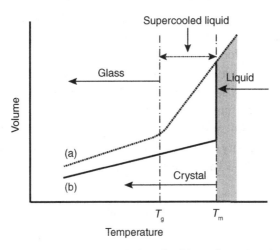

Figure 2.12 Molar volume changes during liquid cooling: the alternative pathways (a) glass formation and (b) crystallisation. T_m is the melting-point, and T_g the glass transition temperature.

actually crystallising); the term *supercooled* is applied to this liquid state. This process is associated with changes in molar volume, or density, which can be measured. With further reduction in temperature, the atoms become unable to accommodate the temperature change, and the term *glass* is applied. The material now has the characteristic properties of a "solid": it is load-bearing up to the point when fracture occurs. It will also have decreased thermal expansivity. These changes of state are reflected in the way the molar volume changes with temperature, shown in Fig. 2.12.

Here the options are for the liquid to crystallise (sharply at the melting-point), or to continue to cool, initially as a supercooled liquid still in *thermal* equilibrium with its environment, then as a glass at lower temperatures. The supercooled liquid and the glass both have lower densities than the crystalline form (α-quartz ~2.65 Mg m^{-3}, silica glass ~2.2 Mg m^{-3}, or ~80%).

2.6.3 Density

The bulk density of siliceous porcelain is usually ~2.45 Mg m^{-3}, corresponding to a true solid density of ~2.55 Mg m^{-3}. This is a very low value in comparison with the values for metals (the low-density metal aluminium has a density of 2.3 Mg m^{-3}; iron has a density of ~8 Mg m^{-3}), and low even for a ceramic. The primary reason is the large volume of aluminosilicate glass in the system, with its own bulk density in the region of 2.3 Mg m^{-3}. The reason in turn for the low density of the glass, besides the low atomic masses of the constituent atoms (A_r Si 28.1,

Na 23.0, K 39.1, O 16.0), is the open structure of the glass network: it is calculated that a silica glass network contains effectively ~46% of interatomic void space. Within this network are located the network modifying cations and their associated O^{2-} ions. Silicate (and aluminosilicate) glasses commonly have densities in the range 2.3–2.4 Mg m^{-3}. The more highly ordered, though still not "close-packed", crystalline α-quartz structure has a density of ~2.65 Mg m^{-3}. In contrast, the close-packed lattice of α-Al$_2$O$_3$, corundum, containing the lower atomic mass Al (27.0), has the much higher density of 3.99 Mg m^{-3}. The other main component, mullite, has a theoretical density of ~3.16 Mg m^{-3}.

2.6.4 Viscosity

The viscosity of a glass provides a measure of its ability to flow and undergo permanent deformation under load (commonly termed *creep*, in a crystalline material). Viscosity is a function of composition and temperature. The influence of temperature on viscosity (η) follows an approximately exponential function, and can be expressed by equations of the form:

$$\eta = AT\exp\left(\frac{B}{T}\right) \tag{2.7}$$

where T is the absolute temperature, and A and B are functions of the composition of the glass. A is related to B, by the expression:

$$A = \exp[-(0.269B + 1.167)]. \tag{2.8}$$

The term B is not itself the viscosity but represents an apparent activation energy relating viscosity to temperature ($B = \Delta E/R$, where R is the gas constant), that is, the sensitivity of viscosity to temperature change. One way of showing the empirical relationships between the term B and composition is shown in Fig. 2.13 (Urbain *et al.*, 1981; Urbain *et al.*, 1982). Using these relationships, the viscosity at 1200 °C of a potassium aluminosilicate liquid of composition corresponding to the analysed mean for a typical siliceous porcelain would be of the order of 1 GPa s. This is fairly high, but nonetheless would permit significant deformation of the material under modest loads (10 N) on a time-scale of a few minutes (of significance for the distortion of a large component during firing). Slow deformation under load is termed creep, and in porcelain it is the consequence of the flow of the viscous feldspathic glass. At temperatures above 1000 °C creep under loads corresponding to a few hundred grams can become severe, as illustrated for a standard siliceous porcelain at 1035 °C in Fig. 2.14 (Porte *et al.*, 2004). Under these conditions 1–2% strain is produced on a time-scale of an hour or so.

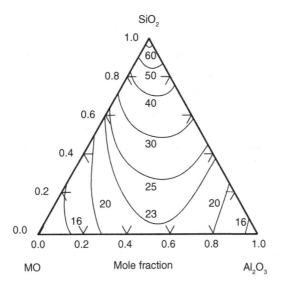

Figure 2.13 Empirical relationship between liquid composition and viscosity, expressed in terms of the coefficient B. (After Urbain *et al.*, 1981.)

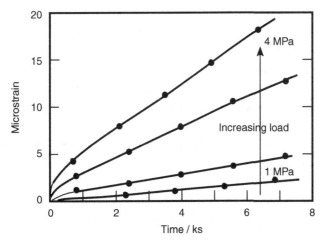

Figure 2.14 The creep behaviour of a standard siliceous porcelain at 1035 °C; strain rate increasing with load. (After Porte *et al.*, 2004.)

Because of the presence of the glass the ceiling temperature for the use of porcelain over a long time-scale is ~ 600 °C. At temperatures of the order of 1500 °C the aluminosilicate liquid is very fluid, and the melting-point and mechanical strength of the two crystalline phases cease to be relevant for the higher-temperature properties of porcelain. At low temperature the intrinsic strength of the glass, as the continuous phase, is important for strength, although the particulate phase has a considerable influence on the properties of the composite. This is true for ceramics containing lesser amounts of intergranular amorphous phase material (though they would not normally be regarded as composites). It is clear that their properties may be controlled more by small amounts of secondary phase, than they are by those of the major phase: this is very evidently so in the case of porcelain, because of the large proportion of glass.

2.6.5 Thermal expansion

The measured thermal expansion coefficient of a porcelain depends on composition (the alkali content in particular) and firing temperature (Rado, 1975). Mean values over the range 25–800 °C of ~7 MK^{-1} are typical, but can be reduced by increasing the proportion of feldspathic material. It is important to appreciate that the overall physical and mechanical properties of what is essentially a composite material may be determined not only by those of the constituent phases themselves, but also by the differences (often small) between the properties of the constituent phases. One consequence of differences in thermal expansivity between the phases is the development of permanent stresses during cooling from production temperature. These stresses can be large enough to cause fracture, either on the microscale, within the microstructure, or over the dimension of the whole piece, when the strength effectively becomes zero (Hasselman *et al.*, 1966). This has been modelled using particles of an inert phase (such as aluminium oxide) in a glass.

It can be shown that a spherical particle in a continuous matrix will be subjected to a pressure P given by:

$$P = \frac{\Delta a \Delta T}{\left(\frac{(1+v_m)}{2E_m} + \frac{(1-2v_p)}{E_p} \right)} \tag{2.9}$$

where Δa is the difference between the two expansion coefficients, ΔT is the cooling range over which matrix plasticity is negligible, and v and E with subscripts m and p are the Poisson ratio and Young modulus of the matrix, or particle (Davidge, 1980). Associated radial and hoop stresses in the glass are $-PR^3/r^3$ and $+PR^3/r^3$ respectively, where R is the particle radius and r the distance from the centre of the particle. The relative decrease in stress (σ/σ_o) with distance for the idealised case of a spherical particle in a homogeneous matrix is shown in

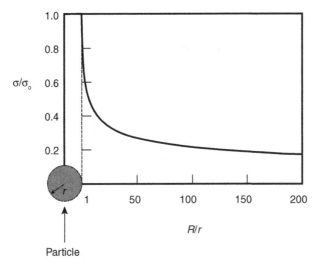

Figure 2.15 The decrease in stress with distance for the idealised case of a spherical particle in a homogeneous matrix.

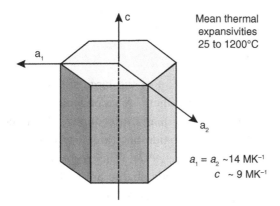

Figure 2.16 The anisotropic thermal expansion of quartz. Values are the means over the temperature range 25–1250 °C.

Fig. 2.15. The mean coefficient of thermal expansion of single-crystal mullite over the temperature range 25–1200 °C is ~5.3 MK^{-1}, but the thermal behaviour of quartz is more complicated, because the quartz crystal is strongly anisotropic with an expansion coefficient parallel to the c-axis of ~9 MK^{-1}, and ~14 MK^{-1} perpendicular to it (illustrated in Fig. 2.16). This large difference would in itself generate large residual stresses within polycrystalline quartz on cooling from high temperature, but in porcelain the glass separating the individual quartz

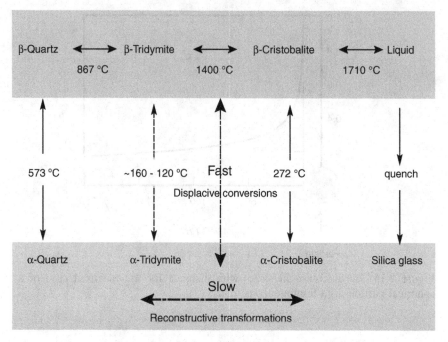

Figure 2.17 Crystalline phases of silicon dioxide: fast conversions and slow reconstructive transformations. The slow reconstructive transformation allows quartz to exist at high temperature.

grains probably prevents the build-up of the large stresses potentially available through direct interactions between them. A second and more important factor is the transformation on cooling of β-quartz to α-quartz which occurs at ~573 °C. Under normal pressures silicon dioxide has three crystalline phases, each with several structures (Stevens *et al.*, 1997). The *reconstructive transformations* between the quartz, tridymite, and cristobalite are slow (and for this reason the low-temperature form, quartz, can exist metastably at high temperature). The *displacive* (α- to β-phase) conversions are fast, as shown in Fig. 2.17. The β-quartz to α-quartz conversion on cooling is associated with a volume contraction of ~4%, and the apparent mean thermal expansion coefficient for quartz over the 1200–25 °C range is therefore around 24 MK^{-1}. This very large volume change equates to an overall linear shrinkage of ~3% (a volume shrinkage of nearly 10%). Because the mean expansion coefficient of the glass matrix in which the quartz crystals are embedded is only ~6 MK^{-1}, the expansion coefficient mismatch is sufficient to induce very large radial tensile stresses (theoretically of the order of 2 GPa) at the interface between the quartz crystals and the glass matrix, sufficient to give rise to circumferential cracks. This behaviour is illustrated in Fig. 2.18 (Davidge, 1980). Although Equation (2.9) shows that the stress should

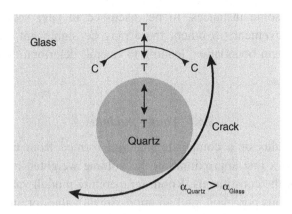

Figure 2.18 Stress distributions and cracks arising from the thermal expansion difference between a particle (quartz) and the matrix (glass). (After Davidge, 1980.)

be independent of particle size, in practice, circumferential cracking around quartz grains is seen mainly with the larger ($>\sim$50 µm) quartz grains. The critical grain dimension for cracking can be estimated by an energy balance criterion: the elastic energy released from the particle and surrounding matrix is equated to the surface energy required to form the observed roughly hemispherical crack. In material containing particles smaller than the critical size, microcracks can still appear at applied stresses lower than the macroscopic failure stress, so that the same weakening effect as in a pre-cracked specimen should occur. This calculation of the residual stress assumes no stress relaxation in the system on cooling; some plastic flow of the glass must be expected at temperatures down to $\sim$800 °C, permitting some stress relief. The large, sharp, cracks forming around the large quartz grains act as important stress concentrators, and contribute towards an overall decrease in strength of the material. In order to maximise the strength of a porcelain one obvious action is to reduce the size of the cracks by using a very fine quartz powder (which also increases the rate of solution of quartz in the feldspathic liquid), or to replace the quartz altogether by particles of material of thermal expansivity closer to that of the glass. By this criterion, aluminium oxide with a mean expansivity of the order of 9 MK^{-1} should be better.

Two general points are raised by these properties of these two components of the porcelain microstructure, which are applicable to all ceramic materials. Firstly, phase changes producing significant volume changes may take place on cooling (from sintering temperature, for example), or simply through thermal cycling. Secondly, it is likely that there will be thermal expansion coefficient differences between major phases, the magnitude of which may also be considerable. In some instances, to be discussed in later sections, there can be strength improvement; in others there may be significant weakening, particularly of the grain boundaries, leading to overall deterioration of mechanical properties.

2.6.6 Young modulus

The Young modulus of a composite material formed from brittle particulate components is to a first approximation the volume weighted mean of the individual moduli of the constituents. Room-temperature moduli values for the main phases of a siliceous porcelain, and an approximate value for an aluminosilicate glass, are given in Table 2.4, together with data for melting-points and densities. The calculated value of ~70 GPa agrees well with the measured value of ~65 GPa, which is sensitive to composition (Vazquez *et al.*, 1998). Because the glass has a reasonably high modulus, porcelain is also a reasonably stiff material, of modulus similar to that of aluminium (80 GPa), but very much stiffer than the thermoplastic and thermoset polymers. The common empirical expression for the influence of pore fraction (p) on modulus ($E = E_o.\exp(-bp)$ where E_o is the modulus of the fully dense material), gives b values in the region of 4. As shown by Fig. 2.19 porosity at the usual level of about 0.05 (that is, 5%) reduces the modulus by about 20%. The stiffness of porcelain decreases gradually with increasing temperature, primarily because of the weakening of the Si–O network of the glass phase. This loss of rigidity becomes severe in siliceous porcelain at temperatures above ~600 °C (at this relatively low temperature the decrease in the modulus of quartz and mullite is very much smaller). Above ~800 °C *creep* (permanent plastic deformation) under load is detectable (Morrell, 1989). However, when low-cost materials able to retain dimensional stability under modest loads at temperatures of a few hundred degrees are needed, the advantages of porcelain over the organoplastic materials become obvious, particularly when stability towards thermal degradation or oxidation is needed.

Differences in elastic modulus between the phases of porcelain can also lead to magnification of applied stress, although this is a much smaller effect than that of thermal expansion mismatch. For a model system consisting of aluminium oxide spheres in a glass of matched thermal expansivity, under uniaxial tension, the stress (S) is typically magnified by a factor of ~1.4, with the maximum occurring at a point on the surface of the sphere. This effect is illustrated in Fig. 2.20. To minimise stress magnification, the elastic constants of the particles and matrix need to be as close as possible. In this respect porcelain is not an ideal material.

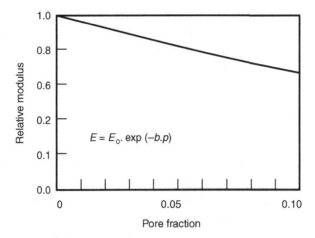

Figure 2.19 The influence of pore fraction (*p*) on relative modulus (*E/E*$_o$) expressed in terms of the standard empirical relationship $E = E_o.\exp(-bp)$.

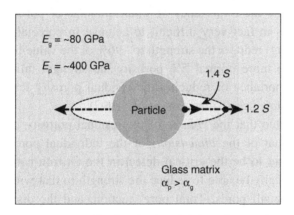

Figure 2.20 Stress (*S*) magnification in the vicinity of a sphere caused by differences in elastic modulus. (After Davidge, 1980.)

2.6.7 Fracture toughness

The fracture toughness (K_{Ic}) is not often used or measured for electrical or domestic porcelains. There is, however, considerable interest in the fracture toughness of dental porcelain which contains ~90% of feldspathic glass (Kelly *et al.*, 1996; Dummond *et al.*, 2000). For siliceous porcelain fracture toughness is in the region of 1–2 MPa m$^{1/2}$ (Soma *et al.*, 1980; Freiman, 1996), which is a little higher than values for standard silicate glass (~0.5 MPa m$^{1/2}$). The values for porcelain in part reflect the higher overall value of the Young modulus, resulting from the presence of mullite and quartz, and in part the increased fracture energy

(γ_i) caused by a more irregular and longer crack path, as the crack tip is diverted by the crystals and fluctuating residual stress fields arising from thermal expansion mismatch. The low values also mean that the material will be defect sensitive: for example, a critical flaw of only 100 μm reduces the strength to ~100 MPa.

2.6.8 Strength

A typical mean strength value of a batch of small bars of fully vitrified, unglazed, siliceous porcelain, measured at room temperature in a three- or four-point bend test, is in the region of 80 MPa. Porcelain is much weaker than the more highly developed polycrystalline structural ceramics, and the high-grade metallic alloys, but its strength is still very respectable compared to those of other low-cost competitive materials such as the thermoplastics and thermoset resins (Braganca *et al.*, 2006). Relationships between strength and porosity of the type expressed by Equation (1.23) in Chapter 1 – and similar to those shown for modulus in Fig. 2.19 – indicate the importance of small amounts of porosity. Only 1% of porosity (which is in fact very difficult to achieve in porcelain because of the problem of bloating) reduces the strength to ~96% of the value for the fully dense material; with the more normal 5% porosity the strength falls to ~80%. This highlights the importance of reducing the residual porosity levels if maximum strength is to be obtained (Kobayashi *et al.*, 1992).

It should be noted that the *volume* of the residual porosity is a quite distinct parameter from that of the *dimensions* of the individual pores, the largest of which may turn out to be the critical defect in the Griffith equation. One large pore may theoretically be able to reduce the strength to that with a large volume fraction of very small pores. Total pore fraction, and the distribution of pore dimensions, while in practice likely to be related, are two separate parameters both needing to be controlled, and if possible minimised, during the powder processing and sintering stages.

The influence of quartz crystals on the strength of porcelain has been examined in the section on thermal expansion. It is clearly necessary to avoid having large crystals in the material, and production is designed to minimise this, through screening of the quartz powders to eliminate very large particles, and in the choice of firing schedule.

2.6.9 Glazing

The application of a thin coating of surface glass, the glaze, can give significant increases in bend strength (Thiess, 1936; Mattyasovszky-Zsolnay, 1957;

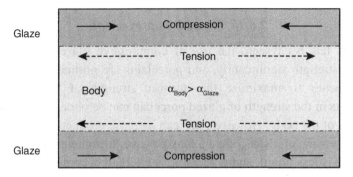

Figure 2.21 Surface stresses developed in a glazed material, as a result of thermal expansivity differences. For surface compression, $\alpha_{body} > \alpha_{glaze}$.

Kobayashi *et al.*, 2003). The reason is that, by use of a glaze of the appropriate thermal expansion coefficient, large compressive stresses can be developed in the glaze on cooling to room temperature (with a counterbalancing tensile stress in the underlying body): this is illustrated in Fig. 2.21. Glaze compositions are chosen so that the mean thermal expansivity of the glaze is around 1 MK^{-1} smaller than that of the body (the *glaze fit*). A typical mean thermal expansivity for the body is 7 MK^{-1} over the glazing temperature range, so that the glaze must have a mean expansivity of ~5.5 MK^{-1}. Compressive stresses as high as 90 MPa can be produced by a glaze of suitable fit to improve bend strength by 40% or more (Kobayashi *et al.*, 2005). This situation is closely similar to that of a "toughened" glass, which also has its surfaces in compression, usually induced in this case by fast surface cooling. Because the body and glaze compositions must be matched to provide the necessary compressive stress level, the choice of body compositions can be restricted by this requirement. Too large a surface compressive stress can lead to cracking and loss of the glaze by *spalling*. For a surface crack to propagate, the applied stress must first be increased by an amount sufficient to neutralize the surface compression. The same principle applies to all structural ceramics; however, there may be practical difficulties in using glazes when the material is used because of its hardness or resistance to wear.

In addition to its strength-improving action, a smooth glaze also has the very useful function of making it easier to keep the surface of a component clean. This is important for high-voltage electrical insulators, when films of dirt on the surface can lead to surface electrical conductivity, and breakdown. Glazes can also have important electrical properties of their own, such as semiconductivity (achieved by incorporating suitable metal ions – hence the glaze colours seen on insulators), but this is outside the present object of focusing on structural applications.

2.6.10 Strength improvement

Although, as has been seen, the use of compressive glazes on porcelain can improve the strength significantly, and porcelains are normally glazed, it would clearly be better to maximise the intrinsic strength of the porcelain first. Improvements in the strength of glazed porcelain can be obtained by reducing the glass content of the body. However, this can increase production costs for several reasons: the raw materials become more expensive, firing temperatures are higher because there is less liquid, and processing is likely to become more difficult, and therefore also expensive.

Two strength-controlling relationships, critical defect size, and porosity, were shown in Equations (1.21) and (1.23). It is difficult to eliminate porosity completely in porcelain, because of the tendency for bloating to occur over long sintering times, or if sintering temperatures are raised, because of the large volume of liquid present. The largest microstructural features (other than occasional large pores) are often those of residual quartz particles. The function of quartz as a strength-controlling factor, and the importance of reducing the size of the residual particles, is well known (Austin *et al.*, 1946; Khandelw and Cook, 1970). This effect of the quartz is intensified by the tendency for the thermal expansion mismatch cracks to be sharp (Ohya *et al.*, 1999). Although quartz powder has the advantage of low cost and density, and the gradual dissolution of the fine quartz particles helps to control glass viscosity, a number of alternatives have been examined. Of these, aluminium oxide has become the most commonly used as a partial or complete replacement for quartz, in the production of what are termed the aluminous porcelains, and now examined.

2.6.11 Aluminous porcelains

Aluminium oxide is a standard additive, partially or completely replacing the silicon dioxide, to generate the subgroup of materials known as *aluminous porcelain*. It can be added as a fine powder to the normal triaxial composition to replace quartz, and the overall system composition moves downwards, towards the Al_2O_3 corner of the diagram. At temperatures in the region of 1200–1400 °C the feldspathic liquid is much less reactive towards aluminium oxide, but some solution of the very fine particle fraction occurs. The result, as shown by the phase diagram, is the formation of additional mullite (termed *tertiary* mullite), as high aspect-ratio needles, nucleated at corundum surfaces (Iqbal and Lee, 1999). The broad ranges of compositions for siliceous and aluminous porcelains, and the equilibrium phases present at firing temperature (assumed here to be 1300 °C), are shown in Fig. 2.22, using the simplified diagram for the subsystem leucite–

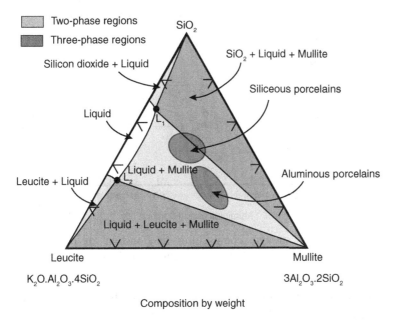

Figure 2.22 The leucite–SiO$_2$–mullite subsystem at 1300 °C, with the compositional regions of the siliceous and aluminous porcelains. (After Kingery *et al.*, 1975. Reprinted with permission of John Wiley & Sons, Inc.)

SiO$_2$–mullite. Although the aluminous porcelains have higher strength than the siliceous, there are drawbacks. There is an increased density, and therefore component weight (because of the much higher density of aluminium oxide than quartz – around 4 Mg m^{-3}, compared with 2.6 Mg m^{-3}) and a higher cost. There are also (perhaps unexpected) additional costs, arising from the hard, and abrasive nature of the aluminium oxide particles in the porcelain body. These result in increased wear rates of extruders and shaping tools used in the manufacture of components, such as electrical insulators. Insulators are often machined before and after firing, to allow fitting to metal components, for example, and the presence of aluminium oxide may necessitate the use of more expensive diamond-loaded grinding wheels.

The mean thermal expansivity of aluminium oxide (~9 MK^{-1}) is higher than that of the feldspathic glass, and mullite, but very much smaller than the overall value for quartz. The critical difference in mean thermal expansivity values between a glass matrix and the particles is around 4 MK^{-1}, and with aluminium oxide the risk of thermal expansion mismatch microcracking is considerably reduced. As the ternary phase diagram (Fig. 2.8) shows, aluminium oxide can be a stable phase once the level of addition is sufficient to move the overall composition of the system into the leucite–mullite–aluminium oxide compatibility

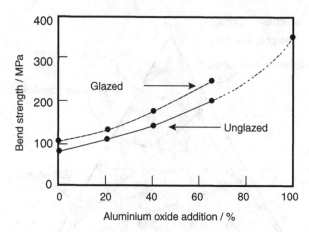

Figure 2.23 Typical strength–composition relationships in porcelain, as a function of aluminium oxide content, for unglazed and glazed materials.

triangle. While a small proportion of the aluminium oxide (particularly the smallest particle size fractions) may react with the liquids formed at the sintering temperature, the reaction rate is slow and most remains as a separate phase. The aluminium oxide notionally replaces some, or all of the quartz, but usually there is a complete reformulation of the porcelain composition. Additions of up to 50% by weight of fine (~5 μm) powder are common (Blodgett, 1961; Warshaw and Seider, 1967).

Strength increases approximately linearly with increasing aluminium oxide content: a typical strength–composition relationship for unglazed and glazed porcelain is shown in Fig. 2.23. However, reduction in internal cracking resulting from the elimination of some of the quartz is only one reason why the strengths of aluminous porcelains are higher. A more important mechanism is believed to be a restriction on internal flaw size, caused by the aluminium oxide particles. The strength of a glass–particulate composite is a function of the volume of dispersed phase at low volume fractions, and the volume and size of the particles at high volume fractions. Strengths measured on model aluminium oxide–glass composite materials are consistent with a Griffith flaw size equal to the mean inter-sphere spacing, which decreases with decreasing particle size and increasing sphere concentration (Hasselman and Fulrath, 1966, 1967). For this reason, the higher the aluminium oxide particle content, the greater should be the strength. The addition of the higher modulus aluminium oxide will also tend to increase strength, through an increase in fracture toughness, and increasing the mullite content (through reactions of the aluminium oxide with the feldspathic liquid) should also increase toughness (Harada *et al.*, 1996; Sugiyama *et al.*, 1997). An early suggestion was that high-aspect-ratio secondary and tertiary mullite needles

Table 2.5 *The compositions of the three standard classifications of porcelain, C110, C120 and C130, expressed in terms of the main phases present (ASTM, 1966).*

Type and phase / volume%	Class	Glass	Mullite	Quartz or cristobalite	Aluminium oxide
Siliceous	C110	60	20	20	0
Aluminous	C120	51	15	2	32
Aluminous	C130	35	20	~0	44

would improve strength, possibly by the interlocking of the needles. Fracture toughness should also be increased through a combination of crack branching and bridging mechanisms, as is the case in sintered silicon nitride containing high aspect-ratio (rod-shaped) β-silicon nitride crystals.

It is more difficult to improve the high-temperature strength of porcelain, because of the high proportion of glass in the microstructure. As seen above, the glass flows under load and creep deformation becomes severe under only modest loads as temperatures exceed 800 °C. Porcelain will not normally be used under load at temperatures much above 600 °C (also see Table 3.6); most applications are at temperatures up to the boiling-points of aqueous solutions.

Table 2.5 shows the compositions of the three standard classifications of porcelain, C110, C120 and C130, expressed in terms of the main phases present (ASTM, 1966). Aluminous porcelain of the C120 type must have bend strengths >110 MPa; for C130 type the value is >160 MPa.

2.6.12 Hardness

Because porcelain contains a large proportion of glass, the hardness of the glass would be likely to control overall hardness (Batista *et al.*, 2001). Vickers indentation hardness values for many glasses are in the region of 6 GPa, which is harder than mild steel and stainless steel, though not as hard as tool steel, with values in the region of 9–10 GPa. Hardness is of importance for a number of reasons. The wear rates of many ceramic materials have been related to indentation hardness (H) by expressions of the form:

$$\text{Wear rate} \propto H^{-a}K^{-b}P^c \tag{2.10}$$

where K is fracture toughness and P is a measure of the load, and a, b and c numerical constants with values in the range 0.5–1.25 (Evans and Wilshire, 1976). This type of expression, with its inverse dependence of wear rate on hardness and toughness, intuitively seems correct. But for a given material other factors strongly influence wear rate; these are grain size and the nature of the grain boundaries. The sliding wear of porcelain under load is also likely to be controlled to a considerable extent by the softening of the glass caused by frictional heat (temperatures can easily exceed 1000 °C), and pull-out of the particulate phases.

2.6.13 Thermal shock

The question of the use of porcelain (and other ceramics) at raised temperatures brings in the questions not only of its strength, but also how quickly a porcelain component can be heated and cooled without being damaged in the process. In practice, a small porcelain component will be severely cracked, or completely fractured by sudden exposure to a source of heat, or by plunging into water or oil from a temperature as low as 200 °C. The general behaviour of a brittle ceramic as a consequence of thermal shock was illustrated in Fig. 1.24, where the residual strength, expressed as a fraction of the original strength (σ/σ_o) was shown as a function of quench temperature (Davidge, 1980). Typically, for a strong material the residual strength does not fall to zero after one shock, but for a quench temperature around the critical value ($\Delta T_{critical}$) to about 20% of the value for undamaged material. The value of α for porcelain varies slightly with composition (and in particular with the quartz content), but the average value over the temperature range 25–1200 °C is in the region of 7 MK^{-1}. Inserting values into Equations (1.30) and (1.33) in Chapter 1 shows that a porcelain bar contracts by ~0.8% on cooling from 1200 °C to room temperature. The calculated value for ΔT_{crit} is ~150 °C, which is in reasonable agreement with the experimentally measured values of ~185 °C (Mora and O'Brien, 1994; Moore, 1996). In these terms porcelain is not particularly good at resisting thermal shock (and in fact only a little worse than sintered alumina). In the very early days of the development of aircraft gas turbine engines, attempts were made to use porcelain turbine blades. The blades survived heating to operating temperature in an engine test rig, but failed spectacularly when the engine was shut down while maintaining a flow of cold air.

2.6.14 Electrical properties

At room temperature porcelain is a DC electrical insulator with resistivity values of the order of 10^{11} Ωm (Eberle and Krönert, 1973). The room temperature dielectric constant is ~6–7, and tan δ ~7–8×10^{-3}, depending on the composition, values similar to those of many common insulators (Oliver and Waye, 1970;

Chaudhuri, 1974; Chaudhuri and Sarkar, 2000). These values are determined by the electrical properties of the constituent phases, glass, mullite, quartz, and aluminium oxide; there are strong, mixed covalent–ionic Si–O and Al–O bonds in the crystalline phases, and in the glass network. These structural units have low mobility even at temperatures of the order of 1000 °C. There are no atomic or ionic species with free electrons (as with metals), or accessible empty electronic conduction bands (as is the case with impure or doped covalent silicon carbide), to permit electronic conductivity. However, the aluminosilicate glass contains the cations Na^+ and K^+ (predominantly), associated with negatively charged O atoms of broken Si–O bonds in the glass network. While these cations are essentially immobile at room temperature, they provide high-temperature ionic conductivity. At temperatures as low as ~300 °C the Na^+ and K^+ ions within the Si–O network of the glass become sufficiently mobile to provide marked ionic electric conductivity – the Nernst effect (Nernst, 1900). Effectively, the glass phase is an ionic semiconductor, for which electrical conductivity increases exponentially with temperature, unlike the case for metallic conductors for which the (electronic) conductivity decreases with increasing temperature.

A second important influence on behaviour is the *polarisability* of the large and "flexible" Na^+ and K^+ ions. The electron clouds of these ions are distorted (*polarised*) by the electric field; the repetitive and rapid polarisation and depolarisation, which the material experiences when it is exposed to kHz to MHz currents, causes significant absorption of power and eventually severe local heating, permitting ionic conduction (Moulson and Herbert, 2003). Because the ionic conduction process also generates internal heating, the temperature rise can become a runaway process, leading to severe breakdown in resistivity. With standard lower-frequency domestic power supplies at 50 or 60 Hz there are no serious problems, but there are important consequences for the use of porcelain as insulators for very high-frequency (kHz, for example) currents and oscillating electrical fields. Under these conditions the presence of the large cations becomes unacceptable, and alternative materials must be used (sintered aluminium oxide, for example). They are also unsuitable, for essentially the same reason, for low-frequency applications requiring sustained exposure to temperatures much above 200 °C.

2.7 Technical applications

Porcelain is used on a very large scale where mechanical stability over long times, in wet or corroding environments, under relatively low loads, is required. Its performance is generally very satisfactory, with deterioration usually being in the glaze, rather than the body. Otherwise the main problem is of slow crack growth in the glass phase under load, aggravated by moisture or water vapour, and which may over several years lead to loss of mechanical strength. Porcelain is a relatively low-cost material because raw material costs are not high, and (for a ceramic) production temperatures are low.

The earliest industrial uses of porcelain were as simple thin-walled vessels for the containment of corrosive or hot fluids, at moderate temperatures (up to ~200 °C), and as low-voltage electrical insulators. Under these conditions the problems of thermal shock are not significant, and rapid heating and cooling are allowed. The porcelain, and its high softening point glazes, resist well attack by concentrated, or hot aqueous acids, and dilute aqueous alkalis. Hydrofluoric acid is an exception: it reacts with all siliceous materials to form silicon tetrafluoride (and its derivatives):

$$4HF_{(aq.)} + SiO_{2(s)} = SiF_{4(g)} + 2H_2O_{(1)} \tag{2.11}$$

$$SiF_{4(g)} + 2HF_{(aq.)} = H_2SiF_{6(aq.)}. \tag{2.12}$$

With the advent of electrical power supplies at the end of the nineteenth century came the need for good, dimensionally stable, high mechanical strength, and low-cost insulators. Dimensional stability, and load-bearing capability of the insulator, are important considerations, particularly when component lives of many decades are required, in humid and contaminated conditions. Porcelain was used for the insulators of internal combustion engine spark plugs, until supplanted in the 1930s by the stronger, and more corrosion resistant, sintered aluminium oxide. Other industrial applications are as grinding media (balls and cylinders) for softer materials.

2.7.1 Electrical insulators

The handling of electricity requires a very wide range of materials, with widely differing properties, according to the operating conditions which they experience. While very many materials can provide adequate insulation for low-voltage DC currents at low temperatures, as was pointed out in the previous section, the requirements become more demanding for high-voltage currents at higher temperature, and at higher frequencies than the domestic 50 or 60 Hz, 115–400 V supplies. Porcelain insulators have very important applications in the DC and 50–60 Hz AC areas (Table 2.6). In many cases the siliceous porcelain (IEC 672 C110) specification is adequate. However, where heavy mechanical loads are experienced over long periods of time the higher strength aluminous porcelains ("normal" strength C120 and "high" strength C130) are used. Glazed aluminous porcelains of these groups are widely used for overhead 275 kV and 440 kV

Table 2.6 *Examples of the applications of porcelain in electrical insulation.*

Category	Voltage
Electricity distribution	
Low	230–400 V
Medium	11–20 kV
High	33–132 kV
Electricity transmission	
Extra high	275–400 kV
Electrified railway	
	750 V
	25 kV

Figure 2.24 High-voltage (400 kV) insulators at a National Grid substation. (Reprinted by kind permission of Allied Insulators Group Ltd.)

electricity transmission power line insulation, under constant applied loads of up to ~500 kN (Liebermann, 2003). They are also used as insulators in medium (to 20 kV) and high (to 132 kV) voltage electricity distribution systems, and for electrified railway 750 V and 25 kV systems. A common example of the use of porcelain insulators on a high-voltage supply is shown in Fig. 2.24. Smaller porcelain insulators for a range of applications are shown in Fig. 2.25 and Fig. 2.26. The size range is large, with some of the larger insulators weighing several hundred kilograms, and requiring proportionately longer processing and

Figure 2.25 Medium-voltage third-rail insulators. (Courtesy of Brecknell, Willis.)

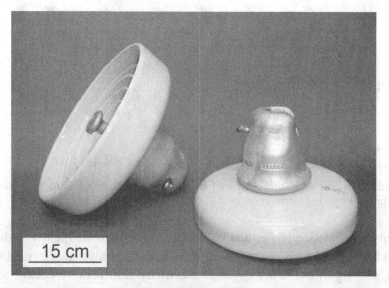

Figure 2.26 Smaller insulators, showing the connection system; the insulator discs are linked together to form an appropriately long column insulator. (Reprinted by kind permission of Allied Insulators Group Ltd.)

firing times. The largest components are usually machined from powder billets, and glazed and fired in a single stage lasting several days.

2.8 Summary

In this first case study, the production and properties of porcelain (which is, strictly, a small family of materials), a wide range of topics has been introduced. While some aspects are of particular relevance for porcelain, others are in varying degrees applicable to all the materials in the structural ceramics class. A picture of what kind of material a structural ceramic is should have emerged: this should take in the chemical and physical processes taking place at high temperature during its production, the way in which the microstructure can develop, and properties determined by the microstructure. This picture should eventually be seen as having some resemblance to those of each of the other structural ceramics, even though each will differ (considerably so in some cases) in detail.

The room-temperature mechanical properties and electrical resistivity of porcelain are strongly controlled by the large volume of aluminosilicate glass phase, in what is essentially an aluminosilicate glass–particulate composite. The glass is responsible for the marked loss of strength at temperatures above ~600 °C, and ~800 °C for aluminous porcelain, and the increasing electrical conductivity at much lower temperatures. Porcelain is a multi-phase material, and physical interactions between the phases also help to determine the properties; differences in thermal expansion coefficient between the glass and crystalline phases, notably the anisotropic quartz, and the internal stresses created have an important influence on strength. Particulate crystalline secondary phases influence strength, through a contribution to fracture toughness by influencing crack propagation, and possibly by limiting the sizes of the Griffith flaws. Large increases in strength can also be obtained by developing surface compressive surfaces. All these aspects can be found in other structural ceramic systems; they are not confined to porcelain.

Further major improvements in mechanical properties require major changes in composition, mainly because of the need to reduce the volume of glass, and perhaps by increasing the proportion of (high aspect ratio) mullite crystals. However, almost by definition, porcelain is a glass-rich material, with the advantage of a relatively low-cost production route using readily available raw materials, and relatively low sintering temperatures. Indeed, it might seem that glass is the one phase to be minimised if at all possible in a structural ceramic when good property values are needed. However, this is to over-simplify the picture, and as will be seen in later chapters, the grain boundary (amorphous) phase can have a very important influence on fracture toughness, and other properties, including resistance to wear.

Questions

2.1. What proportion of the weight loss shown in Table 2.2 is due to the dehydration of the kaolinite?

2.2. Why are porcelain, and a magnesium oxide doped sintered alumina, both translucent in thin section?

2.3. Suggest reasons why it is normal to use a relatively inert "filler", such as silicon dioxide, in porcelain.

2.4. Estimate, using Fig. 2.13, the viscosity at 1000 °C of a liquid of composition (expressed in mole fractions) SiO_2: 0.5; MO: 0.2; Al_2O_3: 0.3. How much lower is the viscosity at 1100 °C?

2.5. Calculate the mean thermal expansion coefficient of quartz over the temperature range 900 to 25 °C.

2.6. Estimate the residual stress developed at the surface of an aluminium oxide particle (mean thermal expansion coefficient 9 MK^{-1}), embedded in an aluminosilicate glass matrix of mean thermal expansion coefficient 6 MK^{-1}. Assume $v = 0.25$ for both materials.

2.7. Explain the steps you would take in order to maximise the strength of a siliceous porcelain.

2.8. Suggest reasons why a C130 glazed aluminous porcelain (prepared with 40 weight% aluminium oxide) would be expected to be significantly stronger than an unglazed siliceous porcelain.

2.9. Why does porcelain become an electrical conductor at high temperature? Is this the same type of conductivity as that seen in a liquid phase sintered alumina? Justify your answer.

2.10. As a possible material of construction for a very large insulator (see Figs. 2.3 and 2.24) what advantages might porcelain have over a standard glass?

Selected reading

Carty, W. M. and Senapati, U. (1998). Porcelain – raw materials, processing, phase evolution and mechanical behavior. *J. Am. Ceram. Soc. Centennial Review*, **81**, 3–20.

Kingery, W. D. (1986). The development of European porcelain. In *Ceramics and Civilisation III: High-Technology Ceramics Past, Present and Future*, ed. W. D. Kingery. Westerville, OH: The American Ceramic Society, pp. 153–80.

Kingery, D. W. (1996). Historical perspective on whiteware science. In *Science of Whitewares*, eds. V. E. Henkes, G. Y. Onada and W. M. Carty. Westerville, OH: The American Ceramic Society, pp. 3–17.

References

Adcock, D. S., Drummond, J. E. and McDowell, I. C. (1959). Pyroplastic index and firing deformation of ceramic bodies. *J. Am. Ceram. Soc.*, **42**, 525–32.

ASTM (1966). Standard definition of terms related to ceramic whitewares and related products. ASTM designation C242. *1966 Annual Book of ASTM Standards*, vol. 15.02. Philadelphia, PA: American Society for Testing and Materials.

Austin, C. R., Schofield, H. Z. and Haldy, N. L. (1946). Aluminium oxide in whiteware. *J. Am. Ceram. Soc.*, **29**, 341–54.

Batista, S. A. F., Messer, P. F. and Hand, R. J. (2001). Fracture toughness of bone china and hard porcelain. *Br. Ceram. Trans.*, **100**, 256–9.

Bergeron, C. G. and Risbud, S. H. (2006). *Introduction to Phase Equilibria in Ceramics*. New York: John Wiley.

Bergstrøm, L. (1994). Rheology of concentrated suspensions. In *Surfactant Science Series, Surface and Colloid Chemistry, Advanced Ceramic Processing*, eds. R. J. Pugh and L. Bergstrøm. New York: Marcel Dekker, pp. 193–244.

Blodgett, W. E. (1961). High–strength aluminium oxide porcelains. *Am. Ceram. Soc. Bull.*, **40**, 74–7.

Bloor, E. C. (1970a). Electrical porcelain: a review. Part I, Production aspects. *J. Br. Ceram. Soc.*, **7**, 77–84.

Bloor, E. C. (1970b). Electrical porcelain: a review. Part II, Technical aspects. *J. Br. Ceram. Soc.*, **7**, 129–34.

Braganca, S. R., Bergmann, C. P. and Hubner, H. (2006). Effect of quartz particle size on the strength of triaxial porcelain. *J. Eur. Ceram. Soc.*, **26**, 3761–8.

Brown, D. R. (1991). Insulators, high voltage. In *Concise Encyclopedia of Advanced Ceramics*, ed. R. J. Brook. Oxford: Pergamon Press, pp. 244–8.

Carty, W. M. and Senapati, U. (1998). Porcelain – raw materials, processing, phase evolution and mechanical behavior. *J. Am. Ceram. Soc.*, **81**, 3–20.

Chaudhuri, S. P. (1974). Ceramic properties of hard porcelain in relation to mineralogical composition and microstructure: III Dielectric behaviour. *Trans. J. Br. Ceram. Soc.*, **73**, 37–41.

Chaudhuri, S. P. and Sarkar, P. (2000). Dielectric behaviour of porcelain in relation to constitution. *Ceram. Int.*, **26**, 865–75.

Davidge, R. W. (1980a). *Mechanical Behaviour of Ceramics*. Cambridge: Cambridge University Press, pp. 81–9.

Davidge, R. W. (1980b). *Mechanical Behaviour of Ceramics*. Cambridge: Cambridge University Press.

Dinsdale, A. (1976). Translucency of tableware bodies. *Am. Ceram. Soc. Bull.*, **55**, 993–5.

Dodd, A. E. and Murfin, D. (2006). *Dictionary of Ceramics [electronic resource]*, 3rd edition. Norwich, NY: Knovel Library.

Drummond, J. L., King, T. J., Bapna, M. S. and Koperski, R. D. (2000). Mechanical property evaluation of pressable restorative ceramics. *Dental Mater.*, **16**, 226–33.

Eberle, H. and Krönert, W. (1973). Electrical conductance and electrical relaxation effects in industrial porcelain bodies. *Trans. J. Br. Ceram. Soc.*, **72**, 323–9.

Einstein, A. (1906). Eine neue bestimmung der molekuldimension. *Ann. Phys.*, **19**, 289–306.

Evans, A. G. and Wilshire, T. R. (1976). Quasi solid state particle damage in brittle solids – I Analysis and implications. *Acta. Metall.*, **24**, 939–56.

Freiman, S. W. (1996). Fracture mechanics: applications for whitewares. In *Science of Whitewares*, eds. V. E. Henkes, G. Y. Onada and W. M. Carty. Westerville, OH: The American Ceramic Society, pp. 293–304.

Funk, J. E. (1982). Designing the optimum firing curve for porcelains. *Am. Ceram. Soc. Bull.*, **62**, 632–5.

Grimshaw, R. W. (1971). *The Chemistry and Physics of Clays and Allied Ceramic Materials*, 4th edition. London: Benn.

Harada, R., Sugiyama, N. and Ishida, H. (1996). Al_2O_3-strengthened feldspathic porcelain bodies: effects of the amount and particle size of aluminium oxide. *Ceram. Eng. Sci. Proc.*, **17**, 88–98.

Hasselman, D. P. H. (1969). Unified theory of thermal shock fracture initiation and crack propagation in brittle ceramics. *J. Am. Ceram. Soc.*, **52**, 600–4.

Hasselman, D. P. H. and Fulrath, R. M. (1966). Proposed fracture theory of a dispersion-strengthened glass matrix. *J. Am. Ceram. Soc.*, **49**, 68–72.

Hasselman, D. P. H. and Fulrath, R. M. (1967). Micromechanical stress concentrations in 2-phase brittle-matrix ceramic composites. *J. Am. Ceram. Soc.*, **50**, 399–404.

International Electrotechnical Commission, Central Office, 3, rue de Varembé, P. O. Box 131, CH – 1211 Geneva 20 Switzerland: <http://www.iec.ch>

Iqbal, Y. and Lee, W. E. (1999). Fired porcelain microstructures revisited. *J. Am. Ceram. Soc.*, **82**, 3584–90.

Iqbal, Y. and Lee, W. E. (2000). Microstructural evolution in triaxial porcelain. *J. Am. Ceram. Soc.*, **83**, 3121–7.

Johnson, P. and Robinson, W. G. (1975). Development of pottery bodies – electrical porcelain. *Trans. J. Brit. Ceram. Soc.*, **74**, 147–52.

Kelly, J. R. (1997). Ceramics in restorative and prosthetic dentistry, *Ann. Rev. Mater. Sci.*, **27**, 443–68.

Kelly, J. R., Nishimura, I. and Campbell, S. D. (1996). Ceramics in dentistry: historical roots and current perspectives, *J. Prosthet. Dent.*, **75**, 18–32.

Khandelw, S. K. and Cook, R. L. (1970). Effect of aluminium oxide additions on crystalline constituents and fired properties of electrical porcelains. *Am. Ceram. Soc. Bull.*, **49**, 74–7.

Kingery, W. D. (1986). The development of European porcelain. In *Ceramics and Civilisation III: High-Technology Ceramics Past, Present and Future*, ed. W. D. Kingery. Westerville, OH: The American Ceramic Society, pp. 153–80.

Kingery, D. W. (1996). Historical perspective on whiteware science. In *Science of Whitewares*, eds. V. E. Henkes, G. Y. Onada and W. M. Carty. Westerville, OH: The American Ceramic Society, pp. 3–17.

Kingery, W. D., Bowen, H. K. and Uhlmann, D. R. (1975). *Introduction to Ceramics*, 2nd edition. New York: Wiley Interscience, pp. 295–310.

Klug, F. J., Prochazka, S. and Doremus, R. H. (1987). Aluminium oxide-silica phase-diagram in the mullite region. *J. Am. Ceram. Soc.*, **70**, 750–9.

Kobayashi, Y., Mukai, M., Mizuno, T., Ohira, O. and Isoyama, H. (2005). Effect of cristobalite formation and glaze on bending strength of α-aluminium oxide reinforced porcelain. *J. Ceram. Soc. Japan*, **113**, 413–18.

Kobayashi, Y., Ohira, O. and Isoyama, H. (2003). Effect of cristobalite formation on bending strength of aluminium oxide-strengthened porcelain bodies. *J. Ceram. Soc. Japan*, **111**, 122–5.

Kobayashi, Y., Ohira, O., Ohashi, Y. and Kato, E. (1992). Effect of firing temperature on bending strength of porcelains for tableware. *J. Am. Ceram. Soc.*, **75**, 1801–6.

Levin, E. M., Robbins, C. R. and McMurdie, H. F. (1979a). *Phase Diagrams for Ceramists*, fourth printing. Columbus, OH: The American Ceramic Society, Fig. 407.

Levin, E. M., Robbins, C. R. and McMurdie, H. F. (1979b). *Phase Diagrams for Ceramists*, fourth printing. Columbus, OH: The American Ceramic Society, Fig. 501.

Levin, E. M., Robbins, C. R. and McMurdie, H. F. (1979c). *Phase Diagrams for Ceramists*, fourth printing. Columbus, OH: The American Ceramic Society, Figs. 407 and 501.

Levin, E. M., Robbins, C. R. and McMurdie, H. F. (1979d). *Phase Diagrams for Ceramists*, fourth printing. Columbus, OH: The American Ceramic Society, Fig. 412.

Liebermann, J. (2001). Avoiding quartz in aluminium oxide porcelain for high-voltage insulators. *Am. Ceram. Soc. Bull.*, **80**, 37–42.

Liebermann, J. (2003). Microstructure and product quality of strength-stressed high-voltage insulators. *Am. Ceram. Soc. Bull.*, **82**, 39–46.

Mattyasovszky-Zsolnay, L. (1957). Mechanical strength of porcelain. *J. Am. Ceram. Soc.*, **40**, 299–306.

McConville, C. J., Lee, W. E. and Sharp, J. H. (1998). Microstructural evolution in fired kaolinite. *Br. Ceram. Trans.*, **97**, 162–8.

Morrell, R. (1989). *Handbook of Properties of Technical & Engineering Ceramics: Part 1: An Introduction for the Engineer and Designer*. National Physical Laboratory, London: HMSO, pp. 77–84.

Moore, R. E. (1996). Thermal shock of triaxial porcelains. In *Science of Whitewares*, eds. V. E. Henkes, G. Y. Onada and W. M. Carty. Westerville, OH: The American Ceramic Society, pp. 281–92.

Mora, G. P. and O'Brien, W. J. (1994). Thermal-shock resistance of core reinforced all-ceramic crown systems. *J. Biomed. Mat. Res.*, **28**, 189–94.

Moulson, A. J. and Herbert, J. (2003). *Electroceramics*, 2nd edition. London: John Wiley.

Nernst, W. (1900). Electrolytic conduction in solid substances at high temperature. *Z. Electrochem.*, **6**, 41.

Ohya, Y., Takahashi, Y., Murata, M., Nakagawa, Z. and Hamano, K. (1999). Acoustic emission from a porcelain body during cooling *J. Am. Ceram. Soc.*, **82**, 445–8.

Oliver, J. and Waye, B. E. (1970). Dielectric strength of electrical porcelain. *Trans. Brit. Ceram. Soc.*, **69**, 121–5.

Porte, F., Brydson, R. D., Rand, B. and Riley, F. L. (2004). Creep viscosity of vitreous china. *J. Am. Ceram. Soc.*, **87**, 923–8.

Powell, P. S. (1996). Ball clay basics. *Am. Ceram. Soc. Bull.*, **75**, 74–6.

Rado, P. (1964). The evolution of porcelain. *J. Brit. Ceram. Soc.*, **1**, 417–25.

Rado, P. (1969). *An Introduction to the Technology of Pottery*. Oxford: Pergamon.

Rado, P. (1971). The strange case of hard porcelain. *Trans. Brit. Ceram. Soc.*, **70**, 131–9.

Rado, P. (1975). Symposium on "Development of pottery bodies": hard porcelain. *Trans. Brit. Ceram. Soc.*, **74**, 153–8.

Schüller, K. H. (1964). Reactions between mullite and glassy phase in porcelains. *Trans. Brit. Ceram. Soc.*, **63**, 102–17.

Singer, F. and Singer, S. S. (1963a). *Industrial Ceramics*. London: Chapman and Hall, pp. 3–26.

Singer, F. and Singer, S. S. (1963b). *Industrial Ceramics*. London: Chapman and Hall, pp. 451–57.

Soma, T., Matsui, M., Oda, I. and Yamamoto, N. (1980). Applicability of crack propagation data to failure prediction in porcelain. *J. Am. Ceram. Soc.*, **63**, 166–9.

Stevens, S. J., Hand, R. J. and Sharp, J. H. (1997). Polymorphism of silica. *J. Mater. Sci.*, **32**, 2929–35.

Sugiyama, N., Harada, R. and Ishida, H. (1997). Effect of aluminium oxide addition on the feldspathic porcelain bodies – strengthening of feldspathic porcelain bodies by aluminium oxide. *J. Ceram. Soc. Japan*, **105**, 126–30.

Thiess, L. E. (1936). Influence of glaze composition on the mechanical strength of electrical porcelain. *J. Am. Ceram. Soc.*, **19**, 70–5.

Urbain, G., Bottinga, Y. and Richet, P. (1982). Viscosity of liquid silica, silicates and alumino-silicates. *Geochim. Cosmochim. Acta*, **46**, 1061–72.

Urbain, G., Cambier, F., Deletter, M. and Anseau, M. R. (1981). Viscosity of silicate melts. *Trans. J. Br. Ceram. Soc.*, **80**, 139–41.

Vasquez, S. B., Velazquez, J. C. M. and Gasga, J. R. (1998). Aluminium oxide additions affect elastic properties of electrical porcelains. *Am. Ceram. Soc. Bull.*, **77**, 81–5.

Warshaw, S. I. and Seider, R. (1967). Comparison of strength of triaxial porcelain containing aluminium oxide and silica. *J. Am. Ceram. Soc.*, **50**, 337–43.

Worrall, W. E. (1986). *Clays and Ceramic Raw Materials*, 2nd edition. London: Elsevier Applied Science.

3

Alumina

3.1 Description and history

Alumina is the name traditionally given to α-aluminium oxide, Al_2O_3, the most stable oxide of trivalent aluminium. The alumina ceramics form a group of materials, in which the major constituent is crystalline α-aluminium oxide. Commercial alumina materials are rarely chemically pure: the range of compositions in this group is very wide – ranging from virtually 100% aluminium oxide with only small traces of additives to control sintering, and accidental impurities, to materials containing a high proportion (10% or more) of liquid phase sintering aids (Dörre and Hübner, 1984; Ryshkewitch and Richerson, 1985a). These materials might seem to be a natural development from the aluminous porcelains, only with a much higher aluminium oxide content; but in fact alumina ceramics were developed over many years as quite a separate industrial activity, and were well-established long before the aluminous porcelains appeared. All the same, the physical and mechanical properties of the aluminous porcelains and the aluminas do generally follow a continuous smooth trend, which can be correlated with the changing silicate content.

The term *alumina* is generally used to describe ceramic materials containing more than 80% by weight of aluminium oxide, the balance usually being a Group IIA metal (magnesium or calcium) aluminosilicate phase, which is often present in the sintered material as a glass; below this level the materials would be regarded as aluminous porcelains (Morrell, 1987; Evans, 1996). These materials have also been referred to as *high-alumina* ceramics, although, with the contrary emphasis, alumina materials containing silicates have been termed *debased aluminas*. Alumina materials have the largest share of the world market for engineering ceramics on a weight basis, "engineering ceramics" in this context being defined as sintered ceramics used in the mechanical engineering and process industry sectors (Briggs, 2007). In one sense they can therefore be regarded as the most important of the structural ceramics. However, other structural ceramics can show better mechanical property values, as will be seen, though they are produced on a much smaller scale. Fine-grain alumina ceramics may appear to be more expensive than the porcelains, but are generally the cheaper of the high-grade structural ceramics. It is, however, not always easy to compare prices across materials on the larger scale, because the cost of the final machining or finishing operations, which can form a large proportion of the total cost, will vary from component to component.

The microstructures of materials containing a high proportion of aluminium oxide consist of small, normally 10–100 μm dimension, aluminium oxide grains, very strongly bonded across the boundaries between the crystal faces. Those of materials prepared with silicates may contain intergranular crystalline aluminate and aluminosilicate phases, and often a substantial proportion of aluminosilicate glass. The glass is for the most part present as pockets in the gaps between irregularly packed and partly directly-bonded aluminium oxide crystals. Small amounts may also be present in the form of very thin films, of the order of 1–2 nm in thickness, at the faces between the grains.

The first commercial application of an alumina ceramic was described in a German patent (Siemens, 1907). Another early German patent (General Electric Company, 1913) describes the production of alumina ceramic dies for drawing wires. But it was not until the 1930s that the serious development of materials of this type began for such diverse applications as container materials for corrosive chemicals, for high-strength metal-cutting tool tips, and as high-voltage electrical insulators. The first large-scale use of alumina ceramics was in spark plug insulators; these materials contained ~90–95% of aluminium oxide, together with liquid-forming sintering aids (McDougal, 1923; Riddle, 1949; Richards, 1981). The rapid development of alumina ceramics was in part driven by the need for improved electrical insulators for use in the more powerful and hotter internal combustion engines in the period immediately preceding the 1939–1945 World War II, related to the requirements of the new aero-engines. This development was aided by the increasing availability of small very high-temperature furnaces able to reach 1600 °C to achieve sintering of alumina materials containing a much smaller proportion of liquid than the porcelains. The need to minimise the quantity of glass in the densified microstructure was in turn determined by the need to achieve high-temperature strength, and electrical insulation qualities. In the 1950s new applications arose for high-grade alumina ceramics: these included thin plates of very high-purity material as substrates in the rapidly developing area of electronic components using kHz and MHz frequency currents, and for wear-resistant thread guides in the new man-made fibre industries.

Aluminium oxide does not occur in nature as a fine powder, unlike the clays which are a major raw material for porcelain. Aluminium is the third most common element (about 8.2% by weight) in the Earth's crust (Table 1.6), and it exists primarily in the form of the aluminosilicate minerals making up the rocks of the crust, and about 25% of their weight. Free aluminium oxide is comparatively rare, but large and workable deposits of aluminium oxide-rich minerals collectively known as *bauxite* (named after Les Baux-en-Provence in France) are found in many parts of the world (Gitzen, 1970). Pure, anhydrous, aluminium oxide Al_2O_3 is the mineral *corundum*, which is not widely found. Even rarer (and consequently costly) are the single-crystal "gem" forms of sapphire and ruby (corundum is derived from the Tamil for ruby – *Kurundam*). The form *emery*, well-known as an abrasive powder (used in *emery paper*), contains small amounts of silicon dioxide and other oxides. Pure single-crystal aluminium oxide is colourless: the colours of mineral sapphire are caused by small amounts of transition metal oxides such as titanium oxide (blue) and nickel oxide (yellow). Ruby contains up to 0.5% of chromium (III) oxide, Cr_2O_3, and the Cr^{3+} ions occupying aluminium sites in the aluminium oxide lattice can be made to emit laser radiation of wavelengths in the region of 700 nm (Orgel, 1957; Maiman, 1960; Arkhagel, 1968).

3.2 Intrinsic properties of α-aluminium oxide

Because the major structural component of the alumina ceramics is α-aluminium oxide in microcrystalline form, it will be useful to start by examining the nature of this phase, and aspects providing the basis for its practical applications.

The crystal lattice of α-aluminium oxide can be regarded as constructed from planes of hexagonal close-packed (relatively large – 140 pm radius) oxygen ions. The small aluminium ions (~50 pm radius) occupy a sublattice of octahedrally coordinated sites, and because of the 2:3 atomic proportions, 1/3 of the sites are empty: the empty aluminium sites can also be considered to occupy a hexagonal close-packed array, which is represented in Fig. 3.1 (Kronberg, 1957). The measured a and c lattice parameters are 476 pm and 1299 pm, respectively, giving a theoretical single-crystal density of ~3.99 Mg m^{-3} (Jan *et al.*, 1960; Dörre and Hübner, 1984). The Al–O bond lengths are short (192 pm), and the bond strength is high (Pauling, 1960; Phillips *et al.*, 1980). Aluminium oxide is usually considered to be an ionic material, consisting of Al^{3+} and O^{2-} ions, and while it can be treated in this way, the bonds have some covalent character (Sousa *et al.*, 1993). Diffusion of the aluminium and oxygen ions is slow at temperatures below ~1000 °C, but the ions have sufficient high-temperature mobility that solid state sintering of fine pure aluminium oxide powder is readily possible at temperatures above ~1300 °C. Both aluminium and oxygen are believed to migrate predominantly by the vacancy diffusion mechanism, the lattice diffusion of Al^{3+} being

Alumina

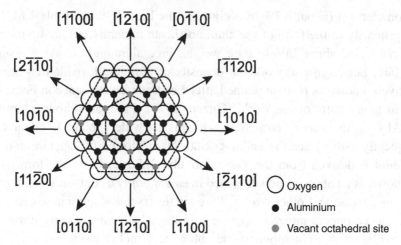

Figure 3.1 Representation of the crystal structure of α-aluminium oxide. The Al^{3+} ions occupy 2/3 of the octahedrally coordinated sites; the other 1/3 of these sites are vacant.

greater than that of O^{2-} by about three orders of magnitude (Cannon and Coble, 1975). Incorporating aliovalent ions such as Mg^{2+} and Ti^{4+} enhances mobility (Freer, 1980). Because of the absence of readily accessible conduction bands and mobile ionic species, the low-temperature electrical conductivity of pure aluminium oxide is extremely small, with a conductivity of 0.01 pS m^{-1} at 400 °C; it has been regarded as a wide-bandgap semiconductor, and the conductivity is still only ~1 μS m^{-1} at 1300 °C (Wills *et al.*, 1992).

Single-crystal α-aluminium oxide has valuable mechanical and physical properties, which are replicated to varying degrees in the polycrystalline alumina ceramics. Artificial single-crystal sapphire is readily (and relatively cheaply) available through the high-temperature growth of aluminium oxide on single-crystal seeds. Probably the foremost of its physical and mechanical properties are hardness, stiffness and high melting-point (Wachtman *et al.*, 1960). The high bond strength in the close-packed crystal structure is reflected by the very high stiffness, and the high melting-point of ~2054 °C (Ryshkewitch and Richerson, 1985). Because it has the rhombohedral crystal structure, α-aluminium oxide (like quartz) is anisotropic, and the single-crystal properties vary along the two major axes, giving room-temperature Young modulus values in the range 335–460 GPa (Wachtman *et al.*, 1961). Values for pure polycrystalline materials are in the region of 400 GPa (Chung and Simmons, 1968).

The single crystal is also very strong: tensile strengths of large crystals can have room-temperature strengths of ~500 MPa, though as with all brittle ceramic materials the degree of surface damage is critical, and scratches and other surface defects are strength-controlling flaws. In contrast, flame polished single-crystal rods have been obtained with tensile strengths of more than 5 GPa (Wachtman and Maxwell, 1959), and fine single-crystal filaments can have strengths in the region of 11 GPa. These values are approaching the calculated theoretical strength for aluminium oxide of ~31 GPa. At high temperature (~1900 °C) plastic deformation through slip on the (0001) basal plane in the {1120} direction becomes possible, and thin rods of single-crystal aluminium oxide have been bent through >360° (Ryshkewitch and Richerson, 1985a).

Single-crystal aluminium oxide is one of the hardest materials known, and in the naturally occurring form of sapphire it provides reference point 9 on the 1 (talc) to 10 (diamond) hardness scale of Friedrich Mohs. Macro- and micro-hardness numbers obtained using diamond indenters depend on the type of test, and on the crystal face. Measurements using the standard Vickers diamond indenter give hardness values of up to ~30 GPa (Ryshkewitch and Richerson, 1985b). Only a handful of ceramic materials have higher values – for example, silicon carbide, boron carbide and cubic boron nitride. The fracture energy is of the order of 6–7 J m^{-2}, depending on the cleavage plane (Wiederhorn, 1969), and the fracture toughness is also very low, at ~2 MPa m$^{1/2}$ (Weiderhorn *et al.*, 1973), which is not much higher than that of a glass.

The mean thermal expansion coefficients over the temperature range 25–1000 °C are ~8.3 MK^{-1} on the *a*-axis, and ~9.0 MK^{-1} parallel to the *c*-axis (Niesen and Leipold, 1963; Grabner, 1978). The difference between the values is not large, but it is sufficient to allow significant stresses (which can be as large as several 100 MPa) to develop in polycrystalline alumina ceramics during cooling from sintering temperature (Ma and Clarke, 1994).

The thermal conductivity of single-crystal aluminium oxide is also slightly anisotropic, with values parallel to the *c*-axis larger than those on the *a*-axis. Thermal conductivity is, as with most materials, temperature dependent. At 100 °C the mean value is ~30 W m^{-1} K^{-1}, making aluminium oxide a reasonably good thermal conductor, though not as good as metals such as aluminium (~210 W m^{-1} K^{-1}) or copper (~380 W m^{-1} K^{-1}) (Nishijima *et al.*, 1965; Fitzer and Weisenberger, 1974), or silicon carbide (Chapter 4).

Aluminium oxide is thermodynamically very stable relative to the constituent elements: in fact it is one of the most stable of all the metal oxides (Brewer, 1953). This is shown by the very large negative Gibbs function of formation ($\Delta G°$):

$$2Al_{(c)} + 1.5O_{2(g)} = Al_2O_{3(a)}; \ \Delta G°_{298\ K} = -1582 \text{ kJ mol}^{-1}. \tag{3.1}$$

At high temperatures and under very low oxygen pressures dissociation into two gaseous suboxides, Al_2O and AlO, is possible (Brewer and Searcy, 1951):

$$Al_2O_{3(a)} = Al_2O_{(g)} + O_{2(g)}; \Delta G^{\circ}_{1500\,K} = 949 \text{ kJ mol}^{-1} \qquad (3.2)$$

$$Al_2O_{3(a)} = 2AlO_{(g)} + \frac{1}{2}O_{2(g)}; \Delta G^{\circ}_{1500\,K} = 1107 \text{ kJ mol}^{-1}. \qquad (3.3)$$

The two vapour pressures are very low, and the reactions are mainly of academic interest.

The use of sintered polycrystalline alumina as a container, or crucible, material depends on its low chemical reactivity. Bulk aluminium oxide is virtually unaffected at low ($<200\,°C$) temperatures by aqueous acids and alkalis, and its resistance to chemically corrosive reagents such as concentrated acids and alkalis can be used to good effect. At temperatures above $1000\,°C$ reactions with molten alkalis, and other mixed oxide systems, such as $CaO–SiO_2$, which have low-melting eutectics, become fast, and the possibility of the solution of aluminium oxide in these, effectively corrosive, liquids must be taken into account (Levin *et al.*, 1979). In fact multi-component silicate systems such as $CaO–SiO_2$, $MgO–SiO_2$, and $CaO–MgO–SiO_2$, are very good liquid phase sintering aids, and are examined in the following sections.

3.3 Ceramic production

3.3.1 Starting materials

Small particle size (of median ~1 μm, or less) high-purity aluminium oxide powders are required for the production of high-strength alumina ceramics. This type of powder is readily (and reasonably cheaply) available: very large quantities of fine *calcined alumina* are produced worldwide for electrolysis to aluminium, of which only a small proportion (~100 000 tonnes p.a., or ~0.2%) is used in the ceramic industries. The starting point is bauxite, good grades of which normally contain ~85–90% of hydrated aluminium oxides, together with oxides such as SiO_2, Fe_2O_3, and TiO_2 (and in small but recoverable amounts the technically important gallium oxide, Ga_2O_3). Depending on its quality, 2–4 tonnes of bauxite ore are required to produce 1 tonne of aluminium oxide. Extraction and purification of the aluminium oxide is achieved using a range of variations of the Bayer process (Bayer, 1888; Gitzen, 1970), the basis of which had been identified in 1855 by Louis LeChatelier (father of Henri LeChatelier, known for *LeChatelier's principle*). Crushed bauxite is dissolved in hot concentrated aqueous sodium hydroxide solution, under several atmospheres pressure, at temperatures up to $250\,°C$. The process can be approximated by the reaction:

$$Al_2O_3.3H_2O_{(s)} + 2NaOH_{(aq.)} = 2NaAlO_{2(aq.)} + 4H_2O_{(l)}. \qquad (3.4)$$

Table 3.1 *Some common mineral precursors for alumina ceramics.*

Mineral	Chemical composition
Gibbsite (hydrargillite)	$Al_2O_3.3H_2O$ or $Al(OH)_3$
Bauxite	$Al_2O_3.2H_2O$ or $Al_2O.(OH)_4$
Diaspore and boehmite	$Al_2O_3.H_2O$ or $AlO.(OH)$
Corundum	α-Al_2O_3

Most of the oxide impurities, including the iron oxide, remain undissolved and are sedimented as "red mud". The remaining supersaturated solution of sodium aluminate is filtered, cooled, and then either seeded, or treated with carbon dioxide to lower the pH, to precipitate hydrated aluminium oxide (sometimes termed *gibbsite*, or *bayerite*), processes which are effectively the reverse of reaction (3.4) above. These hydrated and often poorly crystalline aluminium oxide powders, normally of particle size 60–120 μm, are generally loosely referred to as γ-*alumina*. A large number of individual phases, and complex phase relationships, have been identified. Table 3.1 lists the more common mineral forms of aluminium oxide and its hydrated forms. The precipitated material is washed to remove as much of the soluble sodium compounds (mainly sodium silicate) as possible, and then dehydrated at temperatures in the region of 1000–1200 °C. This converts the metastable γ-alumina phases (*transition aluminas*) more or less completely into the stable α-aluminium oxide. When water is lost during the calcination of hydrated aluminium oxide particles, a sponge-like structure develops. Figure 3.2 shows a typical large particle of this type. At the same time there is partial sintering and considerable crystal growth, so that the calcined *agglomerates* may be ~100 μm in size. However, the sizes of the *primary* crystals are normally <1 μm. The final product is termed *calcined alumina*. Calcination at intermediate temperatures yields a mixture of α-aluminium oxide and transition aluminas, often called *reactive alumina*. Common phases of hydrated aluminium oxide, and the approximate temperatures of their appearance on heating precipitated aluminium hydroxide, are shown in Table 3.2.

For the production of alumina ceramics, powders consisting of small (μm dimension) primary particles are required. In fact a major objective of the calcination process is to form weak agglomerates, which can readily be milled down to uniform sized, small, primary particles. A scanning electron micrograph of a milled calcined alumina powder is shown in Fig. 3.3. The mean *primary* crystal sizes of these powders are in the range 0.3–1.5 μm, giving specific surface areas of the order of 1–5 $m^2 g^{-1}$. Because most calcined alumina powders contain large agglomerates of small primary particles, there may be discrepancies between the

Table 3.2 *Transitional phases formed during the dehydration of aluminium hydroxide minerals. G = Gibbsite; B = boehmite; D = diaspore. (After Gitzen, 1970.)*

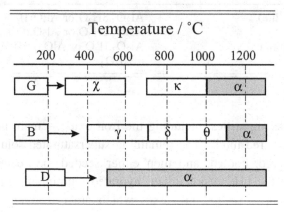

Figure 3.2 A typical calcined alumina powder in the as-calcined state, showing the sponge-like structure of the large crystals. (Courtesy Oras Abdul Kader.)

observed final sintered grain size, and the stated "particle size" of the starting powder. A second important property of calcined alumina is the level of residual sodium oxide impurity. The presence of sodium ions in a sintered alumina has detrimental effects on electrical properties, and mechanical strength at high temperature. Normal calcined aluminas can contain up to 0.6% by weight of sodium oxide (Na_2O). This can be present in the form of crystalline β-alumina,

Table 3.3 *Typical calcined alumina powders.*

Typical analysis / weight%	Normal soda	Low soda
Al_2O_3	99.2	99.8
Na_2O	0.35	0.02
SiO_2	0.05	0.01
Fe_2O_3	0.04	0.02
Primary crystal size / μm	6–10	4.5
Specific surface area / m^2 g^{-1}	0.5	0.75
Particles $> d$ μm	50 μm: 4–20%	45 μm: 0.5%
Powder bulk density / Mg m^{-3}	0.77	–
Compacted density / Mg m^{-3}	1.01	2.38

5 μm

Figure 3.3 Calcined alumina after milling, showing the very much smaller primary particles, each of which is probably an α-aluminium oxide single crystal. (Courtesy Oras Abdul Kader.)

which is not a phase of aluminium oxide at all, but a layer-structured sodium aluminium oxide of composition $Na_2O.xAl_2O_3$, with x in the range 5–12 (Peters *et al.*, 1971). A *low soda* calcined alumina can contain <0.05% of Na_2O. Typical specifications of two types of calcined alumina are shown in Table 3.3.

Very high purity, and submicrometre particle size, aluminium oxide powder is manufactured in much smaller quantities by the decomposition of soluble aluminium compounds, such as ammonium aluminium sulphate, $(NH_4)_2SO_4$. $Al_2(SO_4)_3.24H_2O$. These precursors are readily purified by recrystallisation from water, so that the resulting aluminium oxide is also very pure.

3.3.2 Solid state sintering

Very high-purity aluminium oxide is required for applications of alumina ceramics relying on physical, and high-temperature mechanical, properties. Very fine ($<$ 1 μm) high-purity (99.5%) aluminium oxide powders were first sintered to full density ($\sim$3.95 Mg m^{-3}, $\sim$99% of theoretical) polycrystalline alumina in the 1950s (Cahoon and Cristenson, 1956). Much effort has been put into understanding the solid state sintering process, and controlling grain growth. The key requirement for rapid densification of a pure material is a very small particle size. Aluminium oxide powders with mean particle sizes of $\sim$500 nm are used, and to control grain growth small quantities ($\sim$0.5%) of silicon dioxide or magnesium oxide are incorporated. To further aid the development of a dense alumina, with a minimum of internal porosity, the sintering atmosphere must be hydrogen or water vapour rich, to allow the escape by diffusion of gas trapped in closed pores (Coble, 1961). Hydrogen, unlike nitrogen, is effectively soluble at high temperature in aluminium oxide, and is able to diffuse out of the oxide structure, to allow the closed pores to shrink completely during the final stage of densification. Otherwise the build-up of gas pressure counterbalances the thermodynamic driving force for pore shrinkage outlined in Chapter 1.

The additions of very small amounts of magnesium oxide ($\sim$0.25%) allow aluminium oxide powders to be sintered to full density at 1700–1800 °C under hydrogen, although considerable grain growth occurs. The reason for the action of the magnesium oxide has been the object of considerable speculation (Bennison and Harmer, 1990b), particularly as it appears to be almost unique in its effectiveness. It appears partly, in effect, to homogenise the developing alumina microstructure, and its sintering characteristics. Thin ($\sim$mm) section sintered alumina materials of this type are translucent, with good ($>$80%) in-line transmission of light. Translucency is a consequence of the low level of internal scattering of light at solid–pore, and grain–grain, interfaces (Morrell, 1987a) – as in porcelain. Some grain growth is therefore advantageous: it is controlled (by the magnesium oxide), so that the aluminium oxide grains are uniformly large ($\sim$20–100 μm). In contrast, the presence of very small amounts of liquid (generated by impurities) tends to cause the anisotropic growth of very large aluminium oxide grains at these temperatures (Handwerker *et al.*, 1989). Such materials, sintered at temperatures $\sim$1750 °C, contain up to $\sim$5% of porosity, and are white, and opaque – the normal form of a liquid phase sintered alumina.

3.3.3 *Liquid phase sintering*

The solid state sintering of aluminium oxide powders to full density normally requires very high temperatures, and very fine powders. These materials tend therefore to be expensive. Liquid phase sintering of coarser (and cheaper) powders gives satisfactorily dense and low porosity materials at temperatures between ~1400 and ~1500 °C. By far the largest proportion of alumina used for large-scale applications is therefore made by liquid phase sintering, using a variety of liquid-forming additives. One difference between a liquid phase sintered alumina, and the aluminous porcelains referred to in the previous chapter, is that aluminium oxide is generally sintered using magnesium and calcium silicates, with high eutectic temperatures, while the porcelains tend to be densified with sodium and potassium silicates (though in fact some of the earliest European porcelains were produced using calcium oxide – lime – as the fluxing agent).

As outlined in Chapter 1, liquid phase sintering of a single-phase oxide can be regarded as a solution and crystallisation process, involving a suitable wetting liquid solvent. The smallest particles, and material under pressure at particle contact points, preferentially dissolve, subsequently to recrystallise elsewhere in the system (Kingery, 1959; Kingery and Narasimhan, 1959). This should take place under readily attainable conditions of temperature and pressure (ideally 1 atm – higher pressures can be applied, see below, but this makes the process more expensive). The liquid phase sintering process is illustrated schematically in Figs. 3.4 and 3.5. Figure 3.4 represents the starting powder mixture of α-aluminium oxide particles and additive. During sintering, new phases, including glass, are formed and the aluminium oxide particles partially recrystallise, with grain growth (Fig. 3.4). There is also likely to be a small amount of trapped (mainly intergranular) porosity. Most simple aluminium oxide–metal oxide binary phase equilibrium systems have high eutectic (liquid formation) temperatures (see also Fig. 2.5). However, in the presence of silicon dioxide which forms ternary systems the eutectic temperatures are lowered considerably. With the alkali metal oxides Na_2O and K_2O (as in porcelain) the ternary eutectic temperatures are well below ~1000 °C (Levin *et al.*, 1979a), though with MgO and CaO they are higher, as will be seen. Liquid silicates formed by the Group IA and IIA metal oxides are generally good solvents for aluminium oxide, and are good sintering additives. But the difficulty is to balance the need for faster sintering against the required properties of the alumina ceramic. This is particularly important for high-temperature applications, where softening or melting of intergranular or grain boundary silicate is detrimental to physical and mechanical properties. For this reason the use of sodium and potassium oxides, with

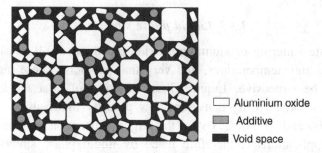

Figure 3.4 Liquid phase sintering of aluminium oxide illustrated schematically: blended powders.

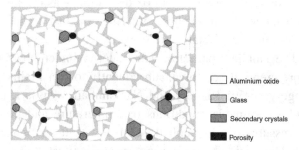

Figure 3.5 Liquid phase sintered alumina, illustrated schematically: the sintered microstructure, illustrating the presence of glass, new crystalline phases, and a small amount of trapped porosity.

very low-melting aluminosilicate phases, is ruled out. Indeed the presence of these oxides at the parts per million impurity level, is undesirable, and *low soda* calcined alumina powders (see Table 3.3) are required when the application requires exposure to high temperatures, or corrosive conditions.

Commonly used liquid phase sintering systems are the Group IIA metal oxides, magnesium oxide and calcium oxide, coupled with a source of silicon dioxide. Naturally occurring silicates are readily available: magnesium monosilicate, $MgSiO_3$, is mined as the mineral *enstatite*, or even more conveniently as a hydroxysilicate, the soft mineral *steatite* (with the alternative names of *soapstone* and *talc*) of composition $Mg_3Si_4O_{10}(OH)_2$. Calcium monosilicate, $CaSiO_3$, is mined as *wollastonite*. Alternatively, mixed systems can be used, for example *chalk* (calcium carbonate, $CaCO_3$), blended with kaolinite as the source of silicon dioxide. The lowest ternary eutectic temperature in the $MgO–Al_2O_3–SiO_2$ system

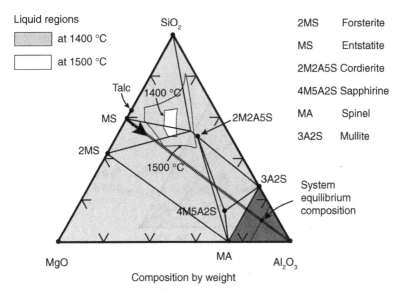

Figure 3.6 The MgO–Al$_2$O$_3$–SiO$_2$ phase equilibrium diagram, showing the compositional regions which are liquid at 1400 and 1500 °C. Standard abbreviations are used for the chemical compositions. The system is assumed to consist initially of aluminium oxide and a small amount (8%) of enstatite (MS). The line joining MS and Al$_2$O$_3$ represents the changing second phase composition as equilibrium is approached. The *equilibrium* composition is a mixture of the three phases forming the corners of the smaller shaded triangle. (After Levin *et al.*, 1979. Reprinted with permission of the American Ceramic Society, 1964. All rights reserved.)

is ~1370 °C, in the enstatite–cordierite–silica compatibility triangle (Fig. 3.6). In this diagram (and in later diagrams), to economise on space, abbreviations (common in the cement science area) for the oxides have been used: A for Al$_2$O$_3$; C for CaO; M for MgO; S for SiO$_2$. The range of compositions which is entirely liquid at 1400 °C is restricted, but by 1500 °C the liquid-forming region is much wider. This diagram also shows that for compositions lying exactly on the MgSiO$_3$–Al$_2$O$_3$ join very little liquid is formed at 1500 °C; in practice extra silicon dioxide (often added as the aluminosilicate mineral kaolinite) is used to tilt the line upwards towards the SiO$_2$ corner of the diagram, so that it crosses the liquid region. The corresponding calcium oxide system is rather less refractory: the lowest melting ternary is at 1170 °C in the wollastonite–anorthite–silica compatibility triangle and the next lowest eutectic is at 1265 °C. By 1400 °C there is extensive liquid formation, as shown in Fig. 3.7. With additions of several per cent by weight of these additive systems (and the combined magnesium–calcium silicate system), normal μm dimension aluminium oxide powders can be sintered to high density at ~1500–1600 °C on a time-scale of a few hours.

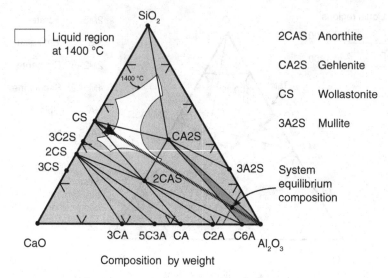

Figure 3.7 The CaO–Al₂O₃–SiO₂ phase equilibrium diagram, showing the compositional region which is liquid at 1400 °C. Again, standard abbreviations are used for chemical compositions. The system is assumed to consist initially of aluminium oxide and a small amount (8%) of wollastonite (CS). The line joining CS with Al₂O₃ represents the changing second phase composition as equilibrium is approached. The *equilibrium* composition is a mixture of the three phases forming the corners of the smaller shaded triangle. (After Levin *et al.*, 1979. Reprinted with permission of the American Ceramic Society, 1964. All rights reserved.)

Sintering is not simply a physical process involving the solution of aluminium oxide in an inert solvent. The whole system (aluminium oxide and silicate) is attempting to reach chemical equilibrium, and reactive liquids are developed. The extent of solid solubility of the magnesium and calcium oxide in the aluminium oxide grains is very small, and the silicate and aluminosilicate reaction products segregate at the grain boundaries, and in gaps between the grains (Kim *et al.*, 1994). Figures 3.6 and 3.7 show (for the two additives) the changes in composition of the systems as chemical equilibrium is approached during heating. The starting composition in each case is assumed to be a mixture of the simple binary silicate and aluminium oxide. The phases which should be seen during progression towards equilibrium are indicated by the lines joining the two starting phases, and the equilibrium phase composition (as defined in terms of the three component oxides) must lie on these lines, at the points indicated within the shaded (*compatibility*) triangles at the Al₂O₃ corner of each diagram. The precise position is determined by the proportion of sintering aid used – 8% in these figures. In practice (as with porcelain) full equilibration does not occur, and crystallisation of the aluminosilicate phases is slow, so although the final chemical composition still corresponds to some point on the line, a proportion of the sintered material consists of aluminosilicate glass derived from the liquid phase. These reactions produce phases (crystalline and glass) in which aluminium oxide has been incorporated. The overall weight (and more importantly volume) of secondary phase material in the sintered product will usually be higher than that of the silicate additive alone. However, the converse may be true when the additive itself contains aluminium and oxygen (kaolinite for example). These aspects have to be taken into account during any quantitative estimation of the phase composition of a sintered alumina, and preferably backed up by direct phase analysis, remembering that small amounts of a crystalline phase may not be readily detectable, or accurately measurable.

The type of liquid phase used has a strong influence on the morphology of the aluminium oxide grains (Kaysser *et al.*, 1987). The use of magnesium oxide as a sintering additive is considered both to homogenise (as it did in solid state sintering), and to inhibit, aluminium oxide grain growth, and the $MgO-SiO_2$ additive system tends to be preferred to that of the $CaO-SiO_2$ (Bateman *et al.*, 1989; Bennison and Harmer, 1990). A liquid containing calcium silicate tends to encourage aluminium oxide grain growth, and lead to the formation of high aspect ratio grains. The rate of sintering densification, besides being roughly proportional to the volume of sintering additive, is also influenced by powder particle size, the powder compaction density, and void size distribution. Fully dense sintered alumina can be obtained in short times at temperatures as low as 1150 °C by seeding nanometre γ-aluminium oxide powder (*reactive alumina*) with small amounts of fine α-aluminium oxide powder (Kumagai and Messing, 1984).

The strengths and Young moduli of all sintered aluminas fall with increasing temperature. This is most marked with materials which have been liquid phase sintered, and therefore contain considerable amounts of intergranular glass which starts to soften at relatively low temperatures (Spriggs *et al.*, 1964).

3.3.4 Hot pressing

Hot pressing, introduced in Section 1.4, is the sintering of a compacted powder shape under applied pressure: it can equally be called *pressure sintering*. Sintering (either solid state, or with liquid additives) is accelerated by enhanced pressure generated at particle contact points by the externally applied pressure, but there is no corresponding acceleration of grain growth, because there is no increase in pressure *gradient across* the boundary (Coble and Ellis, 1963). Sintering to high densities is therefore possible in shorter times, or lower temperatures, and the microstructure retains a small grain size. Pressure can be applied directly to a powder, uniaxially using steel or graphite dies, or, more usually, isostatically using high nitrogen pressures. The component will normally have been pre-sintered to isolate internal porosity, and prevent the penetration of gas into the powder. Applied pressures are of the order of a few MPa, which is an order of magnitude greater than the pressure generated naturally by the interfacial energy (1.5). Although the hot-pressing technique is very effective, it carries a considerable processing cost penalty; it is therefore only used when it is absolutely essential to have a very fine-grain, fully dense, material. The strengths of materials with mean grain sizes of ~ 1 μm can be high – of the order of 400–600 MPa depending on sample size and surface finish (Kirchner *et al.*, 1973).

3.4 Microstructures

3.4.1 Solid state sintered alumina

A scanning electron micrograph of a high-purity solid state sintered alumina is shown in Fig. 3.8. The occasional small holes seen at the grain boundaries, and within large grains, are now genuine pores. This polished surface has been etched thermally (a high-temperature heat treatment in air) to reveal more clearly the grain boundaries; it also shows the tendency for the grain faces to be aligned at 120° to each other, as expected from the equal pulls of grain surface tension at the three-grain edges. There is a striking difference between this microstructure, and those of the liquid phase sintered materials shown below. Small tubes of completely pore-free and translucent alumina are widely used as the envelopes in "high pressure" sodium vapour lamps (though in fact the actual partial pressure of sodium vapour is well below 1 atm).

3.4.2 Liquid phase sintered alumina

Silicate sintering additives are used in amounts of 5–10% by weight, and the theoretical equilibrium composition of the system is well towards the aluminium oxide corner of the phase equilibrium diagram. Corresponding amounts of secondary crystalline, binary and ternary, phases will be present in the sintered material. There will also be an appreciable volume of intergranular aluminosilicate glass. The final phase composition is partly determined by the heating schedule, in particular the time at temperature, and the cooling rate (Powell-Dogan and Heuer, 1990a, b, c). This means that it may not be possible to predict with certainty the phase composition of a sintered alumina on the basis of the starting powder system composition.

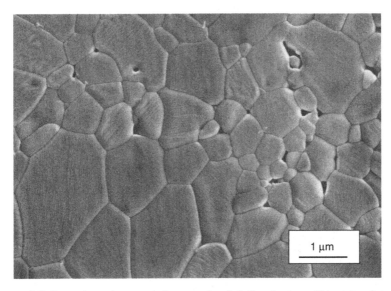

Figure 3.8 Scanning electron micrograph of fully dense solid state sintered alumina. The polished surface has been etched thermally, by heating in air at 1400 °C for a few minutes, to show more clearly the grain boundaries. Note the feature of the common ~120° boundary equilibrium intersection angle. (Courtesy Manuel Miranda-Martinez.)

In the magnesium silicate system, in addition to mullite, there is the binary crystalline phase, magnesium aluminate spinel ($MgO.Al_2O_3$), and the ternary phases cordierite ($2MgO.2Al_2O_3.5SiO_2$), and sapphirine ($2MgO.4Al_2O_3.3SiO_2$). In the calcium silicate system secondary phases are mullite ($3Al_2O_3.2SiO_2$) and calcium hexaluminate ($CaO.6Al_2O_3$), and the ternary phases anorthite ($CaO.Al_2O_3.2SiO_2$) and gehlenite ($2CaO.Al_2O_3.SiO_2$). In practice, with magnesium oxide additions, cordierite and spinel are commonly seen secondary phases; with calcium oxide additions, anorthite and gehlenite, and calcium hexaluminate. These are distributed uniformly as intergranular phases throughout the microstructure, with grain sizes similar to those of the aluminium oxide.

The aluminium oxide grains in liquid phase sintered materials are normally well crystallised and of regular morphology, because they have been partially or completely regenerated during the sintering process from the starting microcrystals of the calcined alumina. Figure 3.9 shows a scanning electron micrograph of the actual microstructure (and to be compared with the schematic diagram in Fig. 3.5) for an alumina sintered with magnesium silicate. The polished specimen has been etched chemically, to remove the aluminosilicate glass, so that only the polished surfaces of the aluminium oxide grains remain. Although some aluminium oxide grains appear to be directly bonded to each other, the presence of

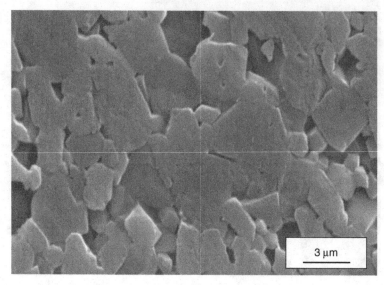

Figure 3.9 Scanning electron micrograph of the polished and chemically etched surface of alumina sintered with magnesium silicate. Etching removes the intergranular glass, to show more clearly the corundum grains. These are generally equiaxed, but with a slight tendency to elongate (actually, in 3-D, plate-like). (Courtesy Dusan Galušek.)

the gaps shows the importance of the glass for structural integrity. Figure 3.10 is a scanning electron micrograph of alumina sintered with calcium silicate: the etched microstructure is similar, but the grains tend to have a slightly elongated (along the c-axis, with a 0001 basal plane), hexagonal, or plate-like, morphology. Grain elongation is generally less marked with magnesium silicate (Park and Yoon, 2002). This may be the result of a smoothing action of the interfacial energies, resulting from the adsorption of Mg^{2+} at grain faces, and the formation, in effect, of a monolayer of isotropic magnesium aluminate spinel. Detailed examinations at higher magnification require transmission electron microscopy (Hansen and Phillips, 1983). Figure 3.11 is a transmission electron micrograph of ion-beam thinned alumina, sintered with 10% addition of calcium monosilicate, which shows clearly the intergranular glass.

High-resolution transmission electron microscopy shows "clean" faces between contacting aluminium oxide grains (Fig. 3.12), and nm dimension films of amorphous aluminosilicate in material sintered with the addition of calcium silicate (Fig. 3.13). These films have compositions, and thicknesses, related to the overall composition of the system. The amorphous material at two-grain faces can have a slightly different composition from material at the three-grain edges (usually known as *triple points* when viewed in section) (Kaplan *et al.*, 1994). This can be shown using high-magnification scanning transmission electron

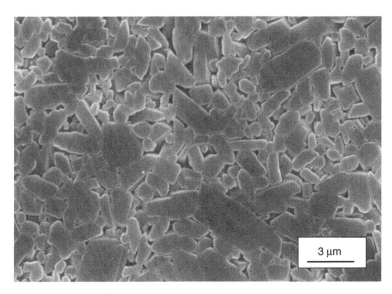

Figure 3.10 Scanning electron micrograph of the polished and chemically etched surface of alumina sintered with calcium silicate. The corundum grains are now distinctly elongated. (Courtesy Shi-Chieh Chen.)

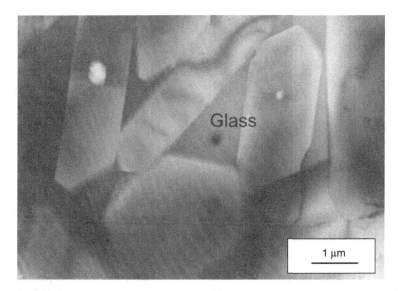

Figure 3.11 Transmission electron micrograph of ion-beam thinned sample of alumina sintered with 10% calcium silicate. The large volume of additive, and long sintering times have allowed grain elongation to occur: the intergranular calcium aluminosilicate glass is clearly shown. (Courtesy Oras Abdul Kader.)

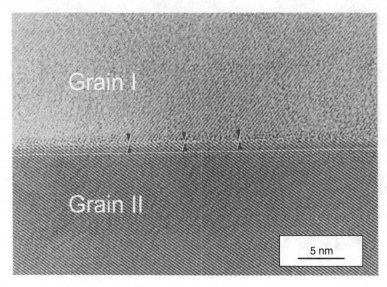

Figure 3.12 A high-resolution transmission electron micrograph of a high purity alumina, showing an essentially "clean" boundary between contacting corundum grains. (Courtesy Rik Brydson.)

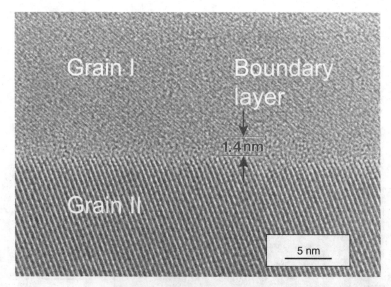

Figure 3.13 A high-resolution scanning transmission electron micrograph of alumina sintered with calcium silicate, showing the presence of a 1–2 nm thick disordered, amorphous, zone at the grain boundary. (Courtesy Rik Brydson.)

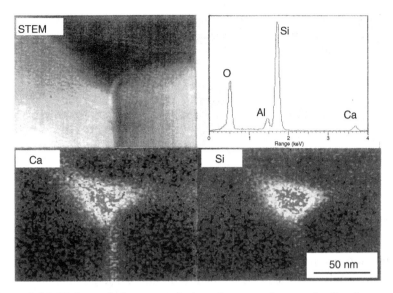

Figure 3.14 High-resolution transmission electron micrograph of a three-grain edge (triple-point) in alumina sintered with calcium silicate. Note the preferential concentration of silicon at the three-grain edge, and its depletion, in favour of calcium, at the two-grain face, indicating variations of system composition on the nm scale. (Courtesy Rik Brydson.)

microscopy, with elemental analysis. In Fig. 3.14 a high concentration of an element is indicated by the density of white dots (the original was a colour image). There is a relatively high calcium content (and low silicon content) at the two-grain face, compared with the grain edges (Brydson *et al.*, 1998). This shows the variations in composition, on the nm scale, within the grain boundary regions of alumina containing glass, and which are likely to influence boundary strength and mechanical properties.

The final grain size is determined by the particle size of the starting powder, and the extent of grain growth (Kaysser *et al.*, 1987; Handwerker *et al.*, 1989). For this reason the mean grain size is usually considerably larger than the mean primary particle size of the starting powder. However, grains may appear to be smaller than the stated particle size of the powder, but this is because the powder particles were actually agglomerates of small primary crystals.

3.4.3 Classification of alumina ceramics

The range of commercial alumina materials is very wide. Materials are commonly categorised by manufacturers in terms of the nominal aluminium oxide content, expressed as per cent by weight, obtained from chemical analyses for

Table 3.4 *Examples of sintered alumina compositions (weight%).*

Oxide	"88"	"95"	"99.7"
Al_2O_3	87.8	95.0	99.7
Na_2O	0.6	0.1	0.01
K_2O	0.3	0.04	0.01
Fe_2O_3	0.3	0.2	0.01
TiO_2	0.2	0.02	0.1
Mn_3O_4	0.01	0.03	0.01
SiO_2	7.0	3.3	0.1
MgO	1.3	1.2	0.04
CaO	1.5	0.1	0.07
Cr_2O_3	1.0	0.01	0.02

aluminium ion, or from analyses of starting powders (Morrell, 1987). These numbers are usually not the same as the aluminium oxide grain content (because aluminium will be present in secondary crystalline phases such as anorthite or spinel, and in the intergranular glass). A "95% alumina" may contain only 90% of aluminium oxide crystal by weight, which is equivalent to about 85% by volume. Slight changes in sintering conditions, or starting powders, can lead to marked changes in final microstructure, for materials of essentially the same nominal aluminium oxide content.

Examples of analyses of sintered alumina materials, taken from manufacturers' data sheets, are shown in Table 3.4. Metal oxides are grouped, depending on whether they are impurity oxides originally present in the calcined alumina powder, or deliberately added. Chromium[III] oxide (Cr_2O_3) is a special case; it is sometimes used in small amounts (usually 1–2%) because it is believed to increase hardness and wear resistance (Bradt, 1967; Ghate *et al.*, 1975). The large Cr^{3+} ions replace smaller Al^{3+} ions in the α-aluminium oxide lattice, creating a compressive lattice strain which may reduce the tendency for microcracking to occur. The presence of the chromium ions is shown, more obviously, by the pink (ruby) colour imparted to the alumina.

3.5 Physical and mechanical properties

A liquid phase sintered alumina can be regarded as a composite material, in which all phases, and their interactions, may determine the properties of the whole. From the composition and microstructural points of view, sintered aluminas are simpler materials than the porcelains, in that they consist predominantly of large α-aluminium oxide grains bonded by intergranular films, but their properties are still influenced by the factors outlined in Chapter 2 for porcelain. Although the major phase is always aluminium oxide with its own (*intrinsic*) properties, the mechanical, and some physical, properties of the sintered material are strongly microstructure dependent (Munro, 1997). Important features are the aluminium oxide grain size and size distribution, the volume and size distribution of the residual porosity, and the nature of the secondary phases. The physical properties of the higher-purity, solid state sintered aluminas are closely related to the intrinsic properties of aluminium oxide, with strength and fracture toughness influenced by the grain size. Intergranular phases in liquid phase sintered materials, their volume, location, and composition, also influence significantly physical and mechanical properties at high temperatures, when strength and creep rate are largely determined by the volume of intergranular glass. Differences between the thermal expansion coefficients of secondary phases and the aluminium oxide crystal generate residual stress of the order of 60–200 MPa, with the larger values in the larger grain size materials (Ma and Clarke, 1994).

3.5.1 Density

The densities of polycrystalline alumina are always lower than the theoretical value for sapphire (about $3.99 \, \mathrm{Mg \, m^{-3}}$) by an amount reflecting the densities of the constituent phase, and the porosity. Most aluminosilicate glasses have densities of less than $3 \, \mathrm{Mg \, m^{-3}}$ (a typical value has been taken to be $2.7 \, \mathrm{Mg \, m^{-3}}$), and there may be several volume per cent of porosity present. With 1% porosity the density of pure alumina is $\sim 3.95 \, \mathrm{Mg \, m^{-3}}$; with 5% porosity $3.79 \, \mathrm{Mg \, m^{-3}}$. Standard commercial sintered aluminas can have density values as low as $3.4 \, \mathrm{Mg \, m^{-3}}$ because of the presence of low-density secondary phases.

3.5.2 Young modulus

Values of the Young modulus (E) for sintered alumina similarly reflects compositions, and particularly the presence of porosity and intergranular glass. Modulus–pore fraction relationships for pure, fine-grained alumina, using the common empirical relationship (Fig. 2.19), give a value of ~410 GPa at zero porosity, and b of ~3.9 (Knudsen, 1962). The influence of porosity on modulus is illustrated in Fig. 3.15 for a pure sintered alumina. This lower maximum value of the Young modulus for the polycrystalline material possibly reflects the degree of atomic disorder at the grain boundaries. Small amounts of porosity have a marked effect on modulus: 2% porosity reduces the modulus by ~8%, 5% porosity by ~18%, showing the need to obtain high sintered densities if the maximum property values are to be achieved.

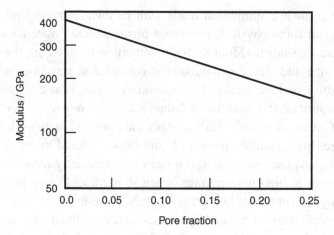

Figure 3.15 The influence of porosity on modulus for pure alumina.

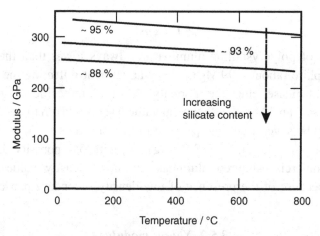

Figure 3.16 The variation of the Young modulus with compositions ranging from 88 to 95% aluminium oxide: collected data from commercial aluminas. This shows the general relaxing action of the less stiff intergranular phases, particularly as the temperature is raised. (After R. Morrell, 1987. © Crown copyright material is reproduced with the permission of the Controller of HMSO and Queen's Printer for Scotland.)

Similar porosity–modulus relationships apply to liquid phase sintered materials. Because the moduli of aluminosilicate glasses and crystalline materials are much lower than those for aluminium oxide, the smaller the aluminium oxide content of the material, the lower is the modulus. A relation between modulus, composition, and temperature, for a range of commercial aluminas, is shown in Fig. 3.16. Young modulus is approximately halved at an aluminium oxide content of ~80%.

The modulus decreases gradually with increasing temperature, more so as the volume of intergranular glass increases. This is partly because of the natural increase of mean interatomic distances and weakening of interatomic bonds with rising temperature. The relationship is approximately linear from room temperature up to ~1100 °C, and to a first approximation is independent of composition (Case *et al.*, 1983). The decrease becomes more severe at temperatures >800 °C, and above ~1200 °C irreversible plastic relaxation begins, which is enhanced in less pure, and liquid phase sintered, materials because of viscous flow of the amorphous intergranular material.

3.5.3 Poisson ratio

The shear modulus (μ) is in the region of 135–160 GPa, with the same general dependence on porosity and composition as the Young modulus (Schreiber and Anderson, 1966). Measurements of E and μ for pure alumina show that the E/μ ratio does not vary up to 1000 °C, so that the value of the Poisson ratio v (which is $E/2\mu-1$) also does not vary significantly. Values of v for pure alumina of zero porosity are in the region of 0.25 (±0.02), and decrease slightly with increasing porosity (Spriggs and Brisette, 1962). The Poisson ratio is insensitive to glass content at room temperature, but rises sharply above ~650–800 °C, approaching 0.5 at ~1200 °C, as the secondary phase becomes increasingly fluid.

3.5.4 Fracture energy and fracture toughness

A large effort over many years has been applied to measuring physical constants associated with failure initiation, and subsequent crack propagation, in ceramic materials. Alumina ceramics have little plasticity at room temperature. There may be some highly localised plasticity in a fracture face, but the materials normally fail under mechanical loading suddenly, and catastrophically, with very fast crack propagation. They can therefore be considered to be macroscopically brittle. For alumina materials with a wide range of compositions, there is a large scatter of data for both fracture energy and fracture toughness. Fracture energy values for dense polycrystalline materials are in the range 20–50 J m^{-2}, which are much larger than the single-crystal value (Rice and Freiman, 1981; Claussen *et al.*, 1982). Porosity reduces the fracture energy, following a relationship similar to those for modulus and strength (Evans and Tappin, 1972; Claussen *et al.*, 1975). There are conflicting trends for the effects of grain size, which depend on test method and the rate of crack propagation. Uncertainty in the interpretation of values arises in part because of the tendency for large (aluminium oxide) grains to undergo microcracking in advance of the main crack tip (Rice, 1977), the result of the thermal expansion anisotropy in the aluminium oxide crystal.

Alumina materials have values for fracture toughness and fracture energy considerably higher than those seen for glasses, and they would be expected to have an improved toleration of small defects, and thus higher strengths. This is borne out by the measured strength values, and resistance to chipping. Room-temperature values for the critical fracture toughness K_{Ic} are similar to those for pure polycrystalline alumina and generally in the range ~3 to ~5 MPa m$^{1/2}$ (Byrne *et al.*, 1982; Dörre and Hübner, 1984, p. 82). There is a tendency for K_{Ic} to decrease with decreasing grain size, and finer grain materials are less resistant to edge chipping in machining and handling. For alumina materials containing up to ~95% aluminium oxide toughness values at temperatures below 600 °C are similar to those of pure alumina. A small peak in toughness at ~800 °C is attributed to the onset of viscous flow of the glass (the same effect of glass softening on strength), but at higher temperatures values fall rapidly (Cheeseman and Groves, 1985).

3.5.5 Mechanical strength

The compressive strength of alumina is often quoted in manufacturers' data sheets as of the order of 1–2 GPa, although this is not a particularly precise parameter, because failure supposedly in compression is the result of local shear, or tensile stresses. However, it does provide an indication of the compressive loads a component can withstand. Normally it is bend strength which is used as the measure of quality. The intrinsic strength of aluminium oxide is very high, but polycrystalline materials tend to have lower strengths. This is because the grains (aluminium oxide and secondary crystalline phases) are effectively defects, so that grain size is an important strength-determining factor. Surface scratches, and other machining damage, can also be important strength-controlling defects, just because they are at the surface where their influence is maximised, and the surface finish of a bend strength bar is important. This also means that the strength of an as-sintered piece can be higher than one which has been ground, even though other factors such as composition, grain size and porosity have clearly not been changed.

In real applications, alumina components are likely to be exposed to stresses over long periods of time under various atmospheric conditions, and possibly of a cyclic nature. Associated aspects of mechanical failure, such as slow crack growth, and stress corrosion, leading to fatigue or delayed failure, will therefore need to be taken into account. This means that the reported mechanical strength values (often obtained from short-term tests on small bars with carefully prepared

Table 3.5 *Examples of composition–property relationships in commercial alumina.*

Al_2O_3 content / weight%	84–86	94–96	97–99	>99
Grain size / μm	3–4	3–6	4–14	1–12
Density / Mg m^{-3}	~3.5	~3.8	~3.8	~3.9
Young modulus / GPa	250	330	335	385
Mean bend strength / MPa	270–350	290–380	300–350	310–500
Thermal expansion coefficient / MK^{-1}	8.2	7.5–8.7	8.0–8.8	8.5–9.0
Limiting temperature of use / °C	1200	1100–1600	~1600	1500–1725

Table 3.6 *Maximum temperatures of use (°C) of standard porcelains and sintered aluminas.*

	Maximum temperature of use / °C		
Material	Short-term use, no load	Detectable creep under load	Long-term use under load
Siliceous porcelain	1000	800	600
Aluminous porcelain	1100	900	800
>95% Alumina	1400	1000	900
>99.5% Alumina	1700	1400	1200

and highly polished surfaces), will be significantly larger than the effective strength (equating to life, under specified loading) of a large piece of material, of uncertain surface finish, under working conditions. These aspects are treated in detail in more specialised treatments of the mechanical properties of ceramic materials (Davidge, 1980; Morrell, 1989).

Table 3.5 shows values for selected physical and mechanical properties for typical commercial alumina materials, classified according to aluminium oxide content. The maximum temperatures at which these materials can be used depend on the load and time of exposure, and the values given are only approximate. A slightly more detailed qualitative assessment (including porcelain for comparison) is in Table 3.6. It must be noted that strength is also a function of time under load, and that, whatever the temperature, in the long term failure may occur as a result of the slow growth of cracks, aided by atmospheric corrosion at the crack tip.

Room-temperature bend strengths of small bars of high-purity commercial alumina materials (>99% aluminium oxide) can be as high as 600 MPa, and as low as 150 MPa. There is a clear association with mean grain size: materials of

Table 3.7 *Influence of powder particle and sintered grain size on strength (Newland, 1989).*

Aluminium oxide content / weight%	99.9	99.9
Powder particle maximum size (d_{99}) / μm	20	5
Density / Mg m^{-3}	3.95	3.97
Mean grain size / μm	3	<2
Maximum grain size / μm	6	3
Bend strength / MPa	400	550

mean grain size ~5 μm normally can be expected to have strengths in the range 300–500 MPa (Binns and Popper, 1966). For commercial materials with nominal additive contents of 0.5–5%, and with mean grain sizes in the region of 20–25 μm, strengths are in the region of 150–250 MPa. These values will be influenced by the quality of surface finish, those for finer grain sizes the more so (Spriggs *et al.*, 1963; Tressler *et al.*, 1974). Only one defect is needed within the zone of maximum stress to cause failure, and this would be the largest defect. Because grains are considered to be defects, grain size-distribution also becomes a strength-controlling factor. Of two similar materials with the same mean grain size, the one with the narrower grain size distribution (prepared from a narrow size-distribution aluminium oxide starting powder, sintered to avoid exaggerated grain growth), would be expected to have the higher strength. It is probably maximum grain size, rather than mean grain size, which is the numerical value which should be used to rank a set of materials (although the two sets of numbers are likely to be related). Data (Table 3.7) usually give the mean grain size (Newland, 1989). Grain orientation developed during the powder pressing stage can also influence strength. Two materials of identical composition but formed using different processing conditions can therefore differ in strength, because of differences in microstructures and responses to surface machining (McNamee and Morrell, 1983; Gonzalez *et al.*, 1984).

Most studies of strength have been made with aluminas of compositions in the range 95–100%. The behaviour of materials with <95% aluminium oxide is more strongly influenced by the presence of silicate. For similar mean grain sizes, strength decreases with decreasing aluminium oxide content (Morrell, 1987d). Water and water vapour are important factors influencing strength (Davidge and Tappin, 1970), particularly with large specimens, and slow loading rates, which give more time for the effects of fatigue. This illustrates the difficulties faced in interpreting and using strength data for alumina (and other ceramics), and indicates the need to allow a substantial safety factor.

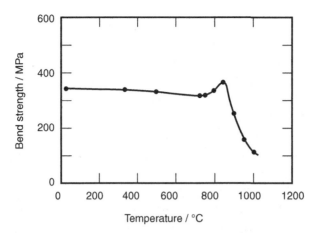

Figure 3.17 Strength of 96%, liquid phase sintered alumina, as a function of temperature; the slight rise at ~850 °C is associated with the softening and viscous relaxation of the intergranular glass. (After R. Morrell, 1987. © Crown copyright material is reproduced with the permission of the Controller of HMSO and Queen's Printer for Scotland.)

Strength decreases with increasing temperature, in part because of the reduced Young modulus of the aluminium oxide, in part because of softening and viscous flow of the intergranular glass leading to cavitation and crack initiation at the grain boundaries. For high-purity materials significant loss of strength starts to occur at around 850 °C, and as the temperature is raised strength decreases further as a result of plastic deformation of the alumina grains, and deformation of grain boundaries (Spriggs *et al.*, 1964). Figure 3.17 illustrates this behaviour. The peak at ~850 °C for constant loading rate is associated with softening and viscous relaxation of the intergranular glass near its glass transition temperature, and may not be so obvious at high strain rates (Davidge and Tappin, 1970). With materials containing larger volumes of intergranular aluminosilicate glass, strengths also decrease sharply at temperatures above the glass transition temperature (~850 °C), and the only way to retain strength at much higher temperatures (up to 1700 °C) is to eliminate the glass content entirely.

Empirical relationships between strength and pore fraction are similar to those for Young modulus (Coble and Kingery, 1956; Rice, 1977). Linear relationships are obtained with semi-logarithmic plots of strength as a function of pore fraction, with a mean value for the constant (*b*) of ~3.6: a porosity of 5% decreases strength by around 20%. The highest strengths can be expected for alumina materials of low or zero porosity, fine grain size and narrow grain size distribution, and with a minimum of intergranular material. Strength values are significantly higher than those for porcelain.

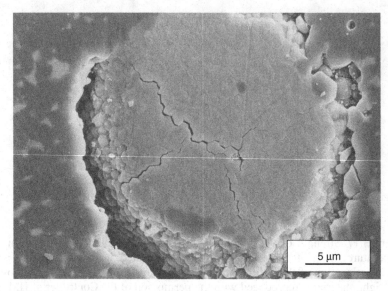

Figure 3.18 A scanning electron micrograph of a severe flaw, generated by the differential shrinkage of an agglomerate of weakly bonded fine particles, during sintering. Microstructural homogeneity is normally essential for strength. (Courtesy Tony Bromley.)

Pores also act as stress concentrating defects, so that pore size and pore fraction are important factors. Because pore size can be related to grain size, interpretations of the influence of grain size on strength in a porous material can be complicated. The critical flaw size can be correlated with the sum of the pore diameter and one grain size. Additional strength-controlling defects are large voids caused by the rapid, preferential, shrinking during sintering of agglomerates of weakly bonded fine particles. This feature is illustrated in Fig. 3.18. Efforts have been made to improve strength by preparing materials under clean-room conditions, that is, where the concentration of air-borne dust particles is kept to an absolute minimum (in practice <100 particles m^{-3}). The view underlying this approach is that particulate organic inclusions in the compacted powder (which on burning out during sintering leave similar size voids) create defects in the sintered article. For fine-grained materials, with a high quality of surface finish, the inclusions or voids then become strength-determining. Their elimination at the powder processing stage is clearly important. Some improvements in strength have been observed through the use of very clean environments for handling powders, but in commercial production it is powder particle agglomeration which is more likely to be of greater importance, and requires the greater attention (Bromley *et al.*, 1996). The wider conclusion is that in practice the strength of a sintered alumina of any composition may just as likely be controlled by the quality of the powder processing and compaction stages, as by grain size or secondary phase content.

3.5.6 Hardness

Single-crystal aluminium oxide is very hard, so it would be expected that polycrystalline aluminas would be almost equally hard materials. Measured hardness is normally a function of load. Typical values for fine-grain, high-density (porosity <1%), high-purity polycrystalline alumina are in the region of 19 GPa, and approximately independent of load in the range 0.1–2.5 kg (Westbrook and Jorgensen, 1965). However, under very low loads (20–50 g) values can be higher and in the range 20–22 GPa. Large-grain materials tend to be less hard under higher loads, because of the tendency for the brittle grains to fragment under the indenter, though for loads of <1 kg, hardness may be higher, reflecting the hardness of the individual aluminium oxide grains. Additional factors controlling the yield resistance of a material under the diamond point (its hardness) are porosity, and second (softer) phase content, and grain size. Liquid phase sintered aluminas with larger volumes of sintering additive are significantly softer. Materials of lower aluminium oxide content (~95%), and possibly higher porosity, can have Vickers hardness values as low as 10 GPa, with considerable load sensitivity (Morrell, 1987c). To achieve materials of the highest hardness, fine grain size, high density and high purity are therefore required.

3.5.7 Thermal conductivity

The thermal conductivity (λ) of sintered alumina is determined primarily by the characteristics of the aluminium oxide crystal, modified by the porosity and the amount and nature of the second phases. Materials with higher aluminium oxide contents tend to show slightly higher thermal conductivities, though the overall patterns of behaviour are closely similar, and not very different from that of aluminium oxide itself. For pure alumina, values peak at low temperatures (~80 K), fall rapidly to ~30 W m^{-1} K^{-1} at room temperature, and reach ~10 W m^{-1} K^{-1} at 500 °C (Charvat and Kingery, 1957; Morrell, 1987b). Averaged data for several types of typical polycrystalline alumina are shown in Fig. 3.19. Thermal conductivity is quite difficult to measure accurately, and differences in values reported for different commercial sintered aluminas are more likely to reflect experimental errors, rather than real differences between the products. Alumina has therefore a slightly higher thermal conductivity (at a specified temperature) than porcelain, but still lower than many metals.

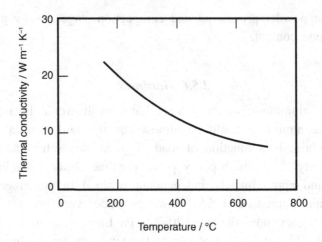

Figure 3.19 Thermal conductivity of liquid phase sintered alumina, as a function of temperature. (After R. Morrell, 1987. © Crown copyright material is reproduced with the permission of the Controller of HMSO and Queen's Printer for Scotland.)

3.5.8 Thermal expansivity

In dense fine-grain alumina the expansivity anisotropy of the aluminium oxide grain tends to be averaged out and there is only a small degree of anisotropy (unless, in the uncommon situation, there is marked grain orientation). The overall expansion is then the weighted average of the two coefficients (a, 8.3 and c, $9.0\,MK^{-1}$). In a dense multi-phase liquid phase sintered alumina the net expansion coefficient is the weighted average of all the phases present. One standard expression for the expansion coefficient (α_m) of a mixture is:

$$\alpha_m = \frac{\sum_i (a_i v_i K_i)}{\sum_i (v_i K_i)} \tag{3.5}$$

where α_i, v_i and K_i are the expansion coefficient, volume fraction, and bulk modulus of the ith component respectively (Morrell, 1989a). A pore, with zero expansion coefficient and zero bulk modulus, does not affect the net expansion coefficient (the high-temperature *bloating* of pores during the sintering of a ceramic containing liquid, and caused by trapped gas bubbles, is an entirely different phenomenon).

Increasing the content of a second phase silicate material (with generally lower expansion coefficients and moduli than aluminium oxide) might be expected to lower the net expansivity. Measured linear expansivity values, while temperature-dependent (Coble and Kingery, 1956; Nielsen and Leipold, 1963), do not vary markedly with composition up to around 10% of additive, and it seems that, unless

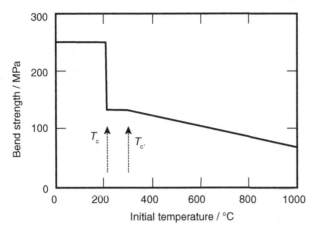

Figure 3.20 Schematic indication of the residual strength of dense sintered alumina, after quenching into water at 20 °C. The critical temperature for severe loss of strength because of surface, and then interior, microcrack development, is ~200 °C. This pattern of behaviour is typical of that of dense, strong, ceramics. (After R. W. Davidge, 1980. Reprinted with permission of Cambridge University Press.)

there is a very high mullite content, the influence of secondary phases is small. Typical mean values of expansivity for liquid phase sintered materials containing 96.5–99.9% aluminium oxide increase from ~6 MK^{-1} over the temperature range 25–300 °C, to ~8 MK^{-1} over the range 25–1000 °C: these values are not far from those of sapphire (Dörre and Hübner, 1984; Morrell, 1987b).

3.5.9 Thermal shock resistance

Values of the critical quench temperature for alumina using a standard water quench are in the region of 160–210 °C, and not significantly dependent on composition. With more severe quenching, strengths fall abruptly to around 20–30% of the initial value (Davidge and Tappin, 1967). The resistance of alumina materials to rapid temperature change is therefore not very different from that of the porcelains. This is largely because of the dominance of the thermal expansion coefficient, which is higher for alumina (~8.5 MK^{-1}) than for porcelain (~7 MK^{-1}). Although alumina materials as a class are much stronger than porcelains, they also have higher Young moduli, and the strength–modulus ratio for the two groups of material remains fairly constant. The expected advantage of the higher strength of alumina is then offset by its higher modulus and thermal expansivity. The general behaviour during quench tests into water at 20 °C of these dense materials (Fig. 3.20) is similar to that for porcelain, shown schematically in Fig. 1.25 (Davidge, 1979). At the first critical quench temperature

Table 3.8 *Calculated values for a dense sintered alumina of three thermal shock resistance parameters.*

Thermal shock parameter	Expression	Approximate numerical value	Unit
R	$\sigma\,(1-v)\,/\,E\alpha$	96–180	K
R'	$\sigma\,(1-v)\,\lambda\,/\,E\alpha$	~2	kW m^{-1}
R'''	$E\,/\,[\sigma^2\,(1-v)]$	~3	MPa^{-1}

(T_c) microcracking is initiated, and the residual strength drops by about half. There is then a slight plateau, associated with the further initiation and spreading of microcracks, before further loss of strength occurs (T'_c).

   Data for the thermal shock behaviour under conditions of an "infinitely fast" quench, of a fine-grain alumina with initial strength in the region of 400 MPa, are shown in Table 3.8. Thermal-shock R-parameters for crack initiation ($\sigma\,(1-v)/E\alpha$) are in the range 96–180 K; values for the thermal shock damage resistance parameter R''', applicable to precracked or damaged materials, are ~3 MPa^{-1} (the units of these parameters may seem odd, but follow simply from combining the units of the individual terms). The parameters are best used simply as guides to likely behaviour, or for comparing different materials. The actual response of an alumina to rapid temperature change will be influenced by the exact conditions of the shock, and the size of the test sample.

3.5.10 Wear

Wear, the removal of surface material through contact with another moving phase (solid or fluid), is a very large and complex subject (Buckley and Miyoshi, 1989). It includes processes such as abrasion, sliding, erosion, cutting, and grinding. No wear mechanism is yet perfectly understood, and wear rates under a given set of circumstances cannot be predicted for alumina ceramics with accuracy. It follows that means for improving resistance to wear of an alumina ceramic cannot be deduced on a fully rational (as opposed to a trial and error) basis. Intuitively, hardness would be a prerequisite for resistance to wear, and while the hardest materials usually also have the best resistance to all types of wear, other factors are clearly involved. One of these is the environment; the presence of water, or lubricants, has a strong influence on wear rate. A commonly applied expression (Evans and Wilshire, 1976) relates the rate of loss of material to the most obvious variables, hardness, fracture toughness, and a measure of the applied load or intensity of the wear treatment (2.10). Aluminium oxide, just because it is a very hard and reasonably tough material, would be expected to resist wear well. Polycrystalline alumina ceramics, as a class, have slow wear rates (by comparison with metals and many aluminosilicate ceramics) under a wide variety of conditions. However, not only wear rates, but wear mechanisms, are influenced by the severity of treatment. For pure alumina materials abrupt transitions can be seen at critical loads, and a transition from initially mild wear to more severe wear can occur after a period of time. The time required for the transition increases with decreasing grain size and decreasing applied load. During the initial stage, surface material is removed by a plastic grooving process, and the accumulation of subsurface damage. With continued sliding, internal stresses associated with the accumulating damage result in grain boundary cracking and grain pull-out, and the onset of fracture-dominated severe wear (Cho *et al.*, 1992). Tribochemical wear becomes important in the presence of water. Sliding wear under wet conditions is much faster than under dry conditions, which appears to be the consequence of the faster dissolution of aluminium oxide under pressure at contact points (Wallbridge *et al.*, 1983).

The rates of wear (under standardised conditions) of different alumina materials can easily vary by an order of magnitude or more. One readily identifiable factor determining wear rate in alumina materials of all types, and under a wide range of wear conditions, is grain size (Liu and Fine, 1993; Davidge and Riley, 1995). In general, fine-grain microstructures are associated with low wear rates. Abrasive wear rates decrease with decreasing grain size down to mean grain sizes of the order of 1 μm. The relationship is very approximately parabolic (that is, wear rate is a function of the square root of the grain size), and therefore follows the same pattern shown by strength – though this should not be taken to mean that strength determines wear rate, but that they are both controlled by the same parameters. However, strength seems to be a good indicator of likely resistance to wear. Figure 3.21 shows the marked influence of grain size on the rates of wear of pure alumina by wet abrasion in a slurry of silicon carbide grit. For larger grain size materials the wear process is that of surface microfracture and grain pull-out, leading to the formation of rough surfaces. The surface of a sintered alumina worn by abrasion using a slurry of silicon carbide grit is shown in Fig. 3.22. Materials of larger grain sizes show predominantly intergranular fracture. For grain sizes <2 μm the influence of grain size vanishes, and surface polishing predominates (Miranda-Martinez *et al.*, 1994). This suggests that either plastic deformation, or a tribochemical process, is now rate-controlling. However, under other wear conditions, microfracture may continue to be a failure mechanism in submicrometre material (Roy *et al.*, 2007). It is clear that much more work remains to be done to clarify wear mechanisms.

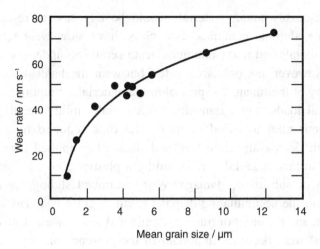

Figure 3.21 Rate of wear of pure alumina by wet abrasion, as a function of grain size. Wear with these grain sizes is associated with surface mechanical damage. For grain sizes $<2\,\mu m$ surface chemical reactions become an important mechanism of material removal. (Miranda-Martinez *et al.*, 1994.)

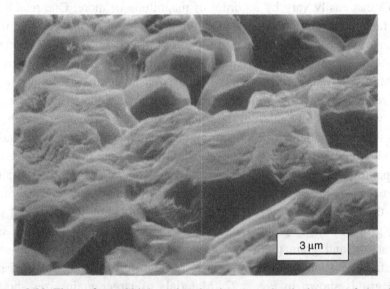

Figure 3.22 The surface of high-purity alumina, worn by the impact of abrasive particles. This illustrates the importance of intergranular fracture for material loss. (Courtesy Manuel Miranda-Martinez.)

Alumina produced by liquid phase sintering, containing intergranular aluminosilicate glass, tends to wear more slowly than pure solid state sintered alumina (by a factor of 50% or more), when comparison is made on the basis of equal grain size and porosity (Brydson *et al.*, 2001). Wear faces for liquid phase

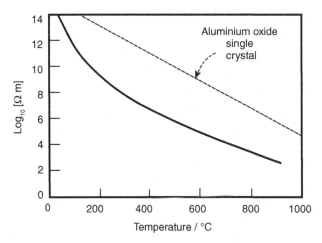

Figure 3.23 DC electrical resistivity of 99.5% alumina, as a function of temperature, compared with data for single-crystal aluminium oxide, and showing the importance of purity. (After R. Morrell, 1987. © Crown copyright material is reproduced with the permission of the Controller of HMSO and Queen's Printer for Scotland.)

sintered materials tend to be smoother, with intragranular fracture predominating. The nm thickness films at the two-grain boundaries appear to have the effect of strengthening the interfaces, and reducing the ease of microcrack propagation. For the best general overall resistance to wear, alumina materials with the smallest grain size, and containing small amounts (5–10%) of intergranular silicates, seem to be required.

3.5.11 Electrical properties

High-purity single-crystal aluminium oxide is a very good electrical insulator up to temperatures of ~1000 °C. This is one of its most useful properties, particularly when it is considered together with mechanical strength and chemical inertness (Peters *et al.*, 1965). While high-purity (>99%) polycrystalline alumina is one of the best high-temperature insulating materials available (only boron nitride being better), it has to be regarded as an electrical conductor at temperatures above 1200 °C (Özkan and Moulson, 1970).

The room-temperature direct current (DC) resistivity of single-crystal aluminium oxide is ~10^{14} Ωm; with increasing temperature resistivity values follow approximately the Arrhenius relationship, as shown by Fig. 3.23 (Binns, 1965). For nominally pure single crystals, high-temperature conductivities are usually only slightly higher than those calculated from aluminium and oxide ion self-diffusion data. At temperatures below ~900 °C conductivity is dependent on purity as a result of ionic and electron–hole conductivity; above this temperature the temperature sensitivity (conduction activation energy) increases, and conductivity values approach those for pure alumina. The electrical conductivity of single-crystal material is dependent on purity and oxygen pressure through their influence on lattice defect concentrations (Brook *et al.*, 1971). Polycrystalline material conductivities tend to be several orders of magnitude higher, at a given temperature, than those for nominally pure alumina (although they may still be considered to be very good insulators). Grain size is now a factor affecting conductivity, because of the increased contributions from diffusion taking place within the grain boundaries.

The relative permittivity (k') of a material reflects the enhancement of its charge storage capacity in an electric field as a result of the electrical polarisation of its crystal structure. Because k' for air is effectively 1, permittivity becomes an approximately linear function of porosity. Permittivity values may also be affected by the secondary crystalline and glass phases present in a liquid phase sintered alumina (Morrell, 1989b). Under AC conditions dielectric loss processes result in an out-of-phase component in the polarisation, and both the real (k') and the imaginary parts (k'') of the permittivity must be considered. Thus

$$k = k' + k'' \tag{3.6}$$

where the resulting phase lag is given by:

$$\tan\delta = \frac{k''}{k'} \tag{3.7}$$

where δ is the phase angle: $\tan\delta$ is the loss tangent, or dissipation factor. In the case of anisotropic materials the value of k' will vary with crystal axis; the two values for single-crystal aluminium oxide differ by ~20% (*c*-axis 11.53, *a*-axis 9.35). A fine-grain polycrystalline alumina may also tend to be anisotropic with respect to permittivity if, as a result of fabrication procedures, there is a non-random orientation of the grains. For high-density and high-purity alumina, relative permittivity values of 9–10 are measured for frequencies of 1 MHz to 1 GHz. These values are not sensitive to composition for aluminium oxide contents of 99%.

When an alternating field is applied to a dielectric there will be a loss of electrical power, and heating. The value of $\tan\delta$, and the frequency of the current, determine the energy loss per cycle. But $\tan\delta$ is also frequency dependent, often in the opposite sense, so that there may be partial compensation to reduce the dependence of energy loss on frequency. Because of the strong bonding in the aluminium oxide lattice, the mean loss for high-purity polycrystalline alumina is normally very low; for frequencies of 1–100 MHz, $\tan\delta$ is typically around 2×10^{-4} (and close to the intrinsic mean value for sapphire), with a temperature coefficient in the region of $120\,MK^{-1}$ over the temperature range 25–400 °C.

Because of the ability of silicate glasses, particularly alkali-containing glasses, to conduct and absorb power, dielectric losses (as indicated by tan δ for example) in impure alumina materials, and liquid phase sintered materials containing glass, can become much larger. Although at room temperature tan δ values for frequencies in the MHz to GHz range can still be between 1×10^{-4} and 1×10^{-3}, they increase rapidly above 300 °C. Alumina materials to be used for high-frequency alternating current insulation must therefore be prepared with very careful control of alkali metal oxide levels (Morrell, 1987d).

DC resistance values show clearly the effect of secondary phase content (reflecting the glass content). It was explained in Chapter 2 that porcelain (like soda-lime glass) is essentially a good electrical (sodium ion) conductor at temperatures above ~300 °C. With a typical ~99% alumina the conductivity is ~100 µS m^{-1} at ~1000 °C; with a ~95% alumina this value is reached at ~700 °C.

3.6 Chemical properties

Aluminium oxide is formally classed as an *amphoteric* oxide. This means that it is able to show both acidic and basic behaviour, depending on the chemical environment. Aluminium oxide can react to give variants of the Al^{3+} cation, or the oxoanion AlO_2^- (which for the hydrated form can be written $Al(OH)_4)^-$:

$$Al_2O_{3(s)} + 6H^+_{(aq.)} = 2Al^{3+}_{(aq.)} + 3H_2O_{(\ell)} \qquad (3.8)$$

$$Al_2O_{3(s)} + 2OH^-_{(aq.)} + 3H_2O_{(\ell)} = 2(Al(OH)_4)^-_{(aq.)}. \qquad (3.9)$$

The ionic species may be able to exist hydrated in aqueous solution (as implied here), or locked within an insoluble three-dimensional lattice. The long-term stability, and life, of a ceramic component is, however, determined more by reaction rate, than by reaction tendency as expressed by thermodynamic factors. In practice the reaction rate of a sintered alumina component is determined partly by the intrinsic surface reactivity, that is the rate at which Al–O bonds are broken and species removed, and by the reaction interfacial area. Aluminium oxide is intrinsically chemically unreactive (because of the strong interatomic bonding and high packing density in the crystal lattice, giving a very high lattice energy). Although a fine (µm dimension) powder has a very large specific surface area (many m^2 g^{-1}), and a reaction rate, measured by the rate of disappearance of solid material, might appear to be reasonably fast, a bulk, dense, sintered alumina has a specific surface area of only a few cm^2 g^{-1}, and its rate of reaction under identical conditions would be some 10^4 or more times slower. Aluminium oxide can be considered to be inert to water and aqueous solutions at low temperatures (~200 °C), but at temperatures above ~500 °C reactions with other oxides, and particularly alkali metal oxides, can become fast. Indications of the likelihood of this type of attack are provided by examinations of the appropriate phase equilibrium diagrams, when the presence of low melting-point eutectics, or new phases, will indicate potential reactivity.

An entirely different kind of factor affecting the chemical stability of an alumina ceramic is the intergranular phase material, and the glass phase in particular.

Most aqueous solutions of acids and alkalis at room temperature are essentially without effect over normal time-scales, and for technical applications sintered alumina materials of all types can be considered to be chemically inert. However, concentrated solutions of alkalis such as sodium hydroxide, and molten alkalis will tend to attack grain boundary phases and cause surface roughening, particularly so at temperatures above 100 °C. In the longer term severe degradation may occur of materials containing large volumes of intergranular silicates. The aluminium oxide crystals themselves are less severely attacked: to completely dissolve a piece of sintered alumina it is necessary to first crush and mill it to a fine powder before treatment with molten alkalis at temperatures in the region of 500 °C. Most mineral acids are without effect unless very high temperatures are used, but hydrofluoric acid is a special case in that it reacts readily with silicon dioxide and silicates at temperatures of ~100 °C. Liquid phase sintered alumina containing silicates is therefore degraded by hot liquid, or gaseous, hydrofluoric acid (2.11) and (2.12). However, at temperatures above ~240 °C (and high pressure) aqueous solutions become corrosive. This is because the super-heated water reacts both with the aluminium oxide grains, and with any intergranular silicate. The presence of impurities or sintering additives is therefore critical and care has to be taken in the choice of material (Oda and Yoshido, 1997; Schacht et al., 2000; Sato et al., 1991). A problem associated with the aqueous corrosion of alumina materials is, as mentioned above, that tribochemical wear rates may increase considerably (Kitaka et al., 1992).

3.7 Applications

The alumina ceramics are the most extensively used of all the technical ceramics, with their applications dependent on mechanical and electrical properties (Richards, 1991; Briggs, 2007). Each application uses an alumina material of a composition which will give the best performance, and cost effectiveness. The cost of production is always a major factor; other things being equal the cost of a component will increase with increasing aluminium oxide content and purity,

Table 3.9 *Main application areas of alumina materials.*

General area	Application	Aluminium oxide content / %
Mechanical	Water pump seals Car engine Washing machine Central heating	95
	Wear resistance – Grinding media Spray nozzles Textile thread guides Cutting tool tips	96
	Body armour	96
Electrical	High-voltage insulators	88–96
Electronic	Substrates, waveguide windows	96–99
Biomedical	Hip joint and other prostheses	>99.5
Windows	High-temperature sodium vapour lamps	99

because of the higher prices of the raw materials and the higher sintering temperatures needed (though factors other than production costs influence the price of a component). Generally, the material will tend to have the lowest aluminium oxide content, compatible with providing the performance and life needed. Accuracy of dimensions, and quality of surface finish or polish, which will require diamond grinding or polishing, are other factors strongly influencing price.

Approximately 50 million tonnes of calcined alumina is produced world-wide each year, of which by far the largest proportion (around 90%) is used for metallurgical processes such as the production of aluminium and aluminium alloys. Of the remaining 4–5 million tonnes of speciality calcined alumina, much is used in the chemical industries, and a smaller proportion is used in the manufacture of coarse-grain refractories, and for technical ceramic products. In these last two categories approximately 25% is used in whiteware ceramics and spark plugs, and 10% in other sintered alumina applications. It can be seen that the overall scale of the alumina technical ceramic market is, in absolute terms, relatively small, although the importance of the components is very high. In most cases the applications of the ceramic are effectively "invisible", in that the alumina component is either itself physically very small, or hidden (as are the spark plugs in the petrol engine, and the circuit boards in electronic equipment) within a much larger component (Briggs, 2007).

In the next sections, examples provide illustrations of the applications, and the relevance of mechanical or electrical properties. Table 3.9 summarises the broad application areas. This table is based on the major property requirements, although in many cases it is a complex set of mechanical, physical, and electrical properties (as well as the prices of competing materials) that determines the choice of alumina for the application.

3.7.1 Electrical insulation

Alumina ceramics were first developed a century ago, but they only came to be more widely used in the 1950s with the major industrial expansion which followed World War II. The petrol engine spark plug provides a good example of materials development to meet a requirement for improved properties; millions of alumina insulators using material of around 90% aluminium oxide content are produced world-wide each week (Ott, 1991). The spark plug insulator used in the earliest petrol engines, carrying the high-voltage (20 000 V) electrode and attached to a steel nut base, was a standard, clay-based, siliceous porcelain. But with the increased temperatures developed in the higher compression engines, the insulators were found to be susceptible to electrical breakdown, and thermal shock. In 1923 anti-knock additives (containing lead tetraethyl) were introduced into the fuel to improve its combustion characteristics, and a further consequence was that the insulator glaze was degraded by attack from the lead oxide produced. Initially, improved (higher clay content) porcelains, and then aluminous porcelains containing 55% of aluminium oxide with mullite as the main crystalline phase, were developed to give the insulators better mechanical and electrical properties. A new insulator, consisting entirely of aluminium oxide, was produced by the Siemens Company in Germany, drawing on their experience with a new range of high-temperature furnaces capable of reaching the 1760 °C required for sintering this material.

Modern spark plug insulators typically contain 88–95% of aluminium oxide of mean grain size in the range 2–5 μm. They are produced from fine calcined alumina powder which is blended with clay and the components of magnesium and calcium oxide sintering aids, such as talc and wollastonite. Sodium and potassium oxide impurities are kept to a minimum. Sintering is carried out at temperatures up to 1650 °C with an 8-hour programme, the exact details of which depend on the composition. The thermal shock resistance of this material is comparable with that of porcelain, but the fine grain size and low glass content meant that high-temperature strengths are considerably higher. Figure 3.24 shows a modern alumina spark plug insulator at successive production stages. The shrinkage from the roughly pressed powder blank is only partly the result of the machining to final shape; a large proportion of the dimensional change is the sintering shrinkage.

With the development of electronic and electrical devices such as radar, and improved radio-communications systems using kHz and MHz frequencies, the

Table 3.10 *International Electrotechnical Commission (IEC 60672) classification of high-alumina materials for electrical insulation.*

Subgroup	Aluminium oxide content / %	Principal applications
C 780	>80 to 86	General purpose, small to medium insulators
C 786	>86 to 95	General purpose, small to medium insulators, substrates
C 795	>95 to 99	Special and low-loss insulators and substrates, metallised parts
C 799	>99	Special and ultra-low loss insulators and substrates, sodium vapour lamp envelopes

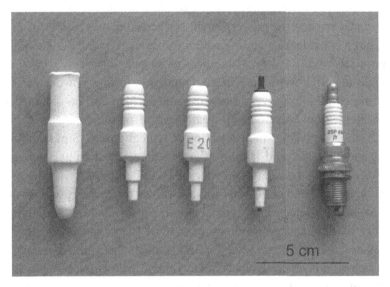

Figure 3.24 Successive production stages in the production of a typical spark plug insulator. The difference in size between the pressed blank on the left, and the sintered insulators, is partly the result of machining to shape, and partly sintering shrinkage.

need arose for insulator materials with lower losses than the magnesium silicates (forsterite), and the porcelains used for low-frequency AC insulators. These low-loss insulators are normally made with >99% aluminium oxide blended with a small quantity of sintering additive (Bailey, 1991). Later, with the arrival of computers and similar devices relying extensively on microelectronic circuitry, thin sheets of high-purity (~99.7%) alumina were needed to form the high dimensional stability, low loss, electrically insulating base on which the circuit components could be constructed. One broad form of classification of electrical insulation materials agreed by the International Electrotechnical Commission (IEC 60672) is shown in Table 3.10. The use of submicrometre *reactive alumina* powders (containing γ-aluminium oxide) enables these materials to be sintered to full density at 1600 °C with a 90-minute cycle.

3.7.2 Wear resistance

One of the early applications for sintered alumina was for dies in the steel and copper wire drawing industries, relying on the material's hardness, and resistance to sliding and abrasive wear. Appreciation of these properties led to the development of alumina thread guides for use in the textile industries, particularly the increasingly important man-made fibre industries where abrasive particles such as titanium dioxide are often incorporated into the filaments as delustrants. A fibre travelling under tension over a very small surface area at velocities of more than 10 m s^{-1} has a considerable cutting action. Alumina, typically containing 98.5% aluminium oxide, with a very wide range of shapes and sizes, is now standard thread-guide material. With care in the choice of starting powder particle size, and sintering conditions, a very smooth guide surface can be obtained without the need for glazing: this reduces wear damage to the filaments themselves. Figure 3.25 shows a (very) small selection of typical guides. Similar applications relying on the abrasive, and sliding wear resistance of alumina, include guides used in paper making, bearings, and water seals in pumps and washing machines, and single-lever hot- and cold-water mixing valves for showers and wash basins. Figure 3.26 shows a selection of typical alumina discs, 2–4 cm in diameter, used in a variety of hot- and cold-water mixer taps. Large quantities of ~95–96% alumina plates and tiles are used for lining dust, grit, and mineral handling systems, such as conveyer channels, and cyclonic collectors, because they have better resistance to erosive and abrasive wear resistant materials than steel, and are cheaper than tungsten carbide-nickel composites. Machining is a very important and specialised aspect of the production and finishing of many metal components. Most machining is carried out with hardened steel, or hard-metals such as tungsten carbide-nickel or cobalt composites. For specific applications, such as the machining of cast iron brake discs and drums, and engine cylinder liners, much higher cutting speeds, up to 10 times those obtainable with tungsten carbide based materials, with the added advantage of a better-quality finish, can be obtained from high-purity alumina (99%) cutting tool tips. Cutting tool tips of this type of dimensions of the order of 1 cm and a range of shapes (and tip holders) are shown in Fig. 3.27. The white tips are of pure alumina; the dark contain oxide pigments or particulate toughening phases such as titanium carbide.

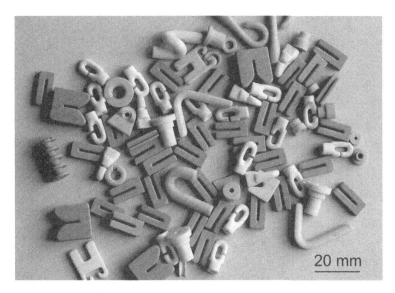

Figure 3.25 A small selection of the very wide range of types of thread guide used in the textiles industries. These are of different colours; some are pink because they contain a small amount of chromium oxide.

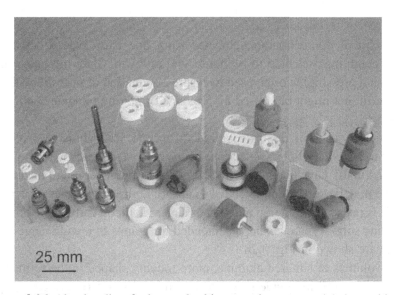

Figure 3.26 Alumina discs for hot- and cold-water mixer taps, and their cartridges. (Reprinted by kind permission of Saint-Gobain Advanced Ceramics, s.r.o.)

Because they are brittle, and susceptible to chipping, alumina tips require continuous cutting routines, with minimisation of impacts of the tool tip caused by discontinuities in the surface of the metal being worked. Strength, and wear resistance, in these components are both maximised by the use of very fine grain

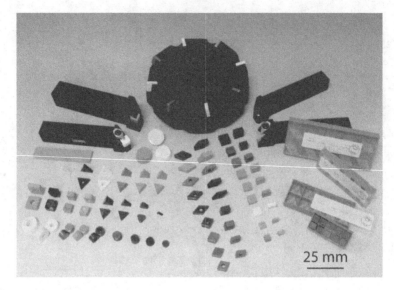

Figure 3.27 Cutting tool tips of a range of shapes and sizes (and their holders) used in high speed metal machining. (Reprinted by kind permission of Saint-Gobain Advanced Ceramics, s.r.o.)

(mean $< 5\,\mu m$) microstructure, obtained from fine particle size powders and sintering conditions selected to minimise grain growth. Sintering under applied pressure (hot pressing) is often used to ensure elimination of porosity without the grain growth that might be caused by prolonged sintering to achieve elimination of porosity.

The internal seals of pumps required to operate with abrasive or corrosive fluids, or to have very long life without maintenance (such as road vehicle water pumps), normally have one mating face produced from a very hard material. The other face may be softer, and intended to absorb abrasive particles (to which the hard face is resistant). Sintered alumina or sintered silicon carbide are commonly used sealing rings: typical examples intended for pumps of a wide range of types and sizes are shown in Fig. 3.28.

Large volumes of sintered alumina, more roughly shaped in the forms of tiles and plates, are used in a wide range of industries, for general protection against wear by abrasive powders and slurries, such as cement, ash, or sand. These can be fastened or used to line metal surfaces, such as in pipes, centrifuges and cyclonic dust collectors, where serious abrasion or erosion of the metal would otherwise occur. In similar types of wet processing involving abrasive particles, sintered alumina components are used in nozzles and seals coming into contact with the slurries. A very large volume of alumina, in spherical and cylindrical form, is

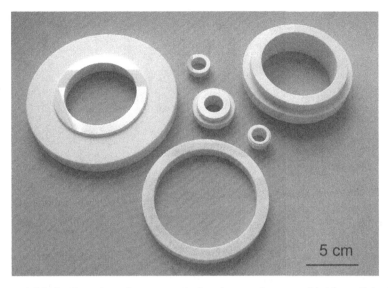

Figure 3.28 Sealing rings for pump shafts: the top faces are highly polished.

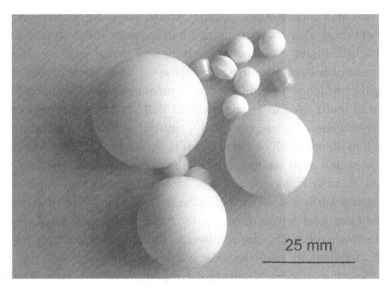

Figure 3.29 A small selection of milling media, for the dry and wet milling of fine powders.

used as grinding media for the production of fine powders – a small sample is shown in Fig. 3.29.

A further extension of the exploitation of the chemical inertness, and sliding wear resistance of alumina, is its use in prosthetic devices, commonly referred to

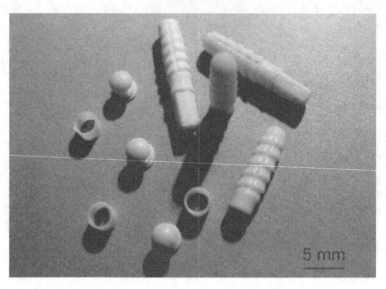

Figure 3.30 Dental implant pegs, made from high-purity sintered alumina. (Reprinted by kind permission of Saint-Gobain Advanced Ceramics, s.r.o.)

as *bioceramic* applications (Daculsi, and Layrolle, 2008). Restorative dental work is carried out using alumina pegs inserted into the gum, to which porcelain teeth can be attached. Figure 3.30 shows a selection of dental implants. More complicated, and much larger, prostheses are those for knee and hip joints, the most widely used of which is the hip-joint replacement unit, with the sintered alumina femoral head (ball) fixed to a stainless steel shaft, rotating in a high molecular weight polyethylene (or, less commonly, sintered alumina) acetabular cup. An example of an alumina ball (femoral head) is shown in Fig. 3.31. Alumina is by far the most widely used ceramic in total hip arthroplasty, and it is estimated that >5 million alumina femoral heads and >500 000 acetabular components have been implanted world-wide since their introduction in the 1970s (Rahaman *et al.*, 2007). The knee joint replacement is a far more complex structure (Fig. 3.32). Because the human body is very sensitive to foreign materials, very high-purity alumina is required to minimise possible leaching by the synovial fluid of impurity ions; the alumina is therefore normally >99.99% pure. The highest possible strength and sliding wear resistance are obtained by keeping the mean grain size as small as possible, in practice <4 µm. Even so, ball fracture is not uncommon when a hip is subjected to abnormally high impact loading, as a consequence of a fall for example. Considerable research effort is being devoted to the extension of alumina bioceramics into other implants and replacement body parts (Park, 1991). This work includes the development of porous aluminas, into which body tissue can grow and bond.

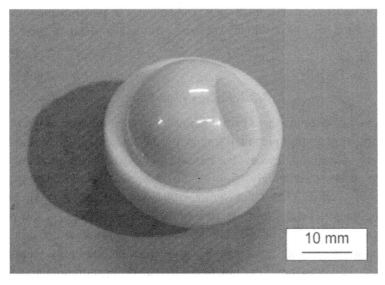

Figure 3.31 A high-purity sintered alumina femoral head, for hip-joint replacement. (Reprinted by kind permission of Saint-Gobain Advanced Ceramics, s.r.o.)

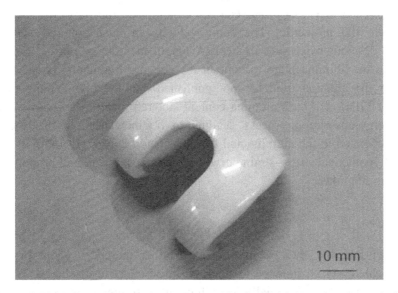

Figure 3.32 A knee joint replacement prosthesis, in high-purity sintered alumina. (Reprinted by kind permission of Saint-Gobain Advanced Ceramics, s.r.o.)

It is difficult to be specific with respect to the optimum alumina compositions
and microstructures required to provide maximised resistance to wear, because
the environmental characteristics, and wear mode, of each situation can vary
so widely. Moreover, in many cases the actual mechanisms of wear are very

imperfectly understood. For each new application, the material may have to be optimised on the basis of previous experience, through an iterative process of trial and error.

3.7.3 Energy absorption

The tendency of alumina, as a brittle material, to fracture with crack branching and the formation of multiple small fragments, has interestingly been exploited to create energy-absorbing devices. For more than 40 years, sintered alumina plates have been used as (relatively) lightweight protective armour plating, and in "bullet-proof" jackets (Viechnicki *et al.*, 1991; Franks *et al.*, 2008). There are two main principles behind this type of application: the hardness, or low plasticity under high local loading of the alumina tends to cause blunting of the more plastic and lower melting-point metallic projectiles or fragments, spreading the impact load. The subsequent fracture under impact of the alumina into very small particles then helps to further absorb the kinetic energy of the projectile, by the generation of alumina surface energy (with work of fracture values for alumina of the order of ~20 to ~50 J m^{-2}). A high sonic velocity is also important, with a large difference in velocity between the armour, and the projectile, helping to disintegrate the projectile. The sonic velocity in 99% sintered alumina is ~10 km s^{-1}, about twice that in steel. A tough, metallic or plastic, backing is applied to the alumina plate to contain the ceramic fragments. Exact compositions for the alumina used are not published, but the high Young modulus (~300–400 GPa) of a high alumina content sintered material will assist the primary energy-absorption process, leading to fragmentation, in that the strain energy generated in the alumina and then dissipated during cracking, is the area under a hypothetical stress–strain curve for the impact event, and proportional to the Young modulus.

3.7.4 Translucency

Many high-intensity fluorescent lights use either mercury vapour or sodium vapour, which can be induced to fluoresce in an electric discharge. While the vapour pressures are not really very high in absolute terms, to achieve the pressures needed high temperatures (~600 °C) are required. Sodium vapour is corrosive towards glass at these temperatures, but vapour-tight envelopes of pure alumina resist attack. Small dimension tubes (~1 to ~10 cm) of either man-made single-crystal sapphire, or (more usually, and lower cost) sintered polycrystalline translucent alumina, are therefore used as containers (Rhodes and Wei, 1991). Attainment of the necessary degree of translucency (typically ~85–90% in-line

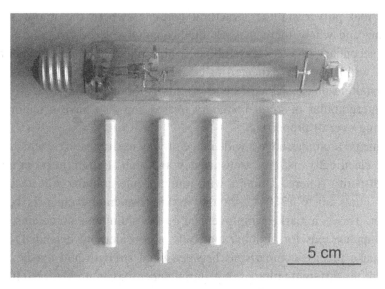

Figure 3.33 An example of a high-pressure sodium vapour lamp (~400 W) and a selection of translucent alumina tubes (with and without end-caps) which are the vapour envelopes. The completely transparent tube on the right is made from single-crystal aluminium oxide.

transmission for 0.75 mm thickness) can be obtained in a fully dense material, with large (20–100 μm) uniform equiaxed grains. These materials are produced by solid state sintering at high temperature (1700–1800 °C), with the incorporation of small amounts (<1%) of magnesium oxide as grain-growth inhibitor. The magnesium oxide is distributed uniformly throughout the grains; other ions may segregate at grain boundaries, with detrimental effects. Because the mean grain size is large, strengths are not high (often ~ 200–300 MPa), but then they do not need to be for this application. Figure 3.33 shows one type of high-pressure sodium vapour lamp (400 W), with a selection of typical tubes, with and without end-plugs. The transparent tube is made from single-crystal aluminium oxide (sapphire).

3.8 Summary

Most of the property differences between an aluminous porcelain and sintered alumina are the result of the continuing replacement of the mullite and quartz by fine-grain aluminium oxide, and the considerable reduction in the volume of glass. An important feature of sintered alumina is that the aluminium oxide crystals are now essentially (if nm films at grain faces are disregarded) the continuous phase. Alumina therefore provides an example of what might be called a real polycrystalline structural ceramic, where the major phase is crystalline, and where there may be small amounts of (or virtually no) intergranular secondary phases – crystalline or amorphous. The properties of these materials would then be expected to be related more strongly to those of the main crystalline phase. As will be seen, this is the case, but even so the secondary intergranular or grain boundary phases can also have a major rôle in determining overall properties.

The sintered aluminas have a wide range of very useful features, which include hardness, electrical resistivity, translucency and high strength at room temperature. High-purity sintered alumina with minimal intergranular silicate is usable at temperatures up to 1700 °C, though at this temperature creep is becoming significant. This is a marked improvement on the aluminous porcelains, and the range of applications for alumina materials is considerably extended. It is not easy to make precise comparisons between bend strength and tensile strength (because of the different volumes of material under maximum load) but if a three- or four-point bend strength of 600 MPa is taken to be approximately equivalent to a tensile strength of ~350 MPa, then a fine-grain sintered alumina is stronger than cast iron and compares favourably with a structural steel. The big difference is in the fracture toughness; this is of the order of 10 times greater in the steel – the result of the ductility and much larger strain to failure.

Strength increases are obtained by reducing the grain size (as is illustrated by Fig. 7.1), and very fine-grain materials can be produced by hot pressing. Grain growth tends to give large, slab-like, chunky grains, which do not seem to have the capacity to increase fracture toughness and strength to a great extent (unlike the rod-like grains developed in silicon nitride). Although sintered alumina is from the microstructural point of view a relatively simple material, because the thermal expansivity of aluminium oxide is anisotropic, there can be large residual stresses in large grains, with an associated potential for self-destruction on local loading. Additional residual stresses may be introduced through thermal expansion mismatch between aluminium oxide and secondary crystalline and glass phases. Reducing the glass content in liquid phase sintered material leads at first to improved strength and modulus, but thin intergranular amorphous films can also have a beneficial effect in reducing the rates of wear by abrasion. The reasons for this are still not fully understood, even though large quantities of alumina are used in applications depending on a resistance to wear. This illustrates the point that the properties of a structural ceramic are not entirely controlled by those of the major phase alone. In aqueous media, alumina (like the transformation toughened zirconia in Chapter 6) tends to be slowly corroded over long periods of time, and because of surface chemical reactions its wear resistance is not as good as it is under dry conditions.

Questions

3.1. Aluminium has a slightly smaller relative atomic mass than silicon (27.0 and 28.1), but aluminium oxide, Al_2O_3, is significantly more dense than quartz. Why is this?

3.2. Explain, on the bases of Figs. 3.6 and 3.7, which you would expect to give faster sintering, all other conditions being equal, for aluminium oxide: talc or wollastonite?

3.3. A crystal forming from solution prefers to develop as a plate, with preferential growth along the *a*- and *c*-axes of the crystal, while growth along the *b*-axis is inhibited. What information does this provide about the relative values of the interfacial energies of the crystal planes normal to these axes?

3.4. Why might small amounts of silicate impurity in a polycrystalline alumina tend to segregate to the grain boundaries?

3.5. Describe what phases you would expect to see develop, and in which sequence, at the surface of an aluminium oxide crystal, heated in contact with calcium monosilicate, $CaSiO_3$, at 1500 °C.

3.6. Why does the Young modulus of a polycrystalline alumina tend to decrease with increasing silicate content?

3.7. Estimate the size of the critical defect in a polycrystalline sintered alumina, with a room-temperature bend strength of 350 MPa, explaining any assumptions.

3.8. Describe the steps you would take to maximise (a) the room-temperature strength, and (b) the strength at 1000 °C, of a sintered alumina.

3.9. Compare the resistance to damage by a quench into water from 200 °C of pure polycrystalline alumina, with aluminous porcelain.

3.10. Why are small alumina insulators, such as those in spark plugs, often glazed, while alumina substrates for microelectrical components are not?

Selected reading

Dörre, E. and Hübner, H. (1984). *Alumina: Processing Properties and Applications.* Berlin: Springer-Verlag.

Evans, K. A. (1996). The manufacture of alumina and its use in ceramics and related applications. In *Key Engineering Materials 122–124, Advanced Ceramic Materials,* ed. H. Mostaghaci. Switzerland: TransTech Publications, pp. 489–526.

Morrell, R. (1987). *Handbook of Properties of Technical & Engineering Ceramics: Part 2 Data Reviews; Section I High-alumina Ceramics.* National Physical Laboratory, London: HMSO.

Richards, G. (1991). Aluminium oxide ceramics. In *Concise Encyclopedia of Advanced Ceramic Materials,* ed. R. J. Brook. Oxford: Pergamon Press, pp. 16–20.

References

Arkhagel, G. E., Morgensh, Z. L. and Neustrue, V. B. (1968). On nature of colour centres of ruby. *Phys. Stat. Solidi,* **29**, 831–56.

Bailey, J. T. (1991). Substrate materials. In *Concise Encyclopedia of Advanced Ceramic Materials*, ed. R. J. Brook. Oxford: Pergamon Press, pp. 466–8.

Bateman, C. A., Bennison, S. J. and Harmer, M. P. (1989). Mechanism for the rôle of magnesia in the sintering of alumina containing small amounts of a liquid phase. *J. Am. Ceram. Soc.*, **72**, 1241–4.

Bates, C. H., Drew, C. J. and Kell, R. C. (1971). A simplified process for preparing translucent alumina tubes from boehmite powder. *Trans. J. Brit. Ceram. Soc.*, **70**, 128–30.

Bayer, K. J. (1888). German Patent. *Verfahren zur Darstellung von Thonerdehydrat und Alkalialuminat*, No. 42977, 3rd Aug. 1888.

Bennison, S. J. and Harmer, M. P. (1990a). Effect of magnesia solute on surface-diffusion in sapphire and the rôle of magnesia in the sintering of alumina. *J. Amer. Ceram. Soc.*, **73**, 833–7.

Bennison, S. J. and Harmer, M. P. (1990b). A history of the rôle of MgO in the sintering of α-Al₂O₃. In *Sintering of Advanced Ceramics, Ceramic Transactions 7*, eds. C. A. Handwerker, J. E. Blundell and W. Kaysser. Westerville, OH: The American Ceramic Society, pp. 13–49.

Binns, D. B. (1965). Results and discussion of testing programme. In *The Use of Ceramics in Valves. Special publication*, No. 46, ed. P. Popper. British Ceramic Research Association, Stoke-on-Trent, UK, pp. 167–89.

Binns, D. B. and Popper, P. (1966). Mechanical properties of some commercial alumina ceramics. *Proc. Brit. Ceram. Soc.*, **6**, 71–82.

Bradt, R. C. (1967). Cr_2O_3 solid solution hardening of Al_2O_3. *J. Am. Ceram. Soc.*, **50**, 54–5.

Brewer, L. (1953). The thermodynamic properties of the oxides and their vaporization processes. *Chem. Rev.*, **52**, 1–75.

Brewer, L. and Searcy, A. W. (1951). The gaseous species of the Al-Al_2O_3 system. *J. Am. Chem. Soc.*, **73**, 5308–14.

Briggs, J. (2007). *Engineering Ceramics in Europe and the USA*. Worcester, UK: Menwith Wood.

Bromley, A. P., Hutson, D. A., Tovey, L. S. and Riley, F. L. (1996). Clean room processing of an alumina ceramic. *Ind. Ceram.*, **16**, 23–6.

Brook, R. J., Yee, J. and Kröger, F. A. (1971). Electrochemical cells and electrical conduction in pure and doped alumina. *J. Am. Ceram. Soc.*, **54**, 444–51.

Brydson, R., Twigg, P. C., Loughran, F. and Riley, F. L. (2001). Influence of CaO-SiO_2 ratio on the chemistry of intergranular films in liquid-phase sintered alumina, and implications for rate of erosive wear. *J. Mater. Res.*, **16**, 652–65.

Brydson, R. M., Chen, S.-C., Milne, S. J., Pan, X. and Rühle, M. (1998). Microstructure and chemistry of intergranular glassy films in liquid-phase sintered alumina. *J. Am. Ceram. Soc.*, **81**, 369–79.

Buckley, D. H. and Miyoshi, K. (1989). Tribological properties of structural ceramics. In *Structural Ceramics*, ed. J. B. Wachtman, Jr., *Treatise on Materials Science and Technology*, Vol. **29**. San Diego, CA: Academic Press, pp. 293–365.

Byrne, W. P., Hanney, M. J. and Morrell, R. (1982). Slow crack growth of oxide ceramics in corrosive environments. *Proc. Brit. Ceram. Soc.*, **32**, 303–14.

Cahoon, H. P. and Cristensen, C. J. (1956). Sintering and grain-growth of alpha-alumina. *J. Amer. Ceram. Soc.*, **39**, 337–44.

Cannon, R. M. and Coble, R. L. (1975). Review of diffusional creep in Al_2O_3. In *Plastic Deformation of Ceramic Materials*, eds. R. C. Bradt and R. E. Tressler. New York: Plenum Press, pp. 61–100.

Case, E. D., Smyth, J. R. and Hunter, O. (1983). Microcrack healing during the temperature cycling of single-phase ceramics. In *Fracture Mechanics of Ceramics*, Vol. **5**, eds. R. C. Bradt, A. G. Evans, D. P. H. Hasselman and F. F. Lange. New York: Plenum Press, pp. 507–30.

Charvat, F. R. and Kingery, W. D. (1957). Thermal conductivity XIII, effect of microstructure on conductivity of single phase ceramics. *J. Am. Ceram. Soc.*, **40**, 306–12.

Cheeseman, C. R. and Groves, G. W. (1985). The mechanism in the peak in strength and fracture toughness at elevated-temperatures in alumina containing a glass phase. *J. Mater. Sci.*, **20**, 2614–22.

Cho, S. J., Moon, H., Hockey, B. J. and Hsu, S. M. (1992). The transition from mild to severe wear in alumina during sliding. *Acta Metall. Mater.*, **40**, 185–92.

Chung, D. H. and Simmons, G. (1968). Pressure and temperature dependence of the isotropic elastic moduli of polycrystalline alumina. *J. Appl. Phys.*, **39**, 5316–26.

Claussen, N., Mussler, B. and Swain, M. V. (1982). Grain-size dependence of fracture energy in ceramics. *J. Am. Ceram. Soc.*, **65**, C14–16.

Claussen, N., Pabst, R. and Lahmann, C. P. (1975). Influence of microstructure of Al_2O_3 and ZrO_2 on K_{Ic}. *Proc. Brit. Ceram. Soc.*, **25**, 139–49.

Coble, R. L. (1961). Sintering crystalline solids. II. Experimental test of diffusion models in powder compacts. *J. Appl. Phys.*, **32**, 793–9.

Coble, R. L. (1994). Science of alumina: a retrospective, *J. Amer. Ceram. Soc.*, **77**, 293–453.

Coble, R. L. and Ellis, J. S. (1963). Hot-pressing alumina – mechanisms of material transport. *J. Am. Ceram. Soc.*, **46**, 438–41.

Coble, R. L. and Kingery, D. W. (1956). Effect of porosity on physical properties of sintered alumina. *J. Am. Ceram. Soc.*, **39**, 377–85.

Daculsi, G. and Layrolle, P., Eds. (2008). *Ceramics in Medicine: Key Engineering Materials*. Stafa-Zurich: TransTech Publications, Vols. **361–363**.

Davidge, R. W. (1980). *Mechanical Behaviour of Ceramics*. Cambridge: Cambridge University Press.

Davidge, R. W. and Riley, F. L. (1995). Grain-size dependence of the wear of alumina. *Wear*, **186**, 445–9.

Davidge, R. W. and Tappin, G. (1967). Thermal shock and fracture in ceramics. *Trans. Brit. Ceram. Soc.*, **66**, 405–22.

Davidge, R. W. and Tappin, G. (1970). The effects of temperature and environment on the strength of two polycrystalline aluminas. *Proc. Brit. Ceram. Soc.*, **15**, 47–60.

Dörre, E. and Hübner, H. (1984). *Alumina: Processing Properties and Applications*. Berlin: Springer-Verlag.

Elässer, C., Heuer, A. H., Rühle, M. and Wiederhorn, S. M. (2003). International workshop on the science and technology of alumina. *J. Amer. Ceram. Soc.*, **86**, 533–700.

Evans, A. G. and Tappin, G. (1972). Effects of microstructure on measurements of fracture energy of Al_2O_3. *Proc. Brit. Ceram. Soc.*, **20**, 275–97.

Evans, A. G. and Wilshire, T. R. (1976). Quasi static solid particle damage in brittle solids – I: Observations, analysis and implications. *Acta Metall.*, **24**, 939–56.

Evans, K. A. (1996). The manufacture of alumina and its use in ceramics and related applications. In *Key Engineering Materials*, Vol. **122–124**, *Advanced Ceramic Materials*, ed. H. Mostaghaci. Switzerland: TransTech Publications, pp. 489–526.

Fitzer, E. and Weisenberger, S. (1974). Cooperative measurement of heat transport properties of tungsten, alumina, and polycrystalline graphite up to 2700 K. In *Advances in Thermal Conductivity*, ed. R. L. Reisbig. Rolla, pp. 42–6.

Franks, L. P., Salem, J. and Zhu, D., Eds. (2008). *Advances in Ceramic Armor III. Ceram. Eng. Sci. Proc.*, Vol. 28. Hoboken, NJ: John Wiley.

Freer, R. (1980). Self-diffusion and impurity diffusion in oxides. *J. Mater. Sci.*, 15, 803–24.

Gereral Electric Co. Berlin (1913). German Patent. *werkzeuge wie Ziehsteine, für die Metallbearbeitung und Verfahren ihrer Herstellung*, No. 284808, 7th Feb. 1913.

Ghate, B. B., Smith, W. C., Kim, C. H., Hasselman, D. P. H. and Kane, G. E. (1975). Effect of chromia alloying on machining performance of alumina cutting tools. *Am. Ceram. Soc. Bull.*, 54, 210–15.

Gitzen, W. H. (1970). Alumina as a ceramic material. *Am. Ceram. Soc. Special Publication; no. 4*. Columbus, OH: The American Ceramic Society.

Gonzalez, A. C., Muthopp, H., Cook, R. F., Lawn, B. R. and Freiman, S. W. (1984). Fatigue properties of ceramics with natural and controlled flaws: a study on alumina. In *Methods for Assessing the Structural Reliability of Brittle Materials*, eds. S. W. Freiman and C. M. Hudson. ASTM Special Technical Publication 844, pp. 43–56.

Grabner, L. (1978). Spectroscopic technique for measurement of residual-stress in sintered Al_2O_3. *J. Appl. Phys.*, 49, 580–3.

Handwerker, C. A., Morris, P. A. and Coble, R. L. (1989). Effects of chemical inhomogeneities on grain-growth and microstructure in Al_2O_3. *J. Am. Ceram. Soc.*, 72, 130–6.

Hanney, M. J. and Morrell, R. (1982). Factors influencing the strength of a 95% alumina ceramic. *Proc. Brit. Ceram. Soc.*, 32, 277–90.

Hansen, S. C. and Phillips, D. S. (1983). Grain-boundary microstructures in a liquid-phase sintered alumina (alpha Al_2O_3). *Phil. Mag. A*, 47, 209–34.

Hing, P. (1976). The influence of some processing parameters on the optical and microstructural properties of sintered aluminas. *Sci. of Ceram.*, 8, 159–72.

Jan, J. P., Steinemann, S. and Dinichert, P. (1960). Density and lattice parameters of ruby. *J. Phys. Chem. Solids*, 12, 349–50.

Jupp, R. S., Stein, D. F. and Smith, D. W. (1980). Observations on the effect of calcium segregation on the fracture behaviour of polycrystalline alumina. *J. Mater. Sci.*, 15, 96–102.

Kaneno, M. and Oda, I. (1979). Effect of the properties of translucent alumina tube on lamp efficiency of high-pressure sodium lamp. In *Energy and Ceramics. Proc. 4th Int. Conf. Modern Ceramic Technologies*, Saint Vincent Italy, 28–31 May, 1979, ed. P. Vincenzini. Amsterdam: Elsevier, pp. 1114–22.

Kaplan, W. D., Mullejans, H., Rühle, M., Rodel, J. and Claussen, N. (1994). Ca-segregation to basal surfaces in alpha-alumina. *J. Am. Ceram. Soc.*, 78, 2841–4.

Kaysser, W. A., Sprissler, M. and Handwerker, C. A. (1987). Effect of a liquid-phase on the morphology of grain-growth in alumina. *J. Am. Ceram. Soc.*, 70, 339–43.

Kim, D. Y., Widerhorn, S. M., Hockey, B. J., Handwerker, C. A. and Blendell, J. E. (1994). Stability and surface energies of wetted grain boundaries in aluminium oxide. *J. Am. Ceram. Soc.*, 77, 444–53.

Kingery, W. D. (1959). Densification during sintering in the presence of a liquid phase: 1 theory. *J. Appl. Phys.*, 30, 301–6.

Kingery, W. D. and Narasimhan, M. D. (1959). Densification during sintering in the presence of a liquid phase: 2 experimental. *J. Appl. Phys.*, 30, 307–10.

Kirchner, H. P., Gruver, R. M. and Walker, R. E. (1973). Strengthening of hot-pressed alumina by quenching. *J. Am. Ceram. Soc.*, **56**, 17–21.

Kitaka, S., Yamaguchi, Y. and Takahashi, Y. (1992). Tribological characteristics of alpha alumina in high-temperature water. *J. Am. Ceram. Soc.*, **75**, 3075–80.

Knudsen, F. P. (1962). Effect of porosity on Young's modulus of alumina. *J. Am. Ceram. Soc.*, **45**, 94–5.

Kronberg, M. L. (1957). Plastic deformation of single crystals of sapphire: basal slip and twinning. *Acta. Met.*, **5**, 507–24.

Kumagai, M. and Messing, G. L. (1984). Enhanced densification of boehmite sol-gels by α-alumina seeding. *J. Am. Ceram. Soc.*, **67**, C230–1.

Levin, E. M., Robbins, C. R. and McMurdie, H. F. (1979a). *Phase Diagrams for Ceramists*. Columbus, OH: The American Ceramic Society, fourth printing, Figs. 407 and 501.

Levin, E. M., Robbins, C. R. and McMurdie, H. F. (1979b). *Phase Diagrams for Ceramists*. Columbus, OH: The American Ceramic Society, fourth printing, Figs. 630, 712 and 880.

Liu, H. Y. and Fine, M. E. (1993). Modelling of grain-size-dependent microfracture-controlled sliding wear in polycrystalline alumina. *J. Am. Ceram. Soc.*, **76**, 2393–6.

Ma, Q. and Clarke, D. R. (1994). Piezospectroscopic determination of residual-stresses in polycrystalline alumina. *J. Am. Ceram. Soc.*, **77**, 298–302.

Maiman, T. H. (1960). Stimulated optical radiation in ruby. *Nature*, **187**, 493–4.

McDougal, T. G. (1923). The story of spark plug progress. *J. Am. Ceram. Soc.*, **6**, 313–16.

McNamme, M. and Morrell, R. (1983). Textural effects in the microstructure of a 95% alumina ceramic and their relationship to strength. *Sci. Ceram.*, **12**, 629–34.

Meredith, H., Newy, C. W. A. and Pratt, P. L. (1972). The influence of texture on some mechanical properties of debased polycrystalline alumina. *Proc. Brit. Ceram. Soc.*, **20**, 299–316.

Miranda-Martinez, M., Davidge, R. W. and Riley, F. L. (1994). Grain size effects on the wet erosive wear of high purity polycrystalline alumina. *Wear*, **172**, 41–8.

Morrell, R. (1987a). *Handbook of Properties of Technical & Engineering Ceramics: Part 2 Data Reviews; Section I High-alumina Ceramics*. National Physical Laboratory, London: HMSO, pp. 37–57.

Morrell, R. (1987b). *Handbook of Properties of Technical & Engineering Ceramics: Part 2 Data Reviews; Section I High-alumina Ceramics*. National Physical Laboratory, London: HMSO, p. 124.

Morrell, R. (1987c). *Handbook of Properties of Technical & Engineering Ceramics: Part 2 Data Reviews; Section I High-alumina Ceramics*. National Physical Laboratory, London: HMSO, pp. 163–4.

Morrell, R. (1987d). *Handbook of Properties of Technical & Engineering Ceramics: Part 2 Data Reviews; Section I High-alumina Ceramics*. National Physical Laboratory, London: HMSO, p. 191.

Morrell, R. (1989a). *Handbook of Properties of Technical & Engineering Ceramics: Part 1: An Introduction for the Engineer and Designer*. National Physical Laboratory, London: HMSO, pp. 77–84.

Morrell, R. (1989b). *Handbook of Properties of Technical & Engineering Ceramics: Part 1: An Introduction for the Engineer and Designer*. National Physical Laboratory, London: HMSO, pp. 160–1.

Munro, R. G. (1997). Evaluated material properties for a sintered alpha-alumina. *J. Am. Ceram. Soc.*, **80**, 1919–28.

Newland, B. G. (1989). The production of high-grade technical ceramics. In *2nd European Symposium of Engineering Ceramics*, ed. F. L. Riley. London: Elsevier Applied Science, pp. 81–97.

Nielsen, T. H. and Leipold, M. H. (1963). Thermal expansion in air of ceramic oxides to 2200 °C. *J. Am. Ceram. Soc.*, **46**, 381–7.

Nishijima, T., Kawada, T. and Ishihata, A. (1965). Thermal conductivity of sintered UO_2 and Al_2O_3 at high temperatures. *J. Amer. Ceram. Soc.*, **48**, 31–4.

Oda, K. and Yoshio, T. (1997). Hydrothermal corrosion of alumina ceramics. *J. Am. Ceram. Soc.*, **80**, 3233–6.

Orgel, L. E. (1957). Ion compression and the colour of ruby. *Nature*, **179**, 1348.

Ott, W. R. (1991). Insulators for spark plugs. In *Concise Encyclopedia of Advanced Ceramic Materials*, ed. R. J. Brook. Oxford: Pergamon Press, pp. 242–8.

Özkan, O. T. and Moulson, A. J. (1970). The electrical conductivity of single-crystal and polycrystalline aluminium oxide. *Brit. J. Appl. Phys.*, **3**, 983–7.

Park, C. W. and Yoon, D. Y. (2002). Abnormal grain growth in alumina with anorthite liquid and the effect of MgO addition. *J. Am. Ceram. Soc.*, **85**, 1585–93.

Park, J. B. (1991). Aluminium oxide: biomedical applications. In *Concise Encyclopedia of Advanced Ceramic Materials*, ed. R. J. Brook. Oxford: Pergamon Press, pp. 16–20.

Pauling, L. (1960). *Nature of the Chemical Bond, and the Structure of Molecules and Crystals*. Ithaca, NY: Cornell University Press.

Peters, C. R., Bettman, M., Moore, J. W. and Glick, M. D. (1971). Refinement of structure of sodium beta-alumina. *Acta Cryst. B*, **27**, 1826–7.

Peters, D. W., Feinstein, L. and Peltzer, C. (1965). On the high-temperature conductivity of alumina. *J. Chem. Phys.*, **42**, 2345–6.

Phillips, D. S., Mitchell, T. E. and Heuer, A. H. (1980). Precipitation in star sapphire 3: chemical effects accompanying precipitation. *Phil. Mag. A*, **42**, 417–32.

Powell-Dogan, C. A. and Heuer, A. H. (1990a). Microstructure of 96% alumina ceramics. 1. Characterization of the as-sintered materials. *J. Am. Ceram. Soc.*, **73**, 3670–6.

Powell-Dogan, C. A. and Heuer, A. H. (1990b). Microstructure of 96% alumina ceramics. 2. Crystallization of high-magnesia grain-boundary glasses. *J. Am. Ceram. Soc.*, **73**, 3677–83.

Powell-Dogan, C. A. and Heuer, A. H. (1990c). Microstructure of 96% alumina ceramics. 3. Crystallization of high-calcia grain-boundary glasses. *J. Am. Ceram. Soc.*, **73**, 3684–9.

Rahaman, M. N., Aihua, Y., Bal, B. S., Garino, J. P. and Ries, M. D. (2007). Ceramics for prosthetic hip and knee joint replacement. *J. Am. Ceram. Soc.*, **90**, 1965–88.

Rhodes, W. H. and Wei, G. C. (1991). Lamp envelopes. In *Concise Encyclopedia of Advanced Ceramic Materials*, ed. R. J. Brook. Oxford: Pergamon Press, pp. 273–6.

Rice, R. W. (1977). Microstructure dependence of mechanical behavior of ceramics. In *Treatise on Materials Science and Technology*, Vol. **11**: *Properties and Microstructure*, ed. R. K. MacCrone. New York: Academic Press, pp. 199–381.

Rice, R. W. and Freiman, S. W. (1981). Grain-size dependence of fracture energy in ceramics: II. A model for non-cubic materials. *J. Am. Ceram. Soc.*, **64**, 350–4.

Richards, G. (1981). Alumina ceramics. *Trans. J. Br. Ceram. Soc.*, **80**, 120–4.

Richards, G. (1991). Aluminium oxide ceramics. In *Concise Encyclopedia of Advanced Ceramic Materials*, ed. R. J. Brook. Oxford: Pergamon Press, pp. 16–20.

Riddle, F. H. (1949). Ceramic spark-plug insulators. *J. Am. Ceram. Soc.*, **32**, 333–46.

Roy, R. S., Basu, D., Chanda, A. and Mitra, M. K. (2007). Distinct wear characteristics of submicrometre-grained alumina in air and distilled water: a brief analysis on experimental observation. *J. Am. Ceram. Soc.*, **90**, 2987–91.

Ryshkewitch, E. and Richerson, D. W. (1985a). *Oxide Ceramics: Physical Chemistry and Technology*, 2nd edition. Orlando, FL: Academic Press, pp. 109–256.

Ryshkewitch, E. and Richerson, D. W. (1985b). *Oxide Ceramics: Physical Chemistry and Technology*, 2nd edition. Orlando, FL: Academic Press, p. 125.

Ryshkewitch, E. and Richerson, D. W. (1985c). *Oxide Ceramics: Physical Chemistry and Technology*, 2nd edition. Orlando, FL: Academic Press, pp. 126–7.

Ryshkewitch, E. and Richerson, D. W. (1985d). *Oxide Ceramics: Physical Chemistry and Technology*, 2nd edition. Orlando, FL: Academic Press, pp. 161–3.

Sato, T., Sato, S., Okuwaki, A. and Tanaka, S. (1991). Corrosion behavior of alumina ceramics in caustic alkaline-solutions at high temperatures. *J. Am. Ceram. Soc.*, **74**, 3081–4.

Schacht, M., Boukis, N. and Dinjus, E. (2000). Corrosion of alumina ceramics in acidic aqueous solutions at high temperatures and pressures. *J. Mater. Sci.*, **35**, 6251–8.

Schreiber, E. and Anderson, O. L. (1966). Pressure derivatives of the sound velocities of polycrystalline alumina. *J. Am. Ceram. Soc.*, **49**, 184–90.

Siemens Brothers & Co. Berlin (1907). German Patent. *Verfahren zur Herstellung hochfeuerfester, geformter Körper aus geschmolzener Tonerde (künstlichem oder natürlichem Korund)*, No. 220394, 28th Sept. 1907.

Sousa, C., Illas, F. and Pacchioni, G. (1993). Can corundum be described as an ionic oxide. *J. Chem. Phys.*, **99**, 6818–23.

Spriggs, R. M. and Brisette, L. A. (1962). Expressions for shear modulus and Poisson's ratio of porous materials, particularly Al_2O_3. *J. Am. Ceram. Soc.*, **45**, 198–9.

Spriggs, R. M. and Vasilos, T. (1963). Effect of grain size on transverse bend strength of alumina and magnesia. *J. Am. Ceram. Soc.*, **46**, 224–8.

Spriggs, R. M., Mitchell, J. B. and Vasilos, T. (1964). Mechanical properties of pure dense aluminium oxide as a function of temperature and grain size. *J. Amer. Ceram. Soc.*, **47**, 323–7.

Tressler, R. E., Langensiepen, R. A. and Bradt, R. C. (1974). Surface-finish effects on strength-vs-grain-size relations in polycrystalline Al_2O_3. *J. Am. Ceram. Soc.*, **57**, 226–7.

Viechnicki, D. J., Slavin, M. J. and Kliman, M. I. (1991). Development and current status of armor ceramics. *Am. Ceram. Soc. Bull.*, **70**, 1035–9.

Wachtman, Jr., J. B. and Maxwell, L. H. (1959). Strength of synthetic single crystal sapphire and ruby as a function of temperature and orientation. *J. Am. Ceram. Soc.*, **42**, 432–53.

Wachtman, Jr., J. B., Tefft, W. E., Lam, Jr., D. G. and Stinchfield, R. P. (1961). Elastic constants of synthetic single crystal corundum from 77 to 850 K. In *Mechanical Properties of Engineering Ceramics*, cds. W. W. Kriegel and H. Palmour III. New York: Interscience, pp. 213–28.

Wachtman, J. B., Tefft, W. E., Lam, D. G. and Stinchfield, R. P. (1960). Elastic constants of synthetic single crystal corundum at room temperature. *J. Res. Natl. Bur. Stds. Section A – Physics and Chemistry*, **64**, 213–28.

Wallbridge, N. C., Dowson, D. and Roberts, E. W. (1983). The wear characteristics of sliding pairs of high density polycrystalline aluminium oxide under both dry and wet conditions. *Proc. Int. Conf. on the Wear of Materials*. Reston, VA: ASME, pp. 202–18.

Westbrook, J. H. and Jorgensen, P. J. (1965). Indentation creep of solids. *Trans. AIME*, **233**, 425–8.

Wiederhorn, S. M. (1969). Fracture of sapphire. *J. Am. Ceram. Soc.*, **52**, 485–91.

Wiederhorn, S. M., Hockey, B. J. and Roberts, D. E. (1973). Effect of temperature on the fracture of sapphire. *Phil. Mag.*, **28**, 783–96.

Will, F. G., Delorenzi, H. G. and Janora, K. H. (1992). Conduction mechanism of single crystal alumina. *J. Am. Ceram. Soc.*, **75**, 295–304.

4

Silicon carbide

4.1 Description and history

Silicon carbide is a member of the group of materials usually termed the *non-oxides*. These are compounds of metals (or semi-metals) with non-metallic elements other than oxygen. This large group of ceramic materials includes many carbides, nitrides, and borides, of which silicon carbide is the most widely used. Silicon carbide is notable for its hardness, and very high melting-point, but it has other extremely useful properties. It has well-established and widespread industrial applications as a refractory (Geiger, 1923; Singer and Singer, 1963), as an abrasive, and, relatively recently, as a high-grade structural ceramic with very good wear resistance.

Silicon carbide does not occur in significant quantities in the strongly oxidising environment of the Earth's crust because it is readily oxidised (that is, it is thermodynamically unstable with respect to oxidation). The detection of traces of silicon carbide in an Arizona iron meteorite was first reported by Henri Moissan in 1904, and the mineral became known as moissanite (Moissan, 1905). Other deposits of silicon carbide subsequently identified were also believed to be exclusively of extraterrestrial origin. However, it is now clear that it can also occur as small crystals trapped in oxide minerals, formed under special conditions in a very early period of the development of the Earth's crust, where high temperatures and extensive carbon deposits are likely to have existed. This origin also explains why silicon carbide is sometimes found in association with diamond, and graphite. Because of the absence of significant natural deposits of silicon carbide, it has to be prepared from other raw materials. Man-made silicon carbide was first produced, and named *carborundum*, by Edward Goodrich Acheson in 1891 – the unexpected outcome of attempts to synthesise diamond (Acheson, 1893; Schröder, 1986). Today the worldwide production of silicon carbide by the Acheson process runs to several hundred thousand tonnes a year, of which a very small proportion (of the order of 10 000–20 000 tonnes) is used for the production of fine-grain structural ceramics (Briggs, 2007).

Silicon carbide can be produced by direct reaction of its constituent elements at high temperature, using carbon and liquid silicon, or silicon vapour:

$$Si_{(l,v)} + C_{(s)} = SiC_{(s)}; \quad \Delta G^{\circ}_{1500K} = -61.5 \text{ kJ mol}^{-1}. \tag{4.1}$$

However, the basis of Acheson's process, which is still used essentially unchanged today for almost all large-scale industrial production of silicon carbide, is the reduction of pure silicon dioxide (as *sand*), by carbon (as coke or coal):

$$SiO_{2(s)} + 3C_{(s)} = SiC_{(s)} + 2CO_{(g)}. \tag{4.2}$$

The usual reaction mixture consists of the readily available high-purity quartz sand, and petroleum coke or anthracite coal, providing the essential chemical ingredients. To this mixture is usually incorporated sawdust (~1–20%) to maintain an open structure to facilitate escape of the carbon monoxide, and a source of chlorine such as sodium chloride (~1–10%) to bring about the volatilisation of common impurity elements such as iron and aluminium as their chlorides. Electrical resistance heating with inserted graphite electrodes is used to generate temperatures of ~1800 °C and above, and because silicon carbide in its normal slightly impure state is a good electrical conductor, electrical heating can easily be continued for the duration of the process. The maximum core temperature of around 2400 °C is reached after about 24 h. At this temperature, considerable recrystallisation of the silicon carbide product occurs, and the product mostly consists of very coarse (cm dimension) plate-like crystals. Figure 4.1 shows a mass of crystalline silicon carbide, typical of that produced in the Acheson process, and showing the characteristic hexagonal plate-like morphology of large crystals. For its applications in the abrasives, refractory, and ceramics industries, the crystals are ground to grits and very fine (μm and smaller) powders. Because large amounts of electricity are required, the commercial production of silicon carbide has often been developed at sites close to cheap sources of hydroelectric power, such as at the Niagara Falls in the USA, and in Norway.

High-purity and very fine silicon carbide powders can also be prepared by a number of routes, involving the decomposition of precursors containing silicon and carbon, and by reactions of vapour phase species containing silicon (Wei, 1983; Schmidt *et al.*, 1991).

Silicon carbide has had long and extensive industrial use (Srinivasan, 1989; Schlichting and Riley, 1991). Silicon carbide coarse powders and grits can be bonded to form moderately strong blocks and more complex shapes, using a range of methods. These materials are routinely used on a very large scale, in a

Figure 4.1 A mass of very coarse silicon carbide crystals, produced by the Acheson process. (Reproduced by kind permission of Kanthal Ltd.)

wide range of high-temperature furnaces, including the linings of blast furnaces used for the production of iron. The use of silicon carbide as a refractory container for liquid iron depends in part on its high melting-point, but also, paradoxically, on a very high thermal conductivity. The explanation for this is that water-cooling of the outside surfaces of silicon carbide blocks allows the freezing of a thin layer of metal on the inner surfaces that then protects the refractory against attack by molten metal, and associated oxides usually known as *slag*. A large quantity of silicon carbide grit and powder with carefully graded particle sizes is used for the production of *carborundum* paper, and the fabrication of cutting and grinding wheels. The usefulness of silicon carbide in these applications is dependent, partly, as might have been expected, on its hardness, and partly on an unexpected aspect of its fracture behaviour – the tendency for single crystals to cleave, and to generate fresh, sharp, crystal edges. In contrast, aluminium oxide crystals (the white *corundum* materials), which are also widely used as abrasives, tend to wear to smooth, rounded, particles, and lose their cutting power more readily.

Interest in the finer grain and much higher strength forms of silicon carbide, which form the basis for this chapter, developed rapidly in the 1960s and 1970s, as part of a systematic search for new ceramic materials (or new forms of old materials), which could be used at much higher temperatures than metals. It was anticipated that these ceramics could be used in the new nuclear power industries, and in new high-temperature internal combustion engines. In the 1970s silicon carbide was selected, together with silicon nitride, as a ceramic material likely to be useful for applications at temperatures of >2500 °F (~1371 °C), and it was included in ambitious, industrial and Government funded programmes of development work intended to demonstrate the practicality of ceramic components in the hot zones of new ranges of high-efficiency gas-turbine, and diesel, engines (Burke *et al.*, 1974; Katz, 1980; Katz, 1983). These expectations were mostly not realised, for a number of reasons unconnected with the high-temperature properties of the ceramics. However, other applications have been found for high-strength, fine-grain silicon carbide materials, and these now have an established range of specialist niche markets, though the scale of applications for the silicon carbide ceramics remains smaller (~10% by weight) than those of the alumina group of ceramics. There is now renewed interest in silicon carbide as a consequence of the needs to filter very fine particles from industrial and automobile hot exhaust gases.

4.2 Basic aspects

4.2.1 Atomic structure

There is only one solid silicon carbide (SiC) with the atomic proportions expected for a compound between two Group IV elements. It is approximately 85% covalent (Phillips, 1970), and the small, tetrahedrally coordinated, Si and C atoms provide a rigid three-dimensional crystal structure and very strong interatomic bonding. However, this apparent simplicity is deceptive: there are well over 100 established, distinct, crystal structures, known as *polytypes*, the differences between which are the result of subtle variations in the stacking arrangements of the "SiC$_4$" and "CSi$_4$" tetrahedra (Jepps *et al.*, 1979; Verma and Krishna, 1996). Fortunately – from the point of view of the use of silicon carbide as a structural material – the polytypes have very similar physical and mechanical properties, and for most practical purposes silicon carbide can safely be treated as one crystal phase, or maybe at the most, two.

The melting-point of silicon carbide is estimated to be at least 2500 °C (Elliott, 1965); this can be attributed to the rigid and mainly covalent tetrahedral bonding, which would also give a highly structured liquid and thus a small entropy of fusion. However, the exact value is of somewhat academic interest because at these very high temperatures evaporation and decomposition are also occurring (Drowart *et al.*, 1958; Price, 1977). There is also the inherent problem of oxidation. Because of its high hardness (Shaffer, 1965; Brookes *et al.*, 1971) – slightly higher than that of corundum – and a refractive index of around 2.7 (Shaffer, 1971), silicon carbide has been regarded as a potential gemstone.

In spite of its hardness and intrinsic strength, milling of coarse crystals is readily achieved using iron rollers, which become silicon carbide coated, and are therefore effectively silicon carbide rollers.

Very pure silicon carbide is (as might be predicted on the basis of its covalent single bonding) practically colourless, and an electrical insulator (like aluminium oxide). In practice, however, silicon carbide is normally coloured – with colours ranging from (predominantly) very dark blue or green, to almost black – and it is an electrical semiconductor with high conductivity at room temperature. The colouration (and the electrical conductivity) depends on traces of different valency (*aliovalent*) impurities occupying Si and C lattice sites. These impurities create impurity energy levels, effectively narrowing the band gap width, and allowing electrons to jump (*electron promotion*) to otherwise empty conduction bands (Patrick and Choyke, 1969; Kim *et al.*, 1995). The photons absorbed during the promotion of electrons have energies corresponding to the visible spectrum, giving rise to the colours seen. Contamination with boron or aluminium at concentrations of ~50 ppm tends to give dark colours, blue, brown or black; nitrogen gives orange-to-red colouration which is less common. The conduction bands permit electrons to flow readily through the crystal lattice, giving rise to the phenomena of *n*- and *p*-type electrical semiconductivity: this is of considerable practical importance (North and Gilchrist, 1981). This electrical conductivity is completely different in character from the conduction at high temperature of porcelain and aluminium oxide, which is mainly ionic. The wide band gap and related electronic properties of silicon carbide are of considerable technical importance. There is a very wide range of application areas for thin-film and single-crystal silicon carbide devices, which includes optoelectronics, high-temperature electronics and high-power/high-frequency devices (Palmour *et al.*, 1993; Nakashima *et al.*, 1996), but this aspect of the behaviour of silicon carbide is outside the scope of this book.

Most commercial silicon carbide (*carborundum*) grits and powders are generally not very pure, and absorb photons of a wide range of energies. The normal very dark grey colour of commercial powders is believed to be caused by relatively high levels of aluminium impurity, introduced with the starting sand as aluminosilicate minerals and not completely eliminated as the volatile aluminium chloride during production. Very high-purity silicon carbide powders are normally pale grey/green (and much more expensive).

4.2.2 Crystal structure

The crystal structure of silicon carbide is not atomically close-packed, unlike the structure of aluminium oxide. It can be regarded as consisting of covalently

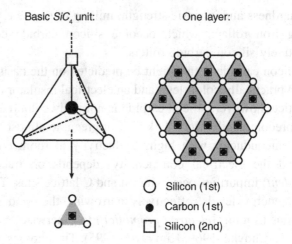

Figure 4.2 The basic tetrahedral SiC$_4$ unit of the silicon carbide crystal structure, and one layer constructed from these units.

bonded tetrahedral units (arbitrarily chosen to be either SiC_4, or CSi_4) each with the expected internal tetrahedral coordination based on carbon and silicon sp$_3$ orbital hybridisation and a bond angle of about 109°. One SiC$_4$ tetrahedral unit, and one layer of the silicon carbide structure built from these units, are shown schematically in Fig. 4.2. The crystal structure can be thought of as constructed by stacking SiC layers on top of each other (and designated A, B, or C, according to the way in which each sits in relation to the first layer), in arrays, which can have overall cubic, hexagonal, or rhombohedral symmetry. If the tetrahedra were simply large spheres, these structures could be regarded as cubic ABC close-packed arrangements, or hexagonal close-packed, depending on the orientation and alignment of each new layer of tetrahedra. Of the hexagonal structures, the ABAB close-packed arrangement is the most simple. The cubic close-packed (or *face-centred cubic*) crystal structure corresponds to the diamond crystal structure in which half the carbon atoms in the diamond have been replaced in a regular way by silicon atoms. This form is termed β-*silicon carbide*, and is unique. Because practically all the physical and other properties of the simple ABAB hexagonal structure and the other more complex structures with hexagonal or rhombohedral symmetry are closely similar, these forms are usually grouped together and described collectively as α-*silicon carbide*. All these structures are known as the silicon carbide *polytypes*.

4.2.3 Polytypism

The crystallography of silicon carbide is dominated by the phenomenon of *polytypism*. This topic is examined briefly for the sake of completeness, though in

Table 4.1 *The most common silicon carbide polytypes: unit cell dimensions and densities. (Snead et al., 2007.)*

Polytype		Unit cell dimensions / pm		Density / Mg m^{-3}
		a^*	c	
β	3C	436	–	3.215
"α"	2H	308	503	3.219
	4H	308	1008	3.215
	6H	308	1512	3.215
	15R	307	3780	–

* Hexagonal cell.

fact its importance for the behaviour of silicon carbide as a structural ceramic is minimal. The subject of polytypism has been reviewed in detail (Verma and Krishna, 1996; Nakashima and Harima, 1997). Polytypism is not unique to silicon carbide, but is shown most spectacularly by it. Through regular repeated alterations in the ways in which the A, B or C sheets of tetrahedra are stacked along the c-axis of the unit cell, slightly different crystal structures of the material can be formed. Several notations have been devised for describing these different stacking arrangements. One widely used notation, the *Ramsdell* (Ramsdell, 1947), uses a numeral to represent the number of layers in the unit cell, to which is added the letter H, R, or C to specify the lattice symmetry, and to indicate a hexagonal, rhombohedral or cubic cell. The simple ABC, β-SiC form, is the cubic 3C polytype. Two other common polytypes are the hexagonal 6H, with the ABCACB repeat pattern of tetrahedra, and the rhombohedral 15R, in which the sheets of tetrahedra have the repeated stacking arrangement ABCBACABACBCACB. The a-axis lattice parameters are approximately constant at ~308 pm; the c-axis parameter varies from ~503 pm for the 2H polytype, through ~1008 pm for the 4H, to claimed values of more than 200 nm for some of the higher order and less certain polytypes (attaining unit cells of μm length). Table 4.1 summarises unit cell data for the more commonly met polytypes. It is of purely academic interest that, on the basis of 50 SiC layers, and using all possible ABC sequences, the existence of more than 93 million polytypes can be predicted (McLarnan, 1981). However, it is, not surprisingly, quite difficult to differentiate the polytypes – about 200 have been identified – on the basis of their X-ray diffraction patterns, and the real existence of many of the less common, and crystallographically more complex, polytypes has often been disputed (Fisher and Barnes, 1990).

The fundamental reasons for the ability of silicon carbide to form such large numbers of polytype structures are still of considerable interest (Bechstedt *et al.*, 1997). However, in discussions of silicon carbide as a structural ceramic,

the more exotic aspects of the polytypes, including the theories for their existence, the mechanisms by which they are formed, and their long-range ordering,
can be safely ignored. The only point perhaps worth noting here is that it is
believed that polytype stability may be influenced by the impurity element type
and content, and the temperature of production or crystallisation. The simplest
hexagonal, AB, or 2H polytype, and the almost as simple 4H variant, ABCB, are
most commonly produced at "lower" temperatures (which means below
~1800 °C); the most stable (and most commonly seen) polytype appears to be the
6H (Jepps and Page, 1983). The cubic 3C polytype is also usually regarded as a
"low-temperature" form, and it readily converts to an α-polytype on heating to
temperatures in the region of 2000 °C: this feature has implications for the
development of microstructure in sintered silicon carbide.

4.3 Physical and mechanical properties

Much of the earlier important literature on silicon carbide (up to about 1986) has
been systematically summarised and reviewed in two Silicon Supplement Volumes, B2 and B3, of the *Gmelin Handbook of Inorganic Chemistry* (Kirschstein,
and Koschel, 1984; Schröder, 1986). The physical and mechanical properties of
silicon carbide in its single and polycrystalline forms are mainly dealt with in
Volume B2. The manufacture of silicon carbide ceramics, special forms of silicon
carbide (single crystals, fibres and coatings), and general applications, are
covered in Volume B3. Other collections of property data have been published
more recently (Pierson, 1996), and reviews carried out (Schwetz, 2000; Roewer
et al., 2002). Precise property values are of particular importance for the applications of silicon carbide in the nuclear fuel industries, and these have been
reviewed in detail (Snead *et al.*, 2007).

   The fundamental, microstructure-insensitive properties (which can be called
the *intrinsic* properties) of silicon carbide, notionally in its single-crystal form,
will be looked at first. As is the case with the alumina group, the properties of
polycrystalline ceramic materials containing silicon carbide as their major phase
are determined by those of the silicon carbide single crystal, modified by the
presence of secondary phases (which includes porosity, and silicon), and the
material's microstructure – of which mean grain size and grain aspect ratio are of
primary importance. Large, high-purity, single crystals of silicon carbide, usually
as thin wafers, can be produced by a range of high-temperature sublimation
techniques (Nakashima *et al.*, 1996), but mechanical and physical properties have
generally been measured either on highly orientated (and often plate-like) single
crystals grown directly during the Acheson process, or on polycrystalline
materials. In the brief review that follows, data are presented where possible for

single-crystal forms. Discussion of the properties of polycrystalline ceramic forms of silicon carbide will be kept to a minimum.

4.3.1 Density

Silicon carbide, like aluminium oxide, is constructed from low atomic mass elements (the total relative mass number is ~40.1); a low physical density, perhaps close to that of corundum, might therefore have been expected. Because of the directional four-coordinate bonding, the silicon carbide structure is not close packed with respect to the atoms (atomic close packing would require 12-coordination). The theoretical density at room temperature calculated from lattice constants is about 3.21 Mg m^{-3}, irrespective of polytype (Taylor and Jones, 1960). This is slightly lower than that of the more strongly bonded diamond allotrope of carbon (which has the same face-centred cubic crystal structure as the ABC polytype), with a density of 3.5 Mg m^{-3}. Polycrystalline materials (which can contain sintering aids) are usually slightly impure (or porous), and tend to have lower densities (Prochazka and Scanlon, 1975). α-Aluminium oxide (corundum), with the true hexagonal close-packed structure, has the slightly higher density of ~3.99 Mg m^{-3}.

4.3.2 Young modulus

The Young modulus values (E) measured using a micropenetration method on pure crystals are widely scattered, but the maximum is in the range 380–450 GPa (Coblenz, 1975). The short (~190 pm), strong, Si–C bonds presumably account for the high stiffness. Values of the room temperature Young modulus for polycrystalline, slightly porous, materials have been fitted to standard empirical equations (Chapter 1, Section 1.4) to give values for fully dense materials in the range 448–460 GPa (Shaffer and Jun, 1962; Carnahan, 1968). Stiffness decreases with increasing levels of impurity (Grow *et al.*, 1994), and gradually falls with temperature to reach ~90% of the room-temperature value at ~1200 °C (Coblenz, 1975).

4.3.3 Poisson ratio

Values of the bulk modulus (K) extrapolated to zero porosity using standard empirical expressions are in the region of 203 GPa (Yean and Riter, 1971; Munro, 1997). The shear modulus (μ) at room temperature is 179 GPa (Munro, 1997). Values of the Poisson ratio are in the range 0.14–0.19 (Carnahan, 1968), which is at the low end of the range for the technical ceramics.

4.3.4 Fracture toughness and fracture energy

Perhaps unexpectedly, given its high hardness and strong bonding, silicon carbide is not a tough material, even by ceramic standards. Measurements of fracture toughness (K_{Ic}) on single crystals of α-silicon carbide (6H) at temperatures between 300 and 1000 K (Henshall *et al.*, 1977; Faber and Evans, 1983), give a value of 3.3 MPa m$^{1/2}$. With polycrystalline sintered silicon carbide, mean values obtained from a range of dense materials are 3.1–3.8 MPa m$^{1/2}$, with a spread of ±1 MPa m$^{1/2}$ (Anstis *et al.*, 1981; Munro, 1997). These values appear to be independent of material purity but there is a clear effect of grain size, with the maximum corresponding to a grain size in the region of 1–5 μm.

The single-crystal value for fracture energy (γ_f) is reported to be 23 J m^{-2} (Henshall *et al.*, 1977). Most values for polycrystalline materials fall in the range 20–30 J m^{-2}, depending on grain size (Faber and Evans, 1983; Morrell, 1989).

4.3.5 Hardness

On the classical Mohs scale of hardness, silicon carbide was placed between corundum (9) and diamond (10). Single-crystal indentation hardness varies with crystal plane, and is a function of load, with smaller loads giving the higher values (Sawyer *et al.*, 1980; Fujita *et al.*, 1986). Standard hardness measurements give room-temperature values of up to ~30 GPa, and even higher values have been reported (Hojo *et al.*, 1983; Munro, 1997). These indicate that silicon carbide is slightly harder than α-aluminium oxide. Values for polycrystalline silicon carbide materials are generally in the region of 20–30 GPa, and small amounts of residual porosity, and impurity levels, must be controlling factors (Lankford, 1983). With increasing temperature, hardness decreases, reaching about 50% of the room-temperature value at 800 °C (Hirai and Niihara, 1979).

4.3.6 Mechanical strength

An estimate of the theoretical tensile strength of silicon carbide single crystal indicates a value in the region of 40 GPa (Veldkamp, 1975). As with single-crystal aluminium oxide, the highest tensile strengths have been measured on very fine whiskers, and room-temperature values have been obtained within the range 9–30 GPa. Bend strengths of small plate-like single crystals are in the region of 1 GPa (Hasseman and Batha, 1963). Very fine-grain (<1 μm) poly-crystalline fibres of silicon carbide with μm diameters have tensile strengths in the region of 1–5 GPa, and the maximum values are obtained with the finest fibres as would be expected. These lower strength values reflect the presence of

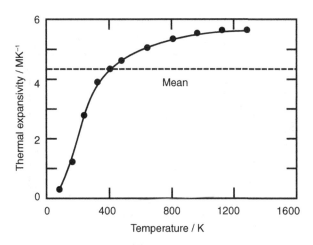

Figure 4.3 Thermal expansivity as a function of thermodynamic temperature: the mean value over 25 to 1250 °C is 4.4 MK^{-1}. (After Slack and Bartram, 1975.)

strength-controlling defects (including grain boundaries) and some oxide, and probably indicate the more realistic target for silicon carbide in its normal polycrystalline ceramic form (Simon and Bunsell, 1984).

4.3.7 Thermal expansivity

The low thermal expansion coefficient is the result of the high Si–C bond strength. There are significant differences in expansion behaviour between the polytypes, and there is anisotropy in the hexagonal and rhombohedral polytypes (Li and Bradt, 1987). Values of expansion coefficient are a function of temperature, and vary smoothly from ~3 MK^{-1} at room temperature to ~5 MK^{-1} at 1000 °C (Popper and Moyuddin, 1964), above which point the expansivity curve levels off (Slack and Bartram, 1975). The typical variation is shown in Fig. 4.3. A value commonly quoted is that of the mean expansivity over the range of room temperature to ~1250 °C. This typically is ~4.4 MK^{-1} (Forrest *et al.*, 1972), which is significantly smaller than the values for other structural ceramics such as alumina (~8.5 MK^{-1}) and porcelain (~7 MK^{-1}), though slightly larger than the mean value for silicon nitride over this range (~3.6 MK^{-1}). The low expansion coefficient was a major factor underlying the selection in the 1960s of silicon carbide ceramics for development as high-temperature internal combustion engine materials, because of their expected better resistance to thermal shock.

4.3.8 Thermal conductivity

Silicon carbide is notable for a very high thermal conductivity. Over temperature ranges of practical interest it is a better thermal conductor than almost all other

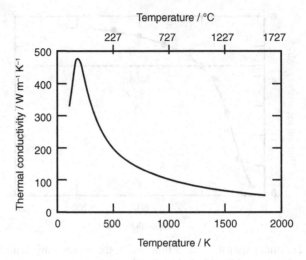

Figure 4.4 Thermal conductivity of pure, dense polycrystalline silicon carbide, as a function of temperature, over a wide range of temperature. (Reproduced by kind permission of Lance Snead.)

ceramic materials, and even better than many metals, which are normally very good thermal conductors (Touloukian, 1977). The thermal conductivity of any material is a function of purity and temperature. The maximum thermal conductivity value is reached between 50 and 200 K; very high-purity single crystals have a thermal conductivity at 50 K of more than ~5000 W m^{-1} K^{-1} (Powell and Tye, 1969; Kingery and Vasilos, 1954). This means that (in the temperature range of more interest for structural applications) single-crystal values can still be high: ~950 W m^{-1} K^{-1} at ~200 K (Slack, 1973; Munro, 1997). Figure 4.4 illustrates the general pattern of behaviour for dense polycrystalline silicon carbide. The conductivity of ~300 W m^{-1} K^{-1} at 300 K can be compared with those of the high thermal conductivity metals such as copper and silver, with values in the region of 400 W m^{-1} K^{-1}, and diamond, with ~2000 W m^{-1} K^{-1} at 300 K. At higher temperatures, the thermal conductivity steadily decreases: by 700 °C values have fallen to ~100 W m^{-1} K^{-1}, though this is still very high for a ceramic. Silicon carbide is particularly sensitive to the concentration of lattice impurity atoms, at levels of much less than 1% (Slack, 1964; Burgermeister *et al.*, 1979). Contamination with low atomic number elements able to occupy silicon carbide lattice sites, such as N and B, can reduce the conductivity by a factor of up to 50. Grain boundaries, and the free carbon in sintered silicon carbide (which may be present at levels as high as 3%), can also reduce the thermal conductivity (Collins *et al.*, 1990). Figure 4.5 shows collected thermal conductivity data for commercial sintered, dense, silicon carbide materials over the temperature range of more interest for structural applications.

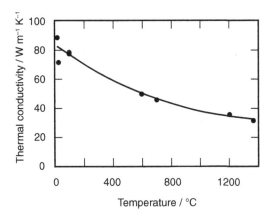

Figure 4.5 Collected thermal conductivity data for dense polycrystalline silicon carbide. (After Kirschstein and Koschel, 1984, pp. 125–9.)

4.3.9 Thermal shock

The high thermal conductivity of silicon carbide is the second factor underpinning an ability to resist thermal shock (Coppela and Bradt, 1973; Weaver *et al.*, 1975). Thermal gradients at the surfaces of rapidly heated or cooled components may (in theory) be much less severe, diminishing the extent of thermal expansion or contraction and hence consequent stresses. Table 4.2 shows values of commonly applied thermal shock, *R*, parameters for several silicon carbide materials, and those for a typical dense sintered alumina for comparison (Edington *et al.*, 1975). The values are slightly arbitrary, in that they depend on the materials' parameters selected, but they indicate that the dense silicon carbides should be significantly better at resisting sudden changes of temperature than alumina, because of their higher thermal conductivities and lower thermal expansivity. Experimental measurements of the loss of strength of these common types of silicon carbide after quenching into water show that the critical quench temperature for significant (>50%) loss of strength is in the range 290–450 °C. The different forms of material respond differently to the quench (as would have been predicted from the *R* parameters). For the porous recrystallised forms the critical temperature ($\Delta T_{\mathrm{critical}}$) is ~300 °C; for the high-density hot-pressed silicon carbide $\Delta T_{\mathrm{critical}}$ is ~425 °C. Reaction-bonded silicon carbide containing the second phase silicon has a higher value, 445 °C, suggesting the fracture energy may be being dissipated by crack branching at the silicon carbide–silicon grain boundaries (Weaver *et al.*, 1975). Of the whole set shown here, the higher-strength sintered silicon carbide and the alumina appear to be the least able to resist propagation of a pre-existing crack (Ainswort and Moore, 1969).

Table 4.2 *Representative mechanical and physical property data for silicon carbide ceramics, and derived thermal shock parameters. Data are shown for pure polycrystalline alumina, for comparison.*

Material	Property/Unit Bend strength / MPa	Modulus / GPa	Poisson ratio	Expansivity / MK^{-1}	Conductivity / W m^{-1} K^{-1}	R / K	R' / kW m^{-1}	R''' / MPa^{-1}
Recrystallised	80	240	0.18	4.8	25	57	1.4	46
Reaction bonded	250	360	0.18	4.1	35	139	4.9	7.0
Sintered	550	410	0.18	4.8	77	229	17.6	1.7
Polycrystalline alumina	400	380	0.25	8.2	20	96	1.9	3.2

Room-temperature strength and modulus values; thermal expansivity over the range 25–1000 °C; thermal conductivity at 700 °C.
$R = \sigma\,(1-\nu)\,/E\alpha$; $R' = \sigma\,(1-\nu)\lambda/E\alpha$; $R''' = E\,/\,[\sigma^2\,(1-\nu)]$.

4.3.10 Electrical conductivity

The electrical behaviour of silicon carbide (like its crystallography) is quite sophisticated (Lambrecht *et al.*, 1997) and is rather remarkable in that it can be controlled to range from being a good insulator to a good conductor. The carbon 2s and 2p and the silicon 3s and 3p hybridised orbitals used for atomic bonding in silicon carbide are completely filled (each element contributes four electrons into the four bonding orbitals). Electron mobility ought therefore to be very low, and silicon carbide would be expected to be a very good electrical insulator. There is, however, an important difference between carbon and silicon, in that silicon also has available to it a set of empty, and relatively low-energy, 3d orbitals. The silicon 3s, 3p and 3d orbitals are able to combine to form higher-energy conduction bands, which can be occupied by electrons, to permit electrical conductivity (Pensl and Choyke, 1993). The energy gap in pure silicon carbide between the completely filled valence and empty conduction bands at room temperature is ~2.4 eV (230 kJ mol^{-1}) in the β-polytype, and ~3 eV (290 kJ mol^{-1}) in the common α-silicon carbide polytypes (Choyke and Patrick, 1957; Kackell *et al.*, 1994). These energy gaps are, under normal circumstances, large enough to prevent the excitation of valence electrons, hence pure silicon carbide is still an electrical insulator. However, silicon carbide is rarely completely pure. The common impurity elements able to occupy lattice sites, nitrogen, aluminium, and boron, provide impurity levels well below the high-energy conduction bands. The transfer of electrons to these levels creates the possibility for electron mobility within the silicon carbide lattice, and thus electronic electrical conductivity. The number of electrons promoted to the conduction bands is a direct (exponential) function of temperature, and silicon carbide is a semiconductor.

The general pattern of behaviour with silicon carbide is that conductivity increases with temperature, to pass through a maximum at ~600 °C, before decreasing again, though with different types and levels of dopant considerable variations on the pattern can be obtained (Hamilton, 1960; Choyke *et al.*, 1996). Most commercial silicon carbides are sufficiently impure that their natural electrical conductivity can be of the order of 10–100 S m^{-1} at room temperature, which makes them very effective electrical conductors (North and Gilchrist, 1981). Figure 4.6 shows the electrical resistivity of dense sintered silicon carbide, expressed relative to the value at 1000 °C.

Because of its electrical conductivity and high-temperature stability, silicon carbide is very widely used as a furnace heating element material; it is particularly useful where temperatures are required well above those normally attainable by lower melting-point metallic conductors such as the nickel-based alloys (which have a ceiling of around 1300 °C).

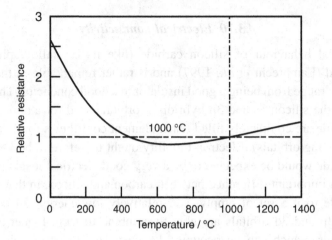

Figure 4.6 The variation of electrical resistivity as a function of temperature. Resistance values are expressed relative to the value at 1000 °C. (Reproduced by kind permission of Kanthal Ltd.)

4.3.11 High-temperature behaviour – thermal stability

Silicon carbide melts with decomposition, at temperatures estimated to be between 2540 °C and 2830 °C. Decomposition gives carbon, and a silicon-rich liquid, in which up to ~20 atomic% of carbon may be dissolved (Scace and Slack, 1959). The vapour over silicon carbide at temperatures above ~2000 °C consists predominantly of low pressures of Silicon vapour together with the gaseous species Si_2C and SiC_2, which also shows the tendency for SiC to decompose (Grieveson and Alcock, 1960). However, from the point of view of structural applications, this high-temperature behaviour is not of practical importance, because useful strength and stiffness will have been lost at much lower temperatures. Furthermore, the difficulty of preventing the oxidation of silicon carbide is also likely to have intervened.

It is therefore perhaps surprising to find, in view of its hardness and very high melting-point, that the thermodynamic stability of silicon carbide with respect to its elements (formally the two solids, graphite and crystalline silicon) is so low: its tendency, chemically speaking, to fall apart, is high. This is partly because the strong bonding and ordered structure within the graphite sheets, and in the diamond lattice of crystalline silicon, make both the enthalpy and the entropy changes in forming silicon carbide very small, so that the Gibbs function (ΔG) of the formation reaction is also small:

$$Si_{(s)} + C_{(graphite)} = SiC_{(\alpha)}; \quad \Delta G^o_{298K} = -69.7 \text{ kJ mol}^{-1}. \quad (4.3)$$

Silicon carbide, therefore, tends to be chemically a very reactive compound (and is used as a source of silicon in metallurgical processing). This behaviour can be contrasted with the extremely high thermodynamic stability of aluminium oxide, in which the oxygen and aluminium can be separated only with difficulty (as in the production of aluminium metal). It also has relevance for the dry sliding wear behaviour of silicon carbide in contact with ferrous alloys for example, where localised very high temperatures can lead to chemical reactions with the formation of metal silicides.

4.3.12 High-temperature behaviour – oxidation

In the context of the use of silicon carbide as a structural material at high temperature, one of the most troubling reactions is its strong tendency to oxidise. The question of the long-term stability of silicon carbide in the Earth's oxidising atmosphere (damp air) has already been referred to. Oxidation occurs in part because of the high thermodynamic stability of the solid oxidation product silicon dioxide, and in part because of the formation of the gaseous reaction products, carbon monoxide or carbon dioxide:

$$2SiC_{(s)} + 3/2O_{2(g)} = SiO_{2(s)} + CO_{(g)}; \ \Delta G^{o}_{298K} = -765 \text{ kJ mol}^{-1} \qquad (4.4)$$

$$2SiC_{(s)} + 2O_{2(g)} = SiO_{2(s)} + CO_{2(g)}; \ \Delta G^{o}_{298K} = -918 \text{ kJ mol}^{-1}. \qquad (4.5)$$

Reaction (4.5) is the normal oxidation reaction under conditions of "high" oxygen pressure, such as the ~0.2 atm of the Earth's crust. Because of the "passivating" (that is, *deactivating*) action of the protective oxide film the reactions yielding silicon dioxide are referred to as *passive* oxidation. Silicon carbide is therefore, like most metals, only able to exist in the oxidising environment of the Earth's crust because of its ability to develop a protective surface layer of an oxide, in which oxygen, silicon, and carbon all have low mobility. The rate of oxidation is determined by the slow inwards diffusion of oxygen (as molecular oxygen or atomic lattice oxygen) through the surface film of silicon dioxide. Platinum marker studies (Costello and Tressler, 1981) show that oxidation occurs at the silicon carbide–silicon dioxide interface, with the release and outwards diffusion of carbon dioxide, or carbon monoxide, as shown schematically in Fig. 4.7. Different crystal faces react at slightly different rates (Ramberg and Worrall, 2001). It is of interest that the rates of oxidation of single-crystal silicon, and single-crystal silicon carbide at temperatures in the range ~1000–1300°C, are virtually identical (Tressler, 1994). This is also strong evidence against the alternative possibility, that the rate-controlling process is the outwards diffusion

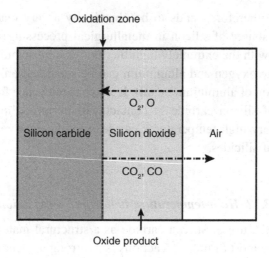

Oxidation zone

Silicon carbide

Silicon dioxide

Air

O_2, O

CO_2, CO

Oxide product

Figure 4.7 The passive oxidation of silicon carbide to silicon dioxide: oxidation is by the inwards diffusion of oxygen through the silicon dioxide surface film. Oxidation occurs at the silicon carbide–silicon dioxide interface. The products carbon monoxide and carbon dioxide then escape by diffusion through the silicon dioxide film.

of silicon and carbon as atomic species, to allow oxidation to take place at the silicon dioxide–air interface (Luthra, 1991; Ramberg and Worrell, 2001).

At this point it will be useful to say a little more about what is meant by reaction rate, and how reaction rate constants are expressed. Reactions of these types between a solid and a gas, in which the rate-controlling process is the diffusion of a reactant (oxygen in this case) through a solid product film (silicon dioxide), have at any moment in time a reaction rate which is an inverse function of the product layer thickness, and follow what is called the *parabolic rate law* (Fig. 4.8). This is because the quantity of oxygen flowing through a plane of unit area normal to the x-axis (the flux, j_x) is determined by the oxygen concentration gradient (dc/dx), by Fick's first law,

$$j_x = -D_x \frac{dc}{dx}. \tag{4.6}$$

Commonly, j_x has units of mole m^{-2} s^{-1}, and D_x units of m^2 s^{-1} (the negative term appears because the flux is positive, although the gradient is formally negative, as shown in Fig. 4.8, which is drawn in the opposite sense to Fig. 4.7 to show this). The rate of production of oxide is proportional to the oxygen flux, and the rate of increase of oxide film thickness (dx/dt) is given by

$$\frac{dx}{dt} = -kD_x \frac{dc}{dx} \tag{4.7}$$

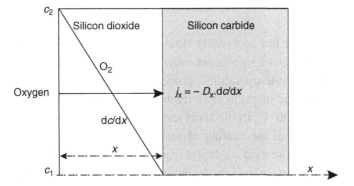

Figure 4.8 An illustration of the diffusion-controlled oxidation of silicon carbide: oxygen diffusion down its concentration gradient. The instantaneous flux is $j_x = -D_x(c_2-c_1)/x$. This diagram is drawn (in the opposite sense to Fig. 4.7) so as to show the negative slope of the concentration gradient, but positive flux.

where k is a constant relating the quantity (moles) of oxygen diffusing to oxide film volume, or, because we can consider volume per unit area, to oxide film thickness (x).

The rate of increase of film thickness at a point in time is

$$\frac{dx}{dt} = -kD_x \frac{c_1 - c_2}{x} \tag{4.8}$$

where c_2 and c_1 are the oxygen concentrations at the outer and inner film surfaces. Both these concentrations are constant: c_2 is fixed by the external oxygen pressure and c_1 by the (very low) pressure of oxygen in equilibrium at the reaction temperature with the reactant (SiC) and product (SiO$_2$). Integrating over $t = 0$ to $t = t$, gives

$$x^2 = -2kD_x(c_1 - c_2)t \tag{4.9}$$

or, combining the constant terms,

$$x^2 = k_p t. \tag{4.10}$$

The reaction rate decreases with time (not because the oxygen has further to diffuse, but because it travels more slowly: the concentration gradient dc/dx driving the diffusion is decreasing with time) and plotting x as a function of t gives a parabola. k_p is the parabolic rate constant, and is used as a standard measure of reaction rate for this type of oxidation mechanism. The units of k_p depend on how the quantity of oxide is expressed. If this is the film thickness, the unit is $m^2 \, s^{-1}$; if it is the weight of oxide formed per unit area (which is what is often measured), it is $(kg \, m^{-2})^2 \, s^{-1}$, or $kg^2 \, m^{-4} \, s^{-1}$. This is the unit used in

Fig. 5.8 in Chapter 5 when the oxidation rates of silicon carbide and silicon nitride are compared.

Pure silicon carbide has acceptably slow oxidation rates in oxygen and air. The rate of oxidation becomes significant only at temperatures above 900 °C and for many industrial "high-temperature" processes materials will survive for long times (many months or years) in air at temperatures up to 1400 °C, and for several thousand hours at 1500 °C. In the short term strengths may even be improved by oxidation, as a result of the healing of surface flaws (Becher, 1983).

However, there is a second oxidation reaction – *active* oxidation – favoured by low oxygen pressures, in which the silicon oxidation product is the gaseous silicon monoxide:

$$SiC_{(s)} + O_{2(g)} = SiO_{(g)} + CO_{(g)}; \ \Delta G^o_{1500K} = -410 \text{ kJ mol}^{-1}. \tag{4.11}$$

Because both reaction products are now gases, the entropy of reaction is large and positive, helping to make the reaction thermodynamically favourable (even though silicon monoxide is not as stable as silicon dioxide – in fact it is very unstable and readily dissociates to silicon and silicon dioxide on cooling). The boundary between "low" and "high" oxygen pressures is a function of temperature, and has to be defined: it is ~100 mbar at ~1000 °C (Schneider *et al.*, 1998).

Both processes show the intrinsic instability of silicon carbide under oxidising conditions. The important difference between passive and active oxidation is that in the first, a layer of silicon dioxide can form, with the possibility of some protection of the underlying silicon carbide against the oxidising atmosphere. In the second, because the silicon-containing product (*silicon monoxide*) is a gas, no protective layer can be formed. It is something of a paradox therefore that, because active oxidation becomes important for low oxygen partial pressures, oxidation of silicon carbide can (at the same temperature) be faster under "low" pressures of oxygen, than under "high" pressures. The rate of oxidation, because it depends on the rate of supply of oxygen, is still a direct function of oxygen pressure and is slower the lower the oxygen pressure. Because oxygen pressures are commonly "high" (in air, or furnace combustion atmospheres) a protective layer of silicon dioxide (usually in the amorphous, or glass, form) will be produced. However, silicon monoxide formation must be taken into account if silicon carbide materials are being considered for long-term use at temperatures in nominally "reducing", or "neutral" atmospheres containing, for example, hydrogen or carbon monoxide, or consisting mainly of water vapour, for which the equilibrium oxygen partial pressure may be reduced below the critical level. This is because the pressure of oxygen gas in a closed system is controlled by the reaction and its associated equilibrium relationship:

Table 4.3 *Maximum surface temperatures for silicon carbide, in different gaseous environments.*

Atmosphere	Maximum surface temperature / °C	Gaseous atmosphere
Oxygen	1625	Normal pressures
Dry air	1500	Lower oxygen pressure
Nitrogen	1350	Silicon nitride formed
Dry hydrogen	1200	Oxidation in wet hydrogen
Dry exothermic gas	1400	Contains $CO_2/H_2/N_2$
Dry endothermic gas	1250	Contains $CO/H_2/N_2$
Vacuum	1200	Short times only possible

$$H_2O_{(g)} = H_{2(g)} + \frac{1}{2} O_{2(g)} \tag{4.12}$$

$$p_{O_2} = k_p \left(\frac{p_{H_2O}}{p_{H_2}} \right)^2, \tag{4.13}$$

so that as the pressure of water vapour drops, for a fixed hydrogen gas pressure, the effective oxygen pressure in the system also falls. In practice the possibility of active oxidation reduces the maximum temperature of long-term use from ~1620 °C (clean air) to ~1200 °C for dry hydrogen or vacuum, and when only very short-term use may be possible. Manufacturers' suggested values for maximum surface temperatures are shown in Table 4.3.

4.3.13 High-temperature behaviour – corrosion

The surface film of silicon dioxide is crucial for the stability of silicon carbide under oxidising conditions. Any substance that can react with silicon dioxide has the potential to compromise the integrity of the film and allow fast oxidation (as *corrosion*) to occur by the inwards transport of oxygen (Singhal, 1976; Lamkin *et al.*, 1992). These reactions can be expressed in the general form, where M_aO_b represents a metal oxide:

$$M_aO_b + nSiO_2 = M_aSi_nO_{(2n+b)}. \tag{4.14}$$

Some product silicates will be solid at reaction temperatures, but those formed from the alkali metal oxides Na_2O and K_2O (as seen in the case of porcelain) can be liquids of variable composition at temperatures as low as 700 °C (which is one reason why the composition of the product silicate is expressed in this general

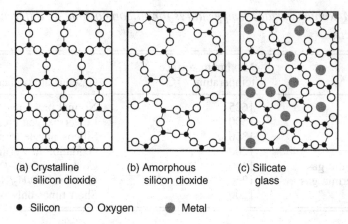

(a) Crystalline   (b) Amorphous   (c) Silicate
    silicon dioxide   silicon dioxide   glass

● Silicon   O Oxygen   ● Metal

Figure 4.9 Illustration of the atomic structures of crystalline silicon dioxide, amorphous silicon dioxide, and an amorphous silicate. Relaxation of the SiO_4 structure is facilitated by the presence of the cations of the metal oxide.

form). Because alkali metal and oxide ions readily break up the silicon dioxide Si–O network, the non-crystalline silicates have much more open structures and looser bonding. The structures of crystalline silicon dioxide, amorphous (glassy) silicon dioxide, and silicate glass are shown schematically in Fig. 4.9. These illustrate the progressive loss of long-range order in the silicon dioxide structure (though short-range order of the tetrahedral SiO_4 unit is retained). Not only do these silicates have very much higher conductivities for both molecular (O_2) and atomic (O) oxygen, but silicon dioxide is in effect soluble (up to a saturation limit determined by the phase equilibrium diagram) in a liquid alkali metal silicate. In the presence of a constant supply of metal oxide, a thick protective silicon dioxide layer may therefore be unable to develop, and continuing oxidation of the underlying silicon carbide at very fast rates is now possible (McKee and Chatterji, 1976). Figure 4.10 shows the corrosion process schematically: at interface I oxidation of silicon carbide to silicon dioxide occurs; at interface II the silicon dioxide reacts with or dissolves into the liquid silicate. The alkali metal oxides can be particularly problematic because of their volatility and mobility at high temperature within a furnace atmosphere; contamination by them can therefore considerably shorten the life of silicon carbide components. Precursors of the oxides such as the halides must also be avoided (Pareek and Shores, 1991). A good (that is, bad) example is sodium chloride; at high temperatures it reacts with oxygen to form sodium oxide, a process that is assisted by any subsequent reaction of the oxide with silicon dioxide:

$$(a) \qquad 2NaCl + \frac{1}{2}O_2 = Na_2O + Cl_2 \qquad (4.15)$$

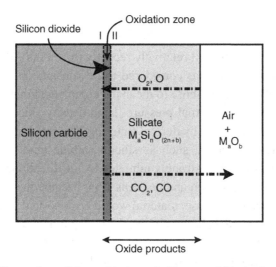

Figure 4.10 Illustration of the oxidation of silicon carbide in the presence of a metal oxide (M_aO_b) able to form a low melting-point silicate. Diffusion of oxygen is much faster through the silicate structure than through the purer silicon dioxide interfacial layer. Oxidation occurs at interface I, between the silicon carbide, and the silicon dioxide. At interface II, conversion of the silicon dioxide to silicate occurs (compare this process with that illustrated in Fig. 5.28).

$$\text{(b)} \qquad Na_2O + nSiO_2 = Na_2SiO_{(2n+1)}. \tag{4.16}$$

This was the basis of the old "salt glazing" process for clay-based cheap earthenware, and pipes, in which "salt" (sodium chloride) was added to the kiln at the ending of the normal firing process. The surface glass film, the *glaze*, then consisted mainly of a sodium potassium aluminosilicate glass. Although silicon carbide is fairly inert in the presence of aqueous acids (apart from hydrofluoric acid) and alkalis, it reacts readily at high temperatures with other oxidising materials, such as chlorine at temperatures above 900 °C (Zheng *et al.*, 1992).

4.4 The development of silicon carbide ceramics

Silicon carbide is an umbrella term (as is *alumina*) covering a wide range of materials. These have differing compositions but are made up predominantly of silicon carbide microcrystals (usually a mixture of α-SiC polytypes), bonded in various ways, and with their distinctive microstructures and properties. Pure silicon carbide can be solid state sintered to full density, although this requires careful control of the conditions. Methods for liquid phase sintering fine silicon carbide powders to full density have also been developed, but they are commercially not yet much used.

4.4.1 Early materials

It has to be appreciated that until relatively recently (given the ~100 year history of the material) silicon carbide was considered to be "unsinterable" (using the term sintering in the slightly special sense to mean the complete elimination of porosity), without the application of external pressure. High-density silicon carbide, of high strength, could not be produced by normal sintering. The rationalisation of this view was that in covalent materials, such as silicon carbide (and silicon and silicon nitride), vacancy diffusion was strongly inhibited by the directional bonding, so that surface diffusion, or evaporation and condensation, became the dominant material transport mechanisms. In this case, material would be deposited preferentially in the neck region where its chemical potential (molar Gibbs function) would be lower, to form interparticle bonds, but the movement of particle centres towards each other (and necessary for densification) would not occur. The possibility of pressing silicon carbide powder to a full density, strong, material using the simultaneous application of heat and pressure (*hot pressing*) in graphite dies with 10–20 MPa applied pressure, had already been explored (Alliegro *et al.*, 1956), as had the use of boron-containing compounds to aid densification (Bind and Biggers, 1975). Workers at the General Electric Company in the USA then identified a method for producing fully dense, fine-grain, sintered silicon carbide without the need to apply pressure, and a better understanding developed of the underlying reasons for the "unsinterability" of silicon carbide (Prochazka, 1975; Greskovich *et al.*, 1976). Up to this time a number of different processes had been developed to bond silicon carbide powder particles, including hot pressing.

These earlier processes provided a range of materials consisting predominantly of fairly coarse grain-size silicon carbide, with different bond phases. Much of this more porous, bonded, material is widely used in the metal processing industries. However, it would be regarded primarily as a refractory, and therefore falls outside the scope of this review. Only the other types of finer grain, and much stronger, silicon carbide material will be examined here. The hot-pressing technique is now not generally used in the commercial production of dense silicon carbide (since it has been superseded by pressureless sintering), but is still useful for the laboratory-scale preparation of dense materials.

In the earlier years of the production of single-phase silicon carbide ceramics and refractories, the best that could be done was to compact and heat reasonably pure silicon carbide powders (10–100 µm) and grits at very high temperatures (normally >2000 °C). The difficulties of attaining temperatures of this order (most conventional oxide refractory materials are melting by ~1700 °C) are considerable, and alternative environments and heating elements (such as carbon

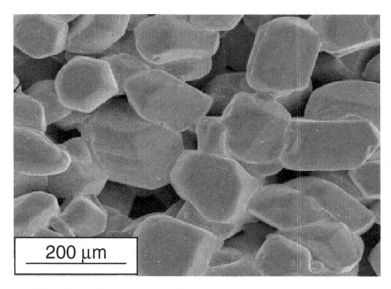

Figure 4.11 The microstructure of recrystallised silicon carbide, shown by scanning electron microscopy. There is only a limited extent of bonding between the silicon carbide grains, and a large volume of intergranular porosity. (Reproduced by kind permission of St. Gobain IndustrieKeramik Rödental.)

tube furnaces) are used. At these temperatures the evaporation and condensation, and surface diffusion, processes create a modest extent of particle shape change and interparticle adhesion. These materials were for this reason often termed *recrystallised* silicon carbide. A scanning electron micrograph of this type of low-density, porous material – in fact, with porosity roughly that of the original compacted powder (15–30%) – is shown in Fig. 4.11.

Recrystallised silicon carbide is an extremely useful, hard, and abrasion-resistant material, and has many important applications at room and high temperature. However, because of the porosity, the potential high strength of silicon carbide cannot be reached.

Alternative ways have been used to bond silicon carbide particles, and fill the interparticle space to create higher-density materials. These include the use of quite large proportions (10–20%) of fusible or more sinterable materials, such as aluminium oxide, or silicates, to produce what can be regarded as silicon carbide–silicate composites. Materials are also produced using silicon nitride and silicon oxynitride as the bonding phases, the nitrides being formed *in situ* by the high-temperature nitridation of silicon powder (described in Chapter 5), blended with the silicon carbide. Most of these oxide and nitride bonded materials are used as coarse-grained refractories and heat-resistant "kiln furniture", or large wear-resistant tiles and protective linings (Singer and Singer, 1963).

One early type of fine-grain, low-porosity, silicon carbide ceramic was termed *siliconised silicon carbide*. Its production makes use of the relatively low melting-point of silicon at ~1412 °C, and the wetability of silicon carbide by liquid silicon. Under an inert gaseous atmosphere, liquid silicon can be drawn by capillary action from a reservoir into the open pores in compacted or lightly sintered silicon carbide powders. The product, though almost fully dense, might be regarded as a two-phase composite, in that it contains a large volume of a low melting-point second phase (silicon) in what would otherwise be accessible porosity. Because it is practically pore-free this material can be produced with considerably higher room-temperature strengths than the standard porous form of recrystallised silicon carbide. It is also impermeable to gas. However, above the melting-point of silicon the material reverts, in effect, to a porous silicon carbide. There is a simultaneous abrupt reduction in mechanical strength, although for many applications the ceiling temperature of just below 1400 °C is quite adequate.

A closely related silicon–silicon carbide two-phase material is *self-bonded silicon carbide*. The type of the material also termed *reaction-bonded* (or *reaction-sintered*) silicon carbide was first produced some 40 years ago (Popper, 1960). It was developed over the following 10 years to meet a requirement for a fully dense, impermeable, silicon carbide (at this time, fully dense pressureless sintered silicon carbide had not been produced). In this material, compacted silicon carbide particles have been bonded by the development of new (bonding) silicon carbide, produced by (4.1) between carbon and silicon. This process can be regarded as a form of liquid phase sintering, in which the crystallising phase is not one which has previously dissolved from the starting powder particles themselves, but one which is formed *in situ* from suitable reactants, hence the term *reaction sintering*. Variations on this process can be used to form a wide range of other ceramic materials that would otherwise be difficult to obtain by straightforward powder sintering, including silicon nitride.

Because of the limiting of the silicon–silicon carbide system to temperatures below ~1400 °C, and the inability to sinter silicon carbide powders to full density, considerable efforts were made in the 1960s to produce a completely dense, high-purity, silicon carbide (Price, 1969). Uses were seen for these materials as fission product barriers in new types of nuclear fuel power generation, operating with higher core temperatures than were possible with metallic fuels and containers (Voice, 1969; Voice and Scott, 1972). The method chosen was to decompose a volatile compound containing the required carbon and silicon, on a very hot surface (*substrate*), which would be a particle of nuclear fuel. Alternatively, an inert former can be used to allow the build up of a specific shape: the thin layer of polycrystalline silicon carbide is then separated from the substrate (or the substrate is chemically removed). For this reason a particularly good substrate material is graphite (graphite is relatively soft and readily machinable, and can be removed by oxidation), though in principle any high melting-point material can be used. Out of this approach evolved a fine-grain polycrystalline material termed *pyrolytic* or *chemical vapour deposition (CVD)* silicon carbide (Popper and Mohyuddin, 1964; Price, 1969). This process can be regarded as another variation on the theme of reaction sintering, in that the silicon carbide is produced by the chemical reaction, with the product (silicon carbide) simultaneously sintering into a dense material. One well-studied readily available source for silicon carbide is the volatile liquid methyltrichlorosilane (it is used for the production of silicone polymers). The decomposition (*pyrolysis*) of methyltrichlorosilane at temperatures above ~1000 °C can be written simplistically as:

$$CH_3SiCl_{3(g)} = SiC_{(s)} + 3HCl_{(g)}. \tag{4.17}$$

This compound appears to be an ideal starting point, in that it contains silicon and carbon in the required 1:1 ratio; in reality the reaction process is complex. It involves very reactive molecules containing unpaired electrons (known as *free radicals*), a small proportion of which finally decompose to release their carbon and silicon on the hot surface (a *heterogeneous* reaction). Under these circumstances the silicon and carbon atoms can readily arrange themselves into the required stable crystalline silicon carbide structure. Figure 4.12 shows the microstructure of a polished surface of fine-grain CVD silicon carbide, which has been chemically etched to reveal the characteristic vapour deposition growth pattern (termed *botryoidal*, from the Greek, meaning a bunch of grapes – though to see the grapes requires a little imagination). Nucleation of the carbide has taken place on a graphite surface, in the bottom right of the picture, and the subsequent development pattern is clear. CVD silicon carbide can have very fine grain sizes (controlled by deposition conditions) and be fully dense, but there are technical difficulties involved with its large-scale production (including detachment from the substrate). A further drawback to this process is that large pieces of CVD materials tend to contain weakening microstructural inhomogeneities, and stresses (which can be as large as 300 MPa) developed during deposition (Airey *et al.*, 1975). This means that the technique has limited applicability for the production of large or thick components. But until the solid state sintering route was developed, the CVD process appeared to offer a flexible route to the production of single-phase, fully dense, silicon carbide, and considerable development work was carried out. This work has led to the use of thin silicon carbide coatings on the nuclear fuel particles used in gas-cooled reactors, and the fluidised bed technique was developed to allow thin, high-strength, coatings to be applied to small (mm dimension) fuel microspheres. In effect the silicon carbide coating is a miniature fission gas "pressure vessel" (Snead *et al.*, 2007).

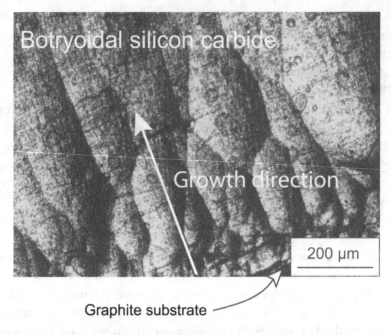

Figure 4.12 The microstructure of a polished and chemically etched surface of fine-grain CVD silicon carbide. Etching reveals the microstructure and the nucleation and outwards growth of the silicon carbide from the graphite substrate (bottom right of the picture).

4.4.2 Sintered silicon carbide

Before discussing the major fine-grain ceramic forms of silicon carbide, and their production methods, a brief account of the underlying difficulty of sintering silicon carbide (and why silicon carbide powder appeared to be "unsinterable") is needed.

A primary factor underlying the difficulty of sintering even very fine silicon carbide powders is the extremely low mobility of silicon and carbon atoms in the crystal lattice, at temperatures that would be more than adequate for the sintering of micrometre dimension aluminium oxide powders. One more consequence of the very strong Si–C bonding, is that the energy of formation of lattice vacancies (*Schottky* defects) will be very high, and because of the directional covalent bonding, the energy of formation of interstitial (*Frenkel*) defects is also very high. The overall result is that, even at very high temperatures, the lattice point defect concentrations, and lattice atomic diffusion coefficients, are very low. Self-diffusion coefficients (D) in the SiC lattice at 2000 °C have been measured for both β-SiC and "α"-SiC; values are of the order of 10^{-17} and 10^{-15} m^2 s^{-1} for Si and C respectively, and insensitive to the polytype. This shows that, perhaps as might be expected, the smaller C atom is the more mobile (Hon *et al.*, 1980; Hong and Davis, 1980; Hong *et al.*, 1981).

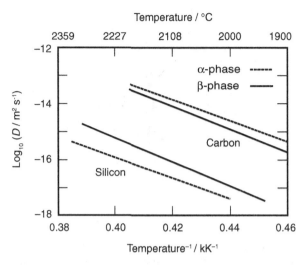

Figure 4.13 Self-diffusion coefficients for carbon and silicon in α- and β-silicon carbide, showing that while all values are small, the larger silicon atom is the less mobile of the two. (After Kirschstein, and Koschel, 1984, p. 108.)

Examples of data for pure silicon carbide for the temperature range 1900–2300 °C, are shown in Fig. 4.13. Atom fluxes in silicon carbide are therefore very low, except at extremely high temperatures, with lattice diffusion of the larger, and slower, Si probably being the rate-controlling process. These values should be compared with those for the lattice self-diffusion coefficients of Al or O in α-aluminium oxide, at a temperature of 1500 °C, of the order of 10^{-10} m^2 s^{-1}. On the basis of the Einstein random walk equation (Section 1.5, Equation (1.13)), some specified extent of sintering which would be achieved in 10 minutes at 1500 °C with aluminium oxide, would with the same particle sized silicon carbide require even at 2000 °C around two centuries.

However, there are other fundamental difficulties. Firstly, the movement – albeit reluctant – of silicon and carbon atoms appears to occur preferentially over free surfaces (*surface diffusion*), rather than within the lattice (*lattice diffusion*) and at grain boundaries (*grain boundary diffusion*). This is not unexpected, because surfaces are regions of atomic disorder and broken bonds, and the degree of adhesion of individual atoms to specific lattice sites would be considerably diminished. A comparison of hypothetical sintering rates devised on the basis of lattice diffusion coefficients has to be treated very carefully: in reality grain-boundary diffusion is likely to be much faster for both materials, but values for the grain-boundary diffusion coefficients for silicon carbide are not easy to measure. Secondly, because of the appreciable volatility (and decomposition) of silicon carbide at the very high temperatures needed (~2400 °C) to drive the

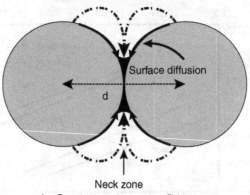

Figure 4.14 The development of a bridge between two silicon carbide particles at their contact point. Material is transported from the surrounding surfaces to the neck zone by surface diffusion, or by evaporation and condensation. There is no loss of material at the contact points, so that the particle–particle centre distance (d) remains constant, and there is no overall shrinkage.

diffusion processes, transport of material by the evaporation and condensation process becomes an important material transport mechanism. One immediate important consequence is that bridges develop across particle contact points in the neck zone, by the movement of material (as atoms) from the surrounding surfaces, as illustrated schematically in Fig. 4.14. The result is that the overall driving force for material flow is reduced, as the neck zone fills up, and radii of curvature increase. To obtain a dense material (by the loss of void space) a reduction in particle centre-to-centre distances (shown as *d*) is needed: this requires the transport of material away from the particle contact points, by grain boundary or lattice diffusion.

There is a further fundamental reason why the normal sintering densification of silicon carbide powders can be difficult. This is a thermodynamic factor, involving the balance of interfacial energies, and the values of local chemical potentials. To see this one must consider a stage in sintering when silicon carbide particles have partially fused, to the point where the microstructure can be considered (in an idealised picture) to contain tubular pores, along the edges shared by three contacting grains. For these tubular pores to shrink and ultimately disappear, atoms must diffuse, either through the lattice, or, (more easily) down the grain boundaries, to the internal surfaces of the pore. The pore surfaces will then move inwards, towards their centres of curvature, and the pore is filled. Figure 4.15 shows a section through a tubular pore at a three-grain boundary. The condition for material transport to the internal surfaces of the pore to occur, is that

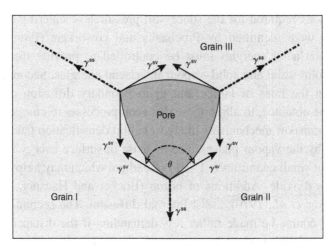

Figure 4.15. Forces at grain–pore interfaces. For an internal angle (θ) of $> 60°$, the solid surfaces become concave, and grain boundary diffusion of material to the pore surface should occur, leading to pore shrinkage.

the surfaces must have a lower chemical potential than the standard value ($\mu°$) for silicon carbide. This in turn requires the pore internal surface to be convex (Section 1.5.3). The internal angle (θ) of the pore is the result of the balance of the interfacial forces (surface tension) such that:

$$\gamma^{ss} = 2\gamma^{sv} \cos(\theta/2). \qquad (4.18)$$

The critical value for θ if pore shrinking is to occur is 60°, for which the ratio γ^{ss}/γ^{sv} is 1.732. However, the solid–solid, grain-boundary, energy (γ^{ss}) for pure silicon carbide is likely to be very high because of misalignment of the carbon and silicon bonds – in the region of 3–5 J m^{-2} (Inomota and Matsumoto, 1971). The pore surface interfacial energy γ^{sv} is likely to be much smaller, because the silicon carbide particle will tend to be coated by a thin film of silicon dioxide (of surface energy less than ~0.3 J m^{-2}, approximately the value for silica glass). Thus the ratio $\gamma^{ss} / \gamma^{sv}$ will be much larger than 1.732, and the internal surfaces of the pore will be concave – the wrong kind. Experimentally, it is observed that pure silicon carbide powder does tend to form pores with concave internal surface curvature. At this stage further densification effectively stops, in accord with the thermodynamic explanation.

All non-oxide powder surfaces tend to be covered with a thin oxide film (possibly only a monolayer thick), through natural oxidation after milling or preparation, although the silicon dioxide films on silicon carbide particles are probably only a few nm thick. However, for small particles this can amount to a considerable volume of oxide, ~6% for a 1 nm film on a 100 nm particle.

The conditions required for the successful pressureless sintering of silicon carbide powders were identified by Prochazka and coworkers (Prochazka, 1975). Firstly, the interfacial energies must be controlled to provide the correct ratio between the solid–solid and solid–vapour interfacial energies. Secondly, a marked acceleration in the rates of lattice and grain boundary diffusion of silicon and carbon must be obtained, to allow these diffusion processes to compete effectively with the two transport mechanisms likely to inhibit densification (surface diffusion and transport by the vapour phase). The increase in surface energy is obtained by the additions of small quantities (~1–3%) of carbon, which may help to remove the surface silicon dioxide. Additions of boron (Böcker and Hausner, 1978) or aluminium (Böcker *et al.*, 1979), assist internal diffusion. The second of these conditions can of course be made rather less demanding if the distances over which diffusion is required to take place are very short. To allow the transport of a sufficient volume of material on a reasonable time-scale of, say, a few hours, the silicon carbide particle size must be very small. In practice, submicrometre powders with median sizes in the region of 100 nm are required, together with the necessary accelerating additives, such as carbon with boron or aluminium.

4.5 Ceramic materials

Table 4.4 provides a summary of the different forms of material based on silicon carbide developed over the last 50 years, using processes designed to try to overcome, or avoid, the fundamental difficulty of sintering standard grades of silicon carbide powder. These materials have widely varying compositions and microstructures, and consequently significantly different properties. It will therefore be more convenient to treat the group as though it consisted of quite distinct materials, though of course each has silicon carbide as the main phase. In this respect the approach is slightly different from that used for alumina materials, which can reasonably be regarded as fitting a smoothly graded spectrum, with changes only in the composition and volume (starting at zero for solid state sintered alumina) of the silicate sintering additive used. In the following sections, the types of silicon carbide material currently produced will be examined briefly in the order of their development, and their properties summarised. Those coarser-grained materials used predominantly for high-temperature applications, and which therefore might be categorised as "refractories", will not be discussed in detail, though it has to be appreciated that the dividing line between a "refractory" and a "structural ceramic" can be rather arbitrary.

The applications of the silicon carbide ceramics will be grouped together in one section. This is because in many cases the choice of material depends on the precise conditions of the application, rather than the type of application itself.

Table 4.4 *Principal forms of silicon carbide ceramic, their main features, and maximum temperatures of use.*

Type	Additive	SiC content / %	Second phase	Porosity	Strength	Maximum temperature of use / °C
Recrystallised	None	100	None	Yes	Low	1600+
Nitride bonded	Silicon	70–80	Silicon nitride, oxynitride	Yes	Low	1500
Reaction bonded	Silicon	80–90	Silicon	No	<1380 °C high	1300
Liquid phase sintered	Al_2O_3/AlN/Y_2O_3	90–95	Silicate	No	High	1500
Solid state sintered	B/Al/C	98–99	Carbon (trace)	No	High	1600+

Different forms of silicon carbide may be used in a particular component, depending on the requirements, or cost restrictions.

4.5.1 Recrystallised silicon carbide

Production

This was one of the earliest forms of structural silicon carbide ceramic, and the earliest high-purity form. Its development took place through the 1950s and early 1960s in a search for materials better able to withstand higher industrial temperatures (Norton, 1958; Torti *et al.*, 1973). Silicon carbide powder, normally of dimensions in the size range 5–100 μm, and with a bimodal distribution of particle size to maximise packing density, is compacted using standard methods to the required shapes. Powders are then heated in a non-oxidising atmosphere to temperatures in the range 2100–2500 °C for several hours. This temperature limit is set in part by the dissociation of the silicon carbide. Very high temperatures are often achieved using an open-ended carbon resistance furnace, in which the gaseous atmosphere is then a "natural" mixture of nitrogen, carbon monoxide, and carbon dioxide. Inorganic additives are not normally used, so that the final product is a nominally 99% purity silicon carbide. Interparticle bonding starts to develop at around 2100 °C by local solid state sintering at particle contact points, and accelerates substantially at 2300–2350 °C. At the highest sintering temperatures the very fine silicon carbide particles may also fuse together with grain growth to form larger crystals, but there is very little, or no, overall sintering shrinkage. Figure 4.16 shows the polished surface of a low-density recrystallised silicon carbide, seen in the scanning electron microscope. The silicon carbide grains (white phase), the relatively small extent of direct, grain-to-grain bonding, and the large volume of residual porosity are obvious.

Properties

No shrinkage takes place during sintering, and recrystallised silicon carbide has virtually the same porosity, and therefore bulk density, as the original compacted powder. The porosity, which is normally in the range 15–30%, is almost all of the open "accessible" type, and the bulk density is of the order of 2.2–2.7 Mg m^{-3}. The strengths and moduli of recrystallised silicon carbides are those to be expected from a rather coarse-grained and porous material, though within which there is an appreciable degree of strong interparticle bonding. Bend strength values are typically in the range 100–130 MPa (depending on the volume of porosity) at room temperature, which is roughly the same as that of glazed porcelain. A typical value for the room-temperature Young modulus would be of

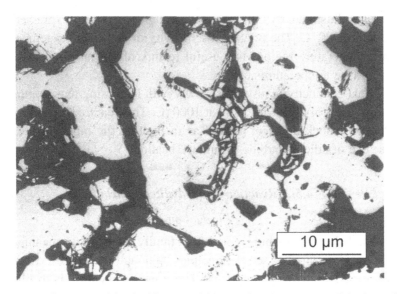

10 µm

Figure 4.16 Recrystallised silicon carbide (light phase); a polished section showing a limited amount of particle–particle bonding.

the order of 200 GPa, for material with 15% porosity. However, a very useful feature is that because there is no intergranular secondary phase (with an absence of glass in particular), the strength and stiffness of recrystallised silicon carbide are maintained up to the highest temperatures of practical interest. Measured values remain unchanged from room temperature to at least 1400 °C. Usable strength and structural rigidity are maintained at temperatures of well over 1500 °C (when porcelain would be largely a viscous fluid, and a sintered alumina would be starting to creep). The maximum operating temperature is often quoted as 1600 °C, and the major problem is likely to be surface oxidation. Oxidation rate, because it is normally controlled by the inwards diffusion of oxygen through silica films, is an exponential function of temperature. Component life depends on the maximum temperature of use, and can be several thousands of hours. The avoidance of contamination is an important factor determining component life; where ceramic articles containing low melting silicates are to be heated, the silicon carbide must be protected by a protective layer of an inert refractory powder, such as aluminium oxide. Under low pressures of oxygen when active oxidation can occur, the temperature must be reduced, or life will be shortened.

Because of the interconnectivity of silicon carbide in the microstructure (and the intrinsic properties of the impure silicon carbide single crystal) both electrical and thermal conductivity are very good. Electrical resistivity first falls quite quickly with increasing temperature, to a minimum in the region of 600–800 °C. Above this point the temperature coefficient of resistivity becomes positive. This behaviour was illustrated by Fig. 4.6, in which the resistivity is expressed relative to its value at 1000 °C. The relative insensitivity of resistance to temperature in the higher temperature regions is a useful feature of the material, and helps to maintain furnace temperature stability.

The thermal conductivity also falls slowly with increasing temperatures, but it is still of the order of 25 W m^{-1} K^{-1} at 1000 °C. This feature helps to reduce the risk of thermal shock damage, as well as facilitating the rapid transfer of heat during the use of silicon carbide in furnace heating.

4.5.2 Reaction-bonded silicon carbide

Production

Reaction-bonded silicon carbide represents a family of materials, manufactured in ways which use a common set of principles, although the precise details of the procedure used in each case may vary. The primary intention is to fill the free space within a compacted silicon carbide powder with new bonding silicon carbide, produced (*in situ*) directly within this space. One method for the production of material on the commercial scale (Forrest *et al.*, 1972) consists of reacting a compacted mixture of silicon carbide powder, and finely dispersed carbon (or an organopolymeric precursor of carbon) with silicon. The starting mixture, in contact with (at this stage, solid) silicon, is slowly heated in an argon atmosphere. Above ~500 °C organic carbon-containing phases decompose (*pyrolyse*) to carbon with the release of gaseous by-products. At ~1412 °C the silicon melts and the liquid silicon is drawn by capillary action into the pores of the silicon carbide/carbon mixture. Carbon is slightly soluble in liquid silicon at high temperature (Fig. 4.17), and rapidly reacts (with the evolution of considerable heat), to form new (*secondary*) silicon carbide (Beckmann, 1963). This is nucleated, and develops epitaxially, on the original silicon carbide particle surfaces (Sawyer and Page, 1978). As the crystals develop and expand, the new bridging silicon carbide forms strong links between particles, as during normal sintering. Effectively, "sintering" of the silicon carbide has been achieved through the transport of silicon carbide through a liquid phase. This process is shown schematically in Figs. 4.18 and 4.19. The reaction zone temperature may be controlled at 1600–1700 °C, but because the silicon–carbon reaction is strongly exothermic, local temperatures within the reacting component can be much higher (up to 400 °C). In some production processes temperatures can reach 2300 °C. The reaction time depends on the temperature and the length over which the liquid silicon must flow, but the process is generally completed within minutes, depending on the distance over which the silicon must travel – another parabolic diffusion process, and the time is a function of (distance)^2.

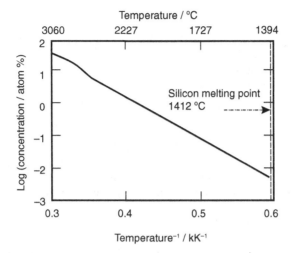

Figure 4.17 The solubility of carbon in liquid silicon (Scace and Slack, 1959. Reprinted with permission from Journal of Chemical Physics. Copyright 1959, American Institute of Physics.)

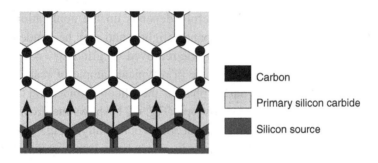

Figure 4.18 The formation of reaction-bonded silicon carbide. Silicon is drawn from a reservoir by capillary action into a porous mixture of carbon and silicon carbide, where reaction takes place.

When carbon is converted to silicon carbide by (4.1) there is a large volume expansion (by a factor of ~2.39); the volume of carbon in the initial mixture must therefore be adjusted to leave some continuous porosity after the reaction, otherwise access of liquid silicon through the remaining pore channels is impeded. The maximum allowable density (ρ) of the green compact is given by $\rho = 3.21/(1+2.33x)$, where x is the volume fraction of graphitised carbon in the starting mixture. In practice a green density ~10% less than the maximum is

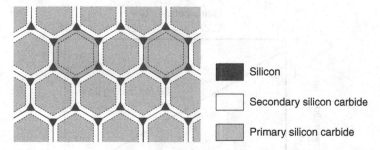

Silicon

Secondary silicon carbide

Primary silicon carbide

Figure 4.19 A schematic illustration of the microstructure of reaction-bonded silicon carbide. The continuous phase is silicon carbide, the darker phase is the original (slightly impure) silicon carbide, the white coating the (purer) secondary silicon carbide, and the darkest regions are residual silicon filling what would be porosity (see also Fig. 4.21).

necessary. However, there is little or no overall dimensional change during the processing and siliconisation stages (slight shrinkage may result from the initial binder burn-out and carbonisation stages). Coarser (100 μm or larger) primary silicon carbide powders are impregnated more rapidly than finer (less than 10 μm) powders, but give products with lower grade mechanical properties; the type of silicon carbide powder used therefore depends on the applications. During cooling the internal liquid silicon expands slightly and may be exuded from the surface of the material: this can easily be removed by grinding.

Figure 4.20 shows the microstructure of a typical sample of reaction-bonded silicon carbide; in this case the darker phase is the silicon carbide, and the light the silicon. More detailed electron microscopy (Fig. 4.21) shows the growth of high-purity (and lighter) α-phase silicon carbide, often epitaxially, on the primary (dark phase), silicon carbide particles, which in effect have then expanded until the crystals merge (Sawyer and Page, 1978). The primary commercial silicon carbide powder particles are invariably slightly impure, and mainly α-phase material; the secondary phase is generally much higher-purity α-phase silicon carbide. This difference in purity readily allows the two types of material to be identified using electron microscopy. There may also be a small amount of cubic β-silicon carbide, which has crystallised within the liquid silicon during cooling. The residual "free" space in the material is occupied by the surplus silicon giving, at temperatures below the melting-point of silicon, essentially a two-phase composite consisting of isolated pockets of silicon within a continuous matrix of silicon carbide.

Reaction-bonded carbide components are normally produced by a preliminary machining of the relatively soft "green" billets of the starting mixture of silicon carbide powder, and the carbon-containing organic phase. The final machining to shape, with the removal of small globules of adhering silicon, is carried out after the reaction-bonding process and cooling to room temperature. The amount of

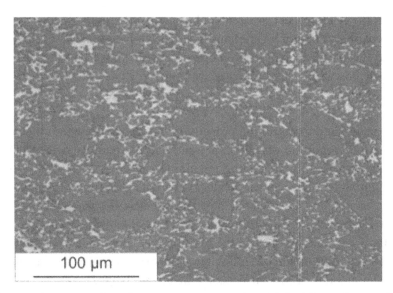

Figure 4.20 A scanning electron micrograph of a polished face of a reaction-bonded silicon carbide showing the two-phase structure; the brighter (more reflective) phase here is silicon. (Reprinted by kind permission of St. Gobain IndustrieKeramik Rödental.)

Primary silicon carbide

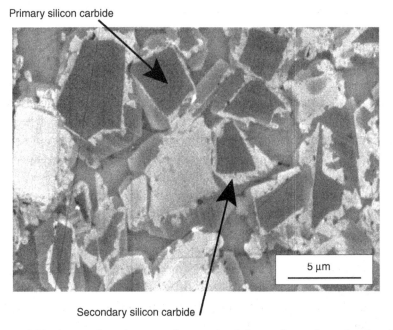

Secondary silicon carbide

Figure 4.21 A scanning electron micrograph of the surface of reaction-bonded silicon carbide, using secondary electron emission to illustrate the two types of silicon carbide. The dark, impure, crystals are primary silicon carbide; the very light phase is the higher purity secondary silicon carbide, which has grown epitaxially on the primary crystals. (Reproduced by kind permission of T. F. Page.)

silicon in the final product is variable, but normally around 8–20% by volume, and there will also be some variations in silicon carbide grain size. These factors influence the physical and other properties of the material, so that reaction-bonded silicon carbide can be treated as a group of materials.

Properties

The two main factors controlling property in the reaction-bonded silicon carbide group are the proportion of silicon present, and the particle size of the silicon carbide powder used in the starting mixture (Kennedy et al., 1973). The silicon carbide starting powders can vary widely in particle size, from means of ~5 μm to ~100 μm, though it is mostly a fine powder, and there is a small increase in grain size as a consequence of the formation and deposition of new carbide. There is very little if any true porosity in the reaction-bonded product, but there will be between 8 and 20% of silicon, filling the void space between the silicon carbide grains. The amount of silicon depends on the packing density of the original compacted silicon carbide, and the amount of carbon incorporated: these will vary with manufacturer.

This type of silicon carbide ceramic shows strikingly the influence of a lower melting-point secondary phase on the properties of a material consisting predominantly of very high melting-point material. The main difference between the liquid phase sintered aluminas (containing intergranular aluminosilicate glass) reviewed in Chapter 3, and the silicon–silicon carbide type of material, is that the glass softens progressively over a wide temperature range, whereas the crystalline silicon melts sharply at a specific temperature. There is therefore an abrupt discontinuity in the mechanical properties of reaction-bonded silicon carbide at ~1412 °C, with, in particular, a sudden loss of strength and modulus. Liquid phase sintered aluminas, in contrast, lose their mechanical integrity more gradually with rising temperature, as the amorphous phase softens. The practical upper limit of use of self-bonded silicon carbide is quoted as 1380 °C.

The Young modulus values follow the standard rule of mixtures, and the overall modulus (E_m) is given approximately by $E_m = V_1 E_1 + V_2 E_2$, in which V_1 and V_2 are the volume fractions of the two components, and E_1 and E_2 their moduli (see Section 1.3). The modulus of silicon at room temperature is 165 GPa, and the composite modulus is therefore in the range 360–420 GPa (similar to that of a high-purity sintered alumina). This stiffness is maintained to ~1200 °C, when the modulus drops to ~100 GPa, close to that for a normal recrystallised silicon carbide (Henshall et al., 1977).

A typical room-temperature three- or four-point bend strength value is between 250 and 450 MPa, depending on crystal size, and silicon content. For a high-quality material containing 8–12% of free silicon (and therefore more silicon

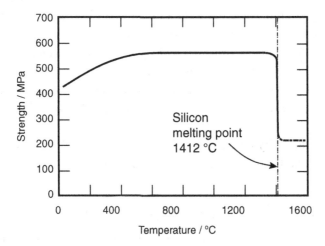

Figure 4.22 Strength of a reaction-bonded silicon carbide as a function of temperature; the loss of strength when the silicon melts at ~1412 °C is abrupt. (After Forrest *et al.*, 1972, reproduced by kind permission of CERAM, Stoke-on-Trent.)

carbide–silicon carbide bonding) the bend strength at room temperature can be as high as 520 MPa. As with many two-phase materials that have been formed at very high temperature, weakening tensile stresses are likely to be developed during subsequent cooling to room temperature, because of differences in the thermal expansion coefficients of the phases. Consistent with this view, bend strength starts to increase at ~500 °C (presumably with stress relaxation), normally reaching a maximum at 1200 °C of ~480 MPa (an increase of ~25%). Very good grade materials can attain 550 MPa at this temperature. Strength invariably falls quite suddenly between 1300 and 1400 °C to ~100 MPa, with the melting of the silicon as shown in Fig. 4.22, and reversion of the material to what is essentially a porous sintered silicon carbide containing between 10–20% volume porosity. As with the modulus, high-temperature strength is similar to that of the porous recrystallised form of silicon carbide.

Fracture toughness (K_{Ic}) values for silicon carbide are in general rather low (Faber and Evans, 1983), in the region of 3–3.5 MPa m$^{1/2}$ for single-crystal silicon carbide, and only slightly higher for polycrystalline and reaction-bonded materials (~4 MPa m$^{1/2}$). The low value means that the strength of a reaction-bonded silicon carbide component is sensitive to defects in general, including surface finish, and damage. However, materials do have good resistance to sliding and abrasive wear, and many successful commercial applications have been developed where low friction and wear resistance are required. In this situation the importance of the silicon content is indicated by the fact that localised cracking and ploughing take place on a larger scale in the vicinity of large areas of silicon, than elsewhere. There is also preferential oxidation of the silicon under the action of frictional heat, followed by loss of the oxide, which can then become a wear controlling process. While in theory, silicon and silicon carbide would be expected to have the same resistance to oxidation and corrosion, in practice reaction-bonded forms of silicon carbide are degraded at a higher rate than pure silicon carbide.

Thermal conductivity remains high, ranging from 210 W m^{-1} K^{-1} at 100 °C to 39 W m^{-1} K^{-1} at 1200 °C (North and Gilchrist, 1981).

4.5.3 Siliconised silicon carbide

This a simpler variant of reaction-bonded silicon carbide produced by filling the porosity in recrystallised silicon carbide with liquid silicon. Infiltration with silicon is carried out under an inert atmosphere at temperatures of up to 2000 °C. Because in this case no new silicon carbide is produced, the porosity and thus final silicon content are normally higher than those of the reaction-bonded silicon carbides. The silicon content of these materials is usually greater than that of the reaction-bonded form, and can reach 20% or more. The microstructure of this form of material consists of interconnected grains of silicon carbide, bonded during the high-temperature recrystallisation stage at 2100–2450 °C, with inter-dispersed silicon. Its properties are similar to those of materials produced by reacting internal carbon with liquid silicon, with due allowance for the higher silicon content.

4.5.4 Solid state sintered silicon carbide

Production

In the early stages of the development of sintered silicon carbide it was believed that only β-phase, fine (and expensive), silicon carbide powder could be densified. However, it was soon realised that the more readily available and slightly lower price α-phase powder would sinter equally satisfactorily provided the particle size was small enough (Coppola *et al.*, 1978). The essential requirement is an extremely fine, high-purity, powder. Particle sizes of less than 500 nm are required, in order to reduce the distances over which atom transport must take place. It is common to express the sizes of such fine powders in terms of specific surface area, rather than particle dimension: values of 5–30 m^2 g^{-1} are standard. The powder is blended with the small amounts of additive necessary to secure densification. Typically, this is around 1% by weight of elemental boron, or a boron compound (Murata and Smoak, 1979), together with 1–3% of carbon, or a carbon precursor material. The use of additives of very fine particle size, and efficient particle blending, are essential to ensure homogeneity of distribution in the very fine silicon carbide powder. Precursors in the form of liquids or solutions are particularly effective at facilitating distribution of the additive elements. After compaction and shaping by conventional methods, the powder is sintered under vacuum or inert gas atmosphere (normally argon – nitrogen would neutralise the boron by reacting to form boron nitride) at 2000–2200 °C for around 30 minutes (additive precursors may require a preliminary calcination stage). Careful control of the sintering conditions, and in particular the temperature, is required: higher temperatures, or longer times, result in excessive grain growth and the development of the large plate-like grains characteristic of the raw Acheson silicon carbide (shown in Fig. 4.1). Even so, it is impossible to avoid grain growth of the 100 nm or so silicon carbide particles, and the mean grain size in the fully dense material is usually of the order of 10–20 µm (Johnson and Prochazka, 1977). Some of the carbon reacts with (and removes as silicon monoxide) the surface silicon dioxide on the silicon carbide particles. The remainder is found in the form of graphite in small pockets at grain edges and corners. The boron normally segregates to the silicon carbide–silicon carbide grain boundaries (Davis *et al.*, 1984). Boron compounds, such as boron carbide, B_4C, can be used as the source of boron. Alternatives to the use of boron were subsequently explored. It was found that aluminium compounds (with carbon) give equally high density silicon carbide. Combinations of aluminium, boron and carbon are used as the sintering aid (Lange, 1975). In this case sintering can involve the formation of liquid phases, and new crystalline phases can be developed at grain boundaries (Yu *et al.*, 2007). Detailed studies have been made of the development of microstructure in all these materials (Suzuki, 1983; Stutz *et al.*, 1985).

Physical and mechanical properties

The important feature of dense sintered silicon carbide is the very low level of low melting-point intergranular phases, and amorphous grain boundary films (Zhang and De Jonghe, 2003). Amorphous phases which do form can be recrystallised by annealing (Chen *et al.*, 2000). Sintered materials are normally >98% dense, with corresponding bulk densities in the region of 3.15 Mg m^{-3}. The microstructure consists of mainly equiaxed grains of α-silicon carbide, although there is a strong tendency for large plate-like grains to develop, particularly if sintering has been carried out at a very high temperature. Commercial materials typically have a mean grain size of <20 µm. A typical microstructure, as shown by transmission electron microscopy, is in Fig. 4.23. There is no apparent grain boundary or secondary phase, though traces of graphite may be present at grain edges and corners. Strength is, as would be expected, influenced by grain size. The sintering

Figure 4.23 A transmission electron micrograph of sintered silicon carbide showing the absence of a second phase. (Courtesy of Steuart Horton.)

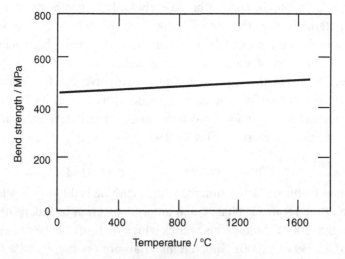

Figure 4.24 The strength of pure sintered silicon carbide as a function of temperature. Because there is no intergranular phase, strength is maintained to very high temperatures. (After Srinivasan, 1989.)

aid is not a strength-controlling factor (Grellner *et al.*, 1981). Typically room-temperature three-point bend strengths are in the region of 400–450 MPa (Munro, 1997; Price and Hopkins, 1982), but strengths as high as 550 MPa have been obtained (Wiederhorn *et al.*, 1999). A valuable feature of sintered silicon carbide is that strength is more-or-less unchanged to the highest temperatures of practical current interest (~1650 °C). The maximum service temperature has been quoted as 1900 °C, though oxidation must be prevented: the strength–temperature relationship is illustrated by Fig. 4.24 (which in other respects lacks any features of interest!). Young modulus values for fully dense materials are close to those for single-crystal material, in the region of 410 GPa (Coblenz, 1975; Schwetz and Lipp, 1980). The adverse result of the absence of intergranular phase material, and the ready cleavage of silicon carbide single crystal to give smooth fracture faces, is that fracture toughness is relatively low for a polycrystalline material. Measured values are in the range 2.3–4.5 MPa m$^{1/2}$ (McHenry and Tressler, 1980; Faber and Evans, 1983). These low values (about 75% of that of a sintered alumina) mean that strength is sensitive to surface defects or damage, and components therefore have to be handled with some care, and impacts avoided. Fracture energy is influenced by grain size and morphology, with values generally ranging from 20–30 J m^{-2} (George, 1973; Mendiratta and Petrović, 1976).

Thermal conductivity decreases almost linearly over the range room-temperature to 400 °C, but is still ~ 77 W m^{-1} K^{-1} at this temperature, making the material a very good thermal conductor (Böder and Heider, 1978). This may convey a certain advantage for the resistance of the material to damage by thermal shock, although the ranking of the three types of silicon carbide reviewed here depends on the parameter being considered (Böder and Heider, 1978). For very violent temperature changes (R), thermal conductivity may not in reality be able to assist very much, but because of the very high strength of the fully dense sintered form it is the best of the three. It is still the best under less violent changes (R'), when it is assumed that there may be time for heat flow to or from the surface to occur. However, if resistance to the propagation of a pre-existing crack (R''') is considered, it is the worst.

Fine-grain sintered silicon carbide resists wear very well, presumably the result of a combination of its hardness, and small grain size (slow wear rates are associated with small grain sizes). The dominant wear mechanism, and the wear rate, depend on the severity of the conditions, and on the existence, or not, of lubrication. Figure 4.25 illustrates typical dry sliding wear rate data, collected from a set of dense sintered silicon carbide and silicon nitride materials with mean grain sizes between 1 and 6 μm, sliding like against like, at constant sliding speed. Both types of material behave similarly, with a discontinuous increase in wear rate at ~10 N load (Horton *et al.*, 1986). At low sliding loads, dry wear is predominantly by a polishing mechanism, with differential wear of different grain facets, as shown in Fig. 4.26. With more severe conditions (and locally high temperatures) oxidation becomes important, with the production and then removal of an amorphous soft oxide film, in combination with microfracture and

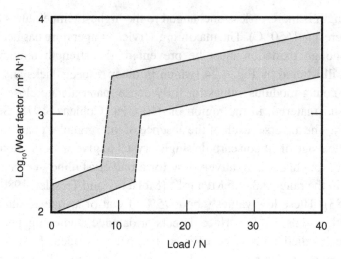

Figure 4.25 Collected wear factor values as a function of load, for the dry sliding wear of silicon carbide against silicon carbide, and silicon nitride against silicon nitride. (After Horton *et al.*, 1986.)

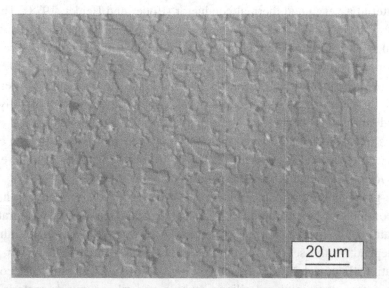

Figure 4.26 The wear surface of sintered silicon carbide, polished by low load dry sliding wear. This shows faint grooving, and the selective wear rates of different crystal faces. (Courtesy of Steuart Horton.)

grain detachment, as shown in Fig. 4.27. Figure 4.28 shows combined data for dry and water lubricated sliding wear. Under wet conditions the wear rates are reduced by factors of between 10^2 and 10^3. Data for dense sintered alumina are shown for comparison: the dry wear rate of the alumina under similar conditions

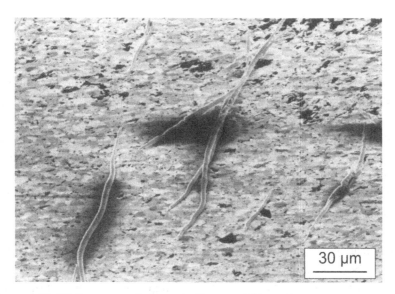

Figure 4.27 Sliding wear of sintered silicon carbide under more severe conditions. This shows the uneven polishing of silicon carbide grains, whole grain detachment, and the rolling up and removal of a soft amorphous silicon dioxide film. (Courtesy of Steuart Horton.)

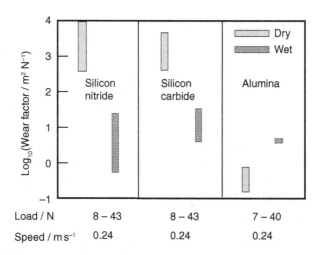

Figure 4.28 Summarised wear factor data for sintered silicon carbide, silicon nitride, and alumina, for dry and wet sliding wear. (After Horton *et al.*, 1986.)

is much lower than those of silicon carbide and silicon nitride, but the wet rates are similar, reflecting the tendency for aluminium oxide to dissolve in water under these conditions (Wallbridge *et al.*, 1983).

4.5.5 *Liquid phase sintered silicon carbide*

Production

To put these materials into perspective it has to be noted that porous and relatively low-strength silicon carbide has been produced over many years with an aluminosilicate glass bond phase, for use as special purpose refractories, and grinding and cutting wheels. Bonding in these materials is typically achieved by incorporating powders such as clay and potash feldspar (as used in porcelain production), in amounts of between 10 and 40%. Sintering is carried out in air at ~1300–1500 °C to melt the additive, which then wets and spreads over the slightly oxidised silicon carbide particles. The product, which has a large volume of accessible porosity, consists of fairly coarse-grained silicon carbide particles, bonded by an aluminosilicate glass. The bend strength and modulus are very low by comparison with the fully dense silicon carbide ceramics discussed above. In a 90% silicon carbide refractory grade for example, the Young modulus would be ~25 GPa, and bend strength ~5 MPa (Singer and Singer, 1963). Stronger materials can be produced from mixtures of silicon and silicon carbide powders, by reacting the silicon with nitrogen at temperatures of 1200–1400 °C (see Chapter 5) to form a silicon nitride bond phase, but these materials still have volumes of porosity that are not much lower than that of the packed starting powders.

It has been known for more than 50 years that fine silicon carbide powder could be hot-pressed to full density at very high temperatures using aluminium oxide. Partial reaction takes place between the natural silicon dioxide film on the silicon carbide particles, with the formation of a wetting aluminosilicate liquid (Alliegro *et al.*, 1956; Lange, 1975). Interest in obtaining fully dense silicon carbide using additive systems similar to those used for silicon nitride (based on aluminium oxide and yttrium oxide) has expanded, and liquid phase sintered silicon carbide of this type is now commercially marketed. This additive, aluminium oxide or aluminium nitride, and a rare earth oxide (usually Y_2O_3), may seem an unusual combination, but the Al_2O_3–Y_2O_3 system is a very refractory one, and the liquids have high viscosity. At temperatures above ~1800 °C it is a good densification aid for silicon carbide (and silicon nitride).

Pressureless sintering of very fine (<1 μm) α- or β-phase silicon carbide powder is readily achieved at ~1850 to 2000 °C under controlled atmospheres (non-oxidising), using wetting liquids formed from the additive oxides and surface silicon dioxide (Mulla and Krstic, 1991; Lee and Kim, 1994). A typical additive combination is ~1.5 weight% aluminium oxide and ~1.5 to 5 weight% yttrium oxide. Densities of >97% of theoretical are readily obtained. A microstructure of material sintered with a mixture of aluminium nitride and yttrium dioxide, is

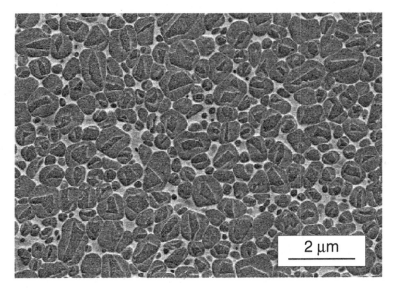

Figure 4.29 A scanning electron micrograph of liquid phase sintered silicon carbide, showing the characteristic silicon carbide growth pattern, and the location at the grain boundaries of the second phase material. (Reproduced by kind permission of the Max-Planck-Institute for Metals Research.)

shown in Fig. 4.29. The secondary phase material is concentrated at grain boundaries, and in intergrain pockets.

Physical and mechanical properties

One advantage of the liquid phase sintering is that hot-pressing or sintering temperatures are reduced. A more important advantage is that the presence of the intergranular glass in the microstructure favours intergranular crack propagation. The fracture of pure silicon carbide tends to be intragranular, with typically a straight line path, giving low fracture toughness. In coarser grain materials the plate-like, higher aspect-ratio causes the crack paths to become tortuous, providing more possibility for crack branching and crack bridging and considerable increases in fracture toughness (Padture, 1994; Padture and Lawn, 1994). In finer grain material the intergranular phase can have an adverse effect on toughness by weakening the grain or interface boundaries, and generating internal stresses from the thermal expansion mismatch between the silicon carbide and the intergranular glass. It is therefore clear that, as with sintered silicon nitride, grain growth and grain size control are important for optimisation of mechanical properties. The microstructure is strongly influenced by the initial silicon carbide powder polytype (α-phase, or β-phase), its particle size and size distribution, and the volume of liquid phase used (Mulla and

Krstic, 1994; Kim *et al.*, 1998). Conversion of the β-polytype to the α-polytypes during sintering gives larger, plate-like grains, and increased fracture toughness (to 8 MPa m$^{1/2}$). In contrast, sintering α-polytype powder produces more equiaxed grains and lower fracture toughness material (to ~5 MPa m$^{1/2}$) (Lee and Kim, 1994). A bimodal particle silicon carbide particle size distribution also gives higher strength and toughness. With fully developed microstructures of this type, toughness values can be doubled from the values typical of solid state sintered silicon carbide, to 6–8 MPa m$^{1/2}$, and strengths can be as high as 1.1 GPa (Keppeler *et al.*, 1998; Rixecker *et al.*, 2001; Hirata *et al.*, 2008). These values are similar to those for high-quality sintered silicon nitride.

4.5.6 CVD silicon carbide

Production

This was the first form of dense, strong, silicon carbide, with the further advantage that sintering densification aids were not involved in its production, so that the purity was very high. However, because the processes involved in the high-temperature gas-phase pyrolytic reactions were complex, early materials tended not to be completely dense, and contaminated with graphite. At a sufficiently high deposition temperature the material is polycrystalline, and fine grained.

The CVD process involves passing the vapour of a volatile species containing silicon and carbon (such as methyltrichlorosilane) at low pressure over a hot substrate (often graphite). Decomposition of the precursor molecule occurs, as is illustrated in Section 4.4.1 and Fig. 4.12, and silicon carbide is nucleated and grows as a thin film on the substrate. Later, because of interest in the material as a coating for microspherical uranium dioxide or uranium carbide fuel particles, the fluidised bed method was used, and deposition of silicon carbide takes place directly on the fuel microsphere (Gulden, 1968). A carrier gas of hydrogen, or a hydrogen/argon mixture, is normally used, giving low concentrations (~1–10%) of reactant vapour. Substrate temperatures are in the range 1200–1800 °C, though good-quality material can be produced at temperatures as low as 1200 °C. Temperature control is critical for deposition rate and efficiency, and for the microstructure of the silicon carbide, and hence its mechanical properties (Popper and Riley, 1967; Su *et al.*, 1995). The process is inefficient (decomposition also takes place in the gas phase, rather than on the surface, which means that potential material is lost) and slow, and it may take many hours to achieve mm thickness layers of silicon carbide. However, large sheets of good-quality, crystalline and fully dense, material can be obtained with thicknesses of several cm, and be deposited with a range of (near net-shape) complex geometries. Nuclear fuel particles are used in the as-coated condition: otherwise the silicon carbide has to be detached from its substrate.

This form of material is fully dense (3.21 Mg m^{-3}), isotropic β-phase (3C polytype), of very high purity (ppm levels of foreign elements). Grain sizes can be in the 1–10 μm range giving room-temperature bend strengths of ~470 MPa, increasing to ~575 MPa at 1400 °C. Room-temperature Young modulus is around 460 GPa, and standard indentation hardness around 25–28 GPa (Niihara, 1984; Price, 1997; Cockeram, 2002). Microhardness values can be higher, of the order of 32–36 GPa. The electrical conductivity is tailored by controlled doping: room-temperature conductivity values cover 3–4 orders of magnitude in the range ~0.1 to ~100 S^{-1} m^{-1}. This means that the material can readily be used for electrical heating during component production (Haigis, 1993). Other advantages in this context are its resistance to corrosion, and an ability to withstand severe temperature cycling without distortion or damage.

4.6 Applications

The low-density, recrystallised or silicon nitride bonded, forms are extensively used as kiln furniture and gas filters. Although some types of dense silicon carbide do have specific applications, there is generally overlapping in the choice of silicon carbide material (sintered or reaction-bonded) for any particular application. For convenience the topic will be treated according to the type of application, rather than material type. The major application areas, and the associated materials' main requirements, are summarised in Table 4.5, and these are the headings which will be used. Many current applications for silicon carbide ceramics are at room or relatively low temperatures (below 1000 °C), and rely for the most part on intrinsic hardness, strength, chemical inertness, and wear resistance. Recent interest has developed in the use of porous silicon carbide as gas filters, with widespread application for the removal of particulate carbon from diesel engine exhaust and industrial waste gases (Briggs, 2007).

4.6.1 General industrial

A longstanding and important application for silicon carbide is the internal support structures, which are usually termed "furniture", widely used in industrial high-temperature ovens and kilns. These structures are used repeatedly, and are subject to mechanical loading, oxidising environments, and rapid temperature changes. While the older type of clay-bonded silicon carbide can be used for firing stacked or separated ceramic components at lower temperatures, single-phase, porous recrystallised silicon carbide is much better for higher temperatures, in the range 1300–1600 °C (although this material is considerably – up to seven times – more expensive than the clay-bonded). The high-purity sintered

Table 4.5 *A summary of the main application areas for silicon carbide ceramics, and the properties required.*

Area	Components	Main materials' requirements
General industrial	Kiln furniture	Thermal stability, thermal shock resistance, chemical inertness, wear resistance
	Furnace heating	Electrical conductivity, thermal conductivity, oxidation resistance, high temperature strength
	Thermocouple sheaths	Thermal conductivity, oxidation resistance, strength
	Cutting and grinding	Hardness, strength, wear resistance, high melting-point
Wear resistance	Pumps and valves	Hardness, wear resistance, chemical inertness
	Brakes	Hardness, low density, thermal conductivity
Filters	Hot metal	Thermal shock resistance, thermal stability, chemical inertness
	Diesel engine exhaust gases	Thermal shock resistance, oxidation resistance
	Industrial waste gases	Thermal shock resistance, oxidation resistance, chemical inertness
Energy absorption	Armour plate	High melting-point, high modulus, low density, high sonic velocity
Automotive	Gas-turbine engine components	Thermal shock resistance, high temperature strength

material is particularly useful where minimum contamination of the material being heated is required, for example, in the sintering of electroceramics and semiconductors. Because the resistance to thermal shock of silicon carbide is good, faster heating times can be used with less risk of damage, and the high thermal conductivity means that heat transport, and temperature smoothing, are fast. The pure forms of silicon carbide can survive several thousand production cycles before they have to be replaced, usually because of distortion or cracking caused by creep and oxidation. Typical examples of supporting kiln furniture made from recrystallised silicon carbide are shown in Figs. 4.30 and 4.31.

4.6.2 Heating

Recrystallised silicon carbide has been widely used over many years as electrical heating elements, in rod or spiral form, in high-temperature furnaces. An example of a small furnace spiral heating element is shown in Fig. 4.32 (Pellisier, 1998).

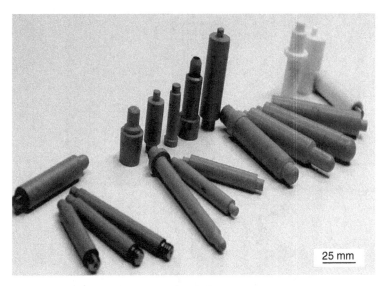

Figure 4.30 Reaction-bonded silicon carbide setters for separating items in a furnace. (Reproduced by kind permission of St. Gobain Advanced Ceramics s.r.o.)

Figure 4.31 Kiln support elements (kiln "furniture") in low-density silicon carbide. (Reproduced by kind permission of St. Gobain IndustrieKeramik Rödental.)

The application depends on the very good electrical conductivity of silicon carbide, and its resistance to oxidation. Silicon carbide can be designed, by suitable doping with other elements, to have a stable electrical conductivity–temperature relationship, as shown by Fig. 4.6. The maximum temperature of use depends on

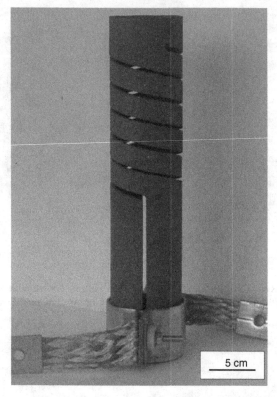

5 cm

Figure 4.32 Spiral electric heating element made in recrystallised silicon carbide. (Reproduced by kind permission of Kanthal Ltd.)

the furnace environment, and the presence of contaminants, but it is usually specified as ~1500 °C for clean air. Oxidation leads to slow loss of material and undesirable increases in resistance. Other valuable features of silicon carbide are its very high thermal conductivity and retention of strength at high temperature. This means that silicon carbide is a very useful material for use in heat exchangers, where the heated gas or fluid must be separated from the source of heat, such as burning gas and combustion products, or impure hot air. It is widely used for burner tubes in gas-heated radiant heaters, where the gas and its combustion products must be separated from the furnace atmosphere. Examples of this type of component tube are shown in Fig. 4.33. Complex units can be constructed using dense sintered silicon carbide as the dividing medium. Figure 4.34 shows heat exchanger plates designed for a unit for use in the chemical processing industries, where fluids are heated and cooled. One type of complete unit is shown in Fig. 4.35. The advantages of dense silicon carbide for these applications are (besides impermeability to gas) a resistance to corrosion by aqueous media and gases, and to wear by abrasive particles. It also has the ability to withstand easily

Figure 4.33 Heat protection: tubular forms of reaction-bonded silicon carbide. (Reproduced by kind permission of St. Gobain IndustrieKeramik Rödental.)

temperatures which would be considered to be high in this context (~1200 °C), and its thermal shock resistance allows rapid thermal cycling. The same types of requirement also apply in principle to thermocouple sheaths, for which a metallic thermocouple, or oxide thermistor, has to be separated from the oxidising or corroding environment, the temperature of which is being measured.

CVD silicon carbide plates are used in the microelectronics industries for processing silicon wafers. The high purity of this form of the material, its high temperature stability and good thermal conductivity, reduce the risk of contamination of the silicon, and other components. It has a wide range of applications relying on these properties in semiconductor processing equipment, and as heat sinks for electronic packaging.

4.6.3 Cutting and grinding

The finer grain, oxide, nitride, or glass-bonded, sintered silicon carbides have been widely used for the production of silicon carbide (carborundum) grinding

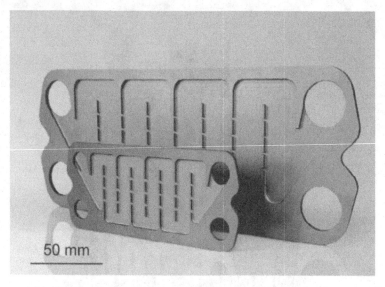

50 mm

Figure 4.34 Heat exchanger plates with internal channels, in sintered silicon carbide. (Reproduced by kind permission of ESK Ceramics GmbH & Co. KG.)

Figure 4.35 Complete heat exchanger unit incorporating silicon carbide heat transfer plates. (Reproduced by kind permission of ESK Ceramics GmbH & Co. KG.)

and cutting blocks and wheels. Many of these systems also use silicon carbide grit, bonded by thermosetting or phenolic resins. Their function depends primarily on the very high hardness, and high melting-point of silicon carbide. A useful feature of silicon carbide in this context is the tendency (mentioned in the Introduction) of individual crystals to cleave, to leave sharp edges. The wear process does not lead therefore to blunting.

4.6.4 Filters – molten metal

There are important applications for ceramic materials as particulate filters at moderate to high temperatures (where metallic materials might dissolve or melt). One well-established use of ceramics, in general, is in the removal of oxide particles from liquid metals, such as molten steel or non-ferrous metals such as aluminium and copper, during casting. Filtering is standard foundry practice, to prevent the trapping of particulate inclusions in the casting. Several types of filter are in use, a common one being the "foam" filter. (How is a ceramic "foam" made? – not too difficult! A conventional polymeric sponge is soaked in a slurry of the ceramic powder, dried, the sponge is burned off, and the residual powder sintered.) While most ceramic foam filters are based on oxides, sintered (that is, pure, recrystallised) silicon carbide foam filters are used in pouring cast iron, and copper alloys. In this form of material, porosity is expressed as pores per linear cm: typical values are in the range 5–25 cm^{-1}. The main advantage of silicon carbide is its resistance to thermal shock, and high-temperature creep resistance. However, because of the potential reactivity of silicon carbide towards metals, and particularly to ferrous alloys, the maximum temperature of use is lower than that of an oxide filter such as partially stabilised zirconia (Chapter 6). Each filter unit is usually quite small (a cm or so thick, and a few cm^2 in area), and the price is a few Euro cents.

4.6.5 Filters – exhaust gases

A market of increasing importance (Fig. 7.2) is in filters for the removal of fine particles ("smoke") from diesel engine exhaust gas, and from industrial hot waste gases. Many industrial and automotive diesel engines incorporate exhaust gas filters in the form of porous honeycomb structures (similar to those used as catalyst supports). These filters can be very efficient at removing the small particles. Recrystallised lightly sintered silicon carbide is among the materials used, with the advantage of thermal stability and resistance to chemical attack by contaminants in the gas stream. Filter materials necessarily have high porosity (~40 to ~60%) and thus low bulk density (~1.9 Mg m^{-3}). Strengths are correspondingly low (~20 to ~50 MPa), but these are quite adequate for the application. The thermal shock resistance of silicon carbide is a further advantage, because the filter temperature is periodically raised to ~450 °C or higher to burn off the trapped carbon. Hot-gas filters are also widely used in industrial processes, such

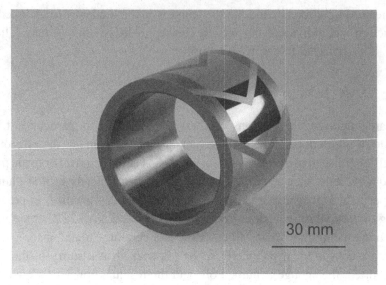

Figure 4.36 Pump shaft sliding bearing for use under high loading. (Reproduced
by kind permission of ESK Ceramics GmbH & Co. KG.)

as power stations and cement and lime production, to trap dusts. These can be
more efficient than electrostatic precipitators in removing submicron particles.

4.6.6 Wear-resistant components – pumps, seals and valves

Important current applications for solid state sintered and reaction-bonded silicon
carbide are for components in mechanical pumps handling liquids, and abrasive
slurries. Here, resistance to sliding and abrasive wear, and possibly corrosion, are
important. The primary material's requirements are hardness, a low coefficient of
friction, and chemical inertness. Many pumps use a hard material for one sealing
face, and a softer (such as a polyamide, or graphite) for the other. Where high
concentrations of abrasive particles are present, both faces must be hard. Components are made from solid state sintered silicon carbide, though reaction-
bonded silicon carbide is slightly cheaper. One type of dense sintered silicon
carbide component, a laser structured (to improve lubrication and reduce leakage)
pump shaft sliding bearing for use under high loading, is shown in Fig. 4.36.
Figure 4.37 shows examples of sintered silicon carbide thrust bearing rings
intended for use in a variety of environments, exposed to corrosive and abrasive
fluids. Automobile water pump seals (at a price of less than €1 each) also provide
an important market for sintered silicon carbide, and for many models silicon
carbide seals are replacing alumina sealing rings. Overall there is an approximately equal split between reaction-bonded silicon carbide, and sintered silicon

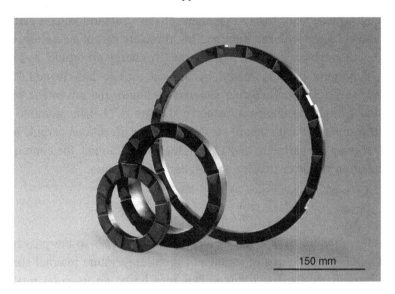

150 mm

Figure 4.37 Thrust bearing rings in sintered silicon carbide. (Reproduced by kind permission of ESK Ceramics GmbH & Co. KG.)

carbide, in pump seal components. The main drawback to the use of reaction-bonded silicon carbide in this type of application is the presence of the silicon, which under aqueous alkaline conditions preferentially reacts and dissolves. Sintered silicon carbide is often produced for applications involving sliding wear with a bimodal grain size distribution, a mixture of >100 μm (up to 1500 μm) platelets in a finer (10 μm) matrix. For this type of application graphite can also be incorporated to improve dry running properties. The differential wear in the microstructure of dense sintered silicon carbide (an example was shown in Fig. 4.26) provides shallow recesses in the polished surfaces, which can trap thin films of fluid to reduce friction during short periods of dry running. Reaction-bonded or siliconised silicon carbide materials are being considered for use as disc brake components in high-speed trains and road vehicles. Besides resisting wear, silicon carbide has the further advantages of its low density (reducing overall weight) and good thermal conductivity, to assist heat dissipation. Liquid phase sintered silicon carbide, with its higher fracture toughness, is now widely used in paper-making machines, and for large gas seals such as those used in compressors in natural-gas pipelines.

4.6.7 Armour plate

Sintered (and hot-pressed) silicon carbide tiles ~5 to ~10 mm thick have over recent years become widely used as protective armour, particularly in the USA

market, and now account for a large share of the structural silicon carbide produced in the USA. As with the alumina tiles, these are part of a carefully designed
system, in which the ceramic tile is supported by energy absorbing, and fragment
trapping, polymeric backing materials. Silicon carbide has, in addition to its
greater hardness (50% higher than sintered alumina), the advantage of its low
density and a higher velocity of sound (~11 km s^{-1}) than alumina. It gives
protection comparable with that of steel, but tile for tile silicon carbide armour is
30% lighter than steel armour, and 15% lighter than alumina, the consequence of
the combination of lower density and higher effectiveness.

4.6.8 Automotive

Considerable attention was given in the 1960s and 1970s to the possible application of silicon carbide in a new range of high-temperature internal combustion
engines, particularly the gas turbine. This was based on its good resistance to
thermal shock, resulting primarily from the low thermal expansivity. A secondary
factor, which in theory should also help materials to withstand thermal shock, is
the very high thermal conductivity, facilitating the smoothing of severe surface
temperature gradients. In practice, however, most temperature changes developed
during engine applications are so fast that the high thermal conductivity is
unlikely to be of real benefit. Reaction-bonded silicon carbide was tested
extensively in early ceramic gas turbine engine programmes as a potential
combustion chamber material. However, while components performed well under
test conditions, difficulties in the commercial mass-production of components of
the required degree of consistency, and at reasonable cost, have for the time being
ruled out such ambitious objectives (Katz, 1980).

4.6.9 Nuclear power

Because of the high strength at high temperature, and low creep rates of fully dense
reaction-bonded silicon carbide, it had originally been thought that it could be used
as relatively large (several cm dimension) containers for the nuclear (uranium
based) fuel, in nuclear power stations working at higher temperatures than those
allowed by metallic containers. The main function of the container is to prevent the
escape of volatile radioactive products into the coolant gas stream. However, this
line of development was supplanted by the use of thin, dense, CVD silicon carbide
coatings for individual fuel microspheres. This type of fuel consists of small particles of uranium oxide or carbide, each of which is coated with multiple, thin, layers
of CVD deposited carbon, and silicon carbide (Snead *et al.*, 2007). The function of
each layer is quite specific, but the key to the containment of the volatile products is

the layer of dense CVD silicon carbide. This in effect acts as a pressure vessel for the reacting fuel particle to contain volatile fission products.

4.7 Summary

Silicon carbide differs from alumina in that more extensive use is made of the solid state sintered form, and solid state sintered silicon carbide provides a very good illustration of a material which has very little if any grain boundary phase. The almost complete absence of grain boundary silicate gives it a very good high-temperature strength, and low oxidation rates, but also a much lower fracture toughness than in the liquid phase sintered form. This illustrates the important influence over properties that even small amounts of intergranular material can have, perhaps even to a greater extent than is shown by alumina. The grain morphology is not dissimilar from that of sintered alumina, with fairly equiaxed microstructures and a tendency for grain growth to give thick plate-like grains, and there is not the same potential for toughening by crack bridging as there might be with higher aspect-ratio grains. Liquid phase sintered silicon carbide has, in contrast, toughness and strength considerably greater than the solid state sintered form (and than alumina), but at the moment the greater expense of its production means that it is not so widely used.

Silicon carbide has a remarkably wide range of useful features: high-temperature stability and an ability to withstand a variety of atmospheres, high hardness, good resistance to wear, and high thermal conductivity. It also has an extremely wide range of electrical conductivity values (from highly insulating to semiconducting). Other advantages are its low thermal expansivity which gives good thermal shock resistance and mechanical stability under conditions of rapid changes of temperature. It is one of the few materials able to maintain strength and creep resistance to over 1500 °C, a property which is widely exploited in applications as kiln furniture. Restrictions on the wider uses of sintered silicon carbide are the higher price of the high purity, very fine, powders needed for its production, and the very high temperatures (1800–2500 °C) required to obtain full density. However, as with all very hard materials, a significant part of the cost of a precision component is the cost of diamond machining to the required shapes and surface finish (see also Fig. 5.32). Because the low fracture toughness of solid state sintered silicon carbide makes it sensitive to surface damage, careful attention is needed in processing and subsequent handling. Nonetheless, where controlled and stable electrical conductivity or good thermal conductivity are required at very high temperatures, it is difficult to find a better material.

The two-phase reaction-bonded form has versatility, in that there is no sintering (or reaction-bonding) shrinkage. Reaction-bonding is useful for the production of large, complex-shaped, components in smaller quantities, where compaction density and normal sintering shrinkage can present problems. This form may therefore be preferred if the component geometry is complex, when slip casting can be used, and for small production runs. However, the silicon infiltration process is not as easy to handle as a dry, solid state, sintering operation, and the general trend now is to use the simpler and easier to manage sintering process, rather than reaction bonding. Reaction-bonded silicon carbide is used for components exposed to sliding or abrasive wear, but the silicon seems to be the weak link and there is a tendency for erosion and corrosion to occur more readily under wet wear conditions. Liquid phase sintered silicon carbide has strength and toughness considerably greater than the solid state sintered form (and alumina), but at the moment the expense of the production means that it is not so widely used. The lesser used CVD process provides a convenient route to the production of thin, dense, barrier layers with applications in the nuclear power industries, the importance of which is likely to increase with decreasing fossil fuel supplies.

Silicon carbide stands out as a material of considerable versatility, and with a set of very good property values, not least its wear and corrosion resistance. It is also of wider interest in that it demonstrates that non-oxides can be sintered to full density (by solid state or liquid phase sintering) and used as high-strength structural ceramics, just as readily as oxides. This opens up considerably the range of materials which might be considered as structural ceramics.

Questions

4.1. The crystal structures of silicon carbide and silicon dioxide are both based on an SiX_4 tetrahedral unit, and carbon (A_r 12) is a little lighter than oxygen (A_r 16.0), but even the most dense form of silicon dioxide (quartz) is less dense ($\sim$2.7 Mg m^{-3}) than silicon carbide ($\sim$3.2 Mg m^{-3}). Why?

4.2. Show that the condition for the ratio of interfacial energies, γ^{ss}/γ^{sv}, for the closure of a pore lying along a three-grain edge in sintering silicon carbide, is $3^{1/2}$. Why is this condition unlikely to be met in normal silicon carbide powders.

4.3. Estimate values for the Young modulus of reaction-bonded silicon carbide containing 15% by volume of silicon (a) at room temperature, (b) at 1420 °C. Assume that the Young modulus of silicon at room temperature is 162 GPa.

4.4. How far could the average carbon and average silicon atom travel in the β-silicon carbide lattice in 1 hour at 2000 °C? How do you explain the difference? What are the implications for sintering silicon carbide powder?

4.5. The parabolic rate constant for the oxidation of dense, solid state sintered, silicon carbide at 1400°C in air at 1 atm pressure is 1×10^{-5} g^4 m^{-2} s^{-1}. Estimate the thickness of silicon dioxide formed on a bar of this material under these conditions in 24 hours, and in 1 year. Assume that the silicon dioxide forms as a continuous amorphous film, of density 2.2 Mg m^{-3}.

4.6. Calculate the fractional volume change in a particle of carbon when it reacts to form silicon carbide. Assume that the density of carbon (as graphite) is 2.3 Mg m^{-3}, and silicon carbide 3.2 Mg m^{-3}.

4.7. Reaction-bonded silicon carbide is to be formed from silicon carbide powder, which has been compacted to a fractional density of 0.60, then infiltrated with a material which will form graphite on heating. What is the maximum allowable proportion (by weight) of graphite in the mixture if complete reaction to silicon carbide is to be possible on exposure to liquid silicon?

4.8. An equation for the rate of wear by abrasion (Evans, 1979) suggests that wear rate can be a function of the material-dependent parameter $\{H^{-1/2} K^{-2/3}\}$, where H is the indentation hardness, and K the critical fracture toughness. On this basis how would dense sintered silicon carbide, dense liquid phase sintered silicon nitride, and a transformation toughened Mg-PSZ zirconia be expected to behave? What other factors should be taken into account in attempting to predict the wear rates of these three materials?

4.9. Why might dense solid state sintered silicon carbide be a better choice of material for an application requiring the ability to withstand a load at 1600 °C, than pure sintered alumina? What would be the arguments against using silicon carbide?

4.10. Suggest explanations for the differences in fracture toughness seen between solid state sintered silicon carbide and liquid phase sintered silicon carbide.

Selected reading

Greskovich, C. and Rosolowski, J. H. (1976). Sintering of covalent solids. *J. Am. Ceram. Soc.*, **59**, 336–43.

Schlichting, J. and Riley. F. L. (1991). Silicon carbide. In *Concise Encyclopedia of Advanced Ceramic Materials*, ed. R. J. Brook. Oxford: Pergamon Press, pp. 426–9.

Schröder, F. A., Ed. (1986). *Gmelin Handbook of Inorganic Chemistry*, 8th edition, *Silicon, Supplement Volume B3 Silicon Carbide*. Berlin: Springer Verlag.

Sōmiya, S. and Inomata, Y., Eds. (1991). *Silicon Carbide Ceramics – 1 Fundamentals and Solid Reactions*. London: Elsevier.

Srinivasan, M. (1989). The silicon carbide family of structural ceramics. In *Structural Ceramics*, ed. J. B. Wachtman Jr., *Treatise on Materials Science and Technology*, Vol. **29**. Boston, MA: Academic Press Inc., pp. 99–159.

References

Acheson, E. G. (1893). On carborundum *Chem. News*, **68**, 179.

Ainswort, J. H. and Moore, R. E. (1969). Fracture behavior of thermally shocked aluminium oxide. *J. Am. Ceram. Soc.*, **52**, 628–9.

Airey, A. G., Cartwright, P. J. and Popper, P. (1975). Stresses developed in pyrolytic silicon carbide coatings during deposition and cooling. In *Special Ceramics 6*, ed. P. Popper. Manchester: The British Ceramic Research Association, pp. 147–60.

Alliegro, R., Coffin, L. B. and Tinkepaugh, J. R. (1956). Pressure-sintered silicon carbide. *J. Am. Ceram. Soc.*, **39**, 386–9.

Anstis, G. R., Chantikul, P., Lawn, B. R. and Marshall, D. B. (1981). A critical evaluation of indentation techniques for measuring fracture toughness: 1 direct crack measurements. *J. Am. Ceram. Soc.*, **64**, 533–8.

Becher, P. (1983). Strength retention in SiC ceramics after long-term oxidation. *Comm. Am. Ceram. Soc.*, C-120–1.

Bechstedt, F., Kackell, P., Zywietz, A., Karch, K., Adolph, B., Tenelsen, K. and Furthmuller, J. (1997). Polytypism and properties of silicon carbide. *Phys. Stat. Solidi B – Basic Research*, **202**, 35–62.

Beckmann, G. E. J. (1963). The growth of silicon carbide from molten silicon. *J. Electrochem. Soc.*, **110**, 84–6.

Bind, J. M. and Biggers, J. K. (1975). Hot pressing of silicon carbide with 1% boron carbide addition. *J. Am. Ceram. Soc.*, **58**, 304–6.

Böcker, W. and Hausner, H. (1978). Influence of boron and carbon additions on microstructure of sintered alpha silicon carbide. *Powder Met. Intern.*, **10**, 87–9.

Böcker, W., Landfermann, H. and Hausner, H. (1979). Sintering of α-silicon carbide with additions of aluminium. *Powder Met. Intern.*, **11**, 83–5.

Böder, H. and Heider, W. (1978). Eigenschaften von reactionsgesintertem Siliciumcarbid und Anwendungsmöglichkeiten in der Fahrzeug-Gasturbinen. In *Keramische Komponenten für Fahrzeug-Gasturbinen*, eds. W. Bunk and M. Böhmer. Berlin: Springer, pp. 235–50. See also: Tiefenbacher, E. (1984). Die Keramische Fahrzeuggasturbine. *Keramische Komponenten für Fahrzeug-Gasturbinen III: Stausseminar: Papers*, eds. W. Bunk, M. Böhmer and H. Kirsler. Bundesministeriums für Forschung und Technologie, pp. 11–26.

Briggs, J. (2007). *Engineering Ceramics in Europe and the USA*. Worcester, UK: Enceram.

Briggs, J. (2008). Private communication.

Brookes, C. A., O'Neill, J. B. and Redfern, B. A. W. (1971). Anisotropy in hardness of single crystals. *Proc. R. Soc. Lond. Series A – Mathematical and Physical Sciences*, **322**, 73–88.

Burgermeister, E. A., Vonmünch, W. and Pettenpaul, E. (1979). Thermal conductivity and electrical properties of 6H silicon carbide. *J. Appl. Phys.*, **50**, 5790–4.

Burke, J. J., Gorum, A. E. and Katz, R. N., Eds. (1974). *Ceramics for High-performance Applications*. Proceedings of the Second Army Materials Technology Conference, Hyannis, MA, November 13–16th 1973. Chestnut Hill, MA: Brook Hill.

Carnahan, R. D. (1968). Elastic properties of silicon carbide. *J. Am. Ceram. Soc.*, **52**, 223–4.

Chen, D., Sixta, M. E., Zhang, X. F., De Jonghe, L. C. and Richie, R. O. (2000). Rôle of the grain-boundary phase on the elevated-temperature strength, toughness, fatigue, and creep resistance of silicon carbide sintered with Al, B and C. *Acta Mater.*, **48**, 4599–608.

Choyke, W. J. and Patrick, L. (1957). Absorption of light in alpha SiC near the band edge. *Phys. Rev.*, **105**, 1721–3.

Choyke, W. J., Devaty, R. P., Clemen, L. L., Macmillan, M. F. and Yoganathan, Y. (1996). Optical properties and characterization of SiC and III-V nitrides. In *Silicon Carbide and Related Materials*, 1995, Institute of Physics Conference Series, **142**, pp. 257–62.

Coblenz, W. S. (1975). Elastic moduli of boron-doped silicon carbide. *J. Am. Ceram. Soc.*, **58**, 530–1.

Cockeram, B. V. (2002). Fracture strength of plate and tubular forms of monolithic silicon carbide produced by chemical vapour deposition. *J. Am. Ceram. Soc.*, **85**, 603–10.

Collins, A. K., Pickering, M. A. and Taylor, R. L. (1990). Grain-size dependence of the thermal conductivity of polycrystalline chemical vapor-deposited beta-SiC at low temperatures. *J. Appl. Phys.*, **68**, 6510–12.

Coppola, J. A. and Bradt, R. C. (1973). Thermal shock damage in SiC. *J. Am. Ceram. Soc.*, **56**, 214–18.

Coppola, J. A., Hailey, L. N. and Mcmurtry, C. H. (1978). *Process for Producing Sintered Silicon Carbide Body.* US Patent 4,124,667, The Carborundum Co., November 7th 1978.

Costello, J. A. and Tressler, R. E. (1981). Oxidation kinetics of hot-pressed and sintered α-SiC. *J. Am. Ceram. Soc.*, **64**, 327–31.

Davis, R. F., Lane, J. E., Carter, J. E., Bentley, J., Waldin, W. H., Griffiths, D. P., Linton, R. W. and More, K. L. (1984). Microanalytical and microstructural analyses of boron and aluminium regions in sintered a-silicon carbide. *Scanning Electron Microsc.*, **3**, 1161–7.

Drowart, J., Demaria, G. and Ingram, M. G. (1958). Thermodynamic study of SiC utilizing a mass spectrometer. *J. Chem. Phys.*, **29**, 1015–21.

Edington, J. W., Rowcliffe, D. J. and Henshall, J. (1975). *Powder Met. Int.*, **7**, 136–47.

Elliott, R. P. (1965). *Constitution of Binary Alloys, 1st Supplement.* New York: McGraw-Hill, pp. 1551–5.

Faber, K. T. and Evans, A. G. (1983). Intergranular crack-deflection toughening in silicon carbide. *J. Am. Ceram. Soc.*, **66**, C94–6.

Fisher, G. R. and Barnes, P. (1990). Towards a unified view of polytypism in silicon carbide. *Phil. Mag. B*, **61**, 217–36.

Forrest, C. W., Kennedy, P. and Shennan, I. V. (1972). The fabrication and properties of self-bonded silicon carbide bodies. In *Special Ceramics 5*, ed. P. Popper. Stoke-on-Trent: The British Ceramic Research Association, pp. 99–123.

Fujita, S., Maedo, K. and Hyodo, S. (1986). Anisotropy of high-temperature hardness in 6H silicon carbide. *J. Mater. Sci. Letters*, **5**, 450–2.

Geiger, C. F. (1923). The history of the development of silicon carbide refractories. *J. Am. Ceram. Soc.*, **6**, 301–6.

George, W. (1973). Thermal property measurements on silicon nitride and silicon carbide ceramics between 290 and 700 K. In *Ceramics for High-Temperature Engineering*, ed. D. J. Godfrey, *Proc. Br. Ceram. Soc.*, **22**. Stoke-on-Trent: The British Ceramic Society, pp. 147–67.

Grellner, W., Schwetz, K. A. and Lipp, A. (1981). Fracture phenomena of sintered alpha-SiC. In *Special Ceramics 7, Proc. Brit. Ceram. Soc.* No. 31, ed. D. Taylor and P. Popper. Stoke-on-Trent: The British Ceramic Society, pp. 27–36.

Greskovich, C. and Rosolowski, J. H. (1976). Sintering of covalent solids. *J. Am. Ceram. Soc.*, **59**, 336–43.

Grieveson, P. and Alcock, C. B. (1960). The thermodynamics of metal silicides and silicon carbide. In *Special Ceramics,* ed. P. Popper. London: Heywood, pp. 183–208.

Grow, J. M. and Levy, R. A. (1994). Micromechanical characterization of chemically vapor-deposited ceramic films. *J. Mater. Res.*, **9**, 2072–8.

Gulden, T. D. (1968). Deposition and microstructure of vapor-deposited silicon carbide. *J. Am. Ceram. Soc.*, **51**, 424–7.

Haigis, W. R. and Pickering, M. A. (1993). CVD scaled up for commercial production of bulk SiC. *Am. Ceram. Soc. Bull.*, **72**, 74–8.

Hamilton, D. R. (1960). The growth of silicon carbide by sublimation. In *Silicon Carbide High Temperature Semiconductors*, ed. J. R. O'Connor and J. Smiltens. Proc. first Conf., Boston, MA, 1959. Oxford: Pergamon, pp. 43–52.

Hase, T., Suzuki, H. and Iseki, T. (1976). Formation of beta-SiC during reaction-sintering. *J. Nucl. Mater.*, **59**, 42–8.

Hasselman, D. P. H. and Batha, H. D. (1963). Strength of single crystal silicon carbide. *Appl. Phys. Lett.*, **2**, 111–13.

Henshall, J. L., Rowcliffe, D. J. and Eddington, J. W. (1977). Fracture toughness of single crystal silicon carbide. *J. Am. Ceram. Soc.*, **60**, 373–4.

Hirai, T. and Niihara, K. (1979). Hot hardness of SiC single crystal. *J. Mater. Sci.*, **14**, 2253–5.

Hirata, Y., Matsunaga, N., Hidaka, N., Tabata, S. and Sameshima, S. (2008). Processing of high performance silicon carbide. *J. Ceram. Soc. Japan*, **116**, 665–73.

Hojo, J., Miyachi, M., Okabe, Y. and Kato, A. (1983). Effect of chemical composition on the sinterability of ultrafine SiC powders. *J. Am. Ceram. Soc.*, **66**, C114–15.

Hon, M. H., Davis, R. F. and Newbury, D. E. (1980). Self-diffusion of Si^{30} in α-SiC single crystals. *J. Mater. Sci.*, **15**, 2073–80.

Hong, J. D. and Davis, R. F. (1980). Self-diffusion of C^{14} in high purity and N-doped α-SiC single crystals. *J. Am. Ceram. Soc.*, **63**, 546–52.

Hong, J. D., Davis, R. F. and Newbury, D. E. (1981). Self-diffusion of Si^{30} in β-SiC single crystals. *J. Mater. Sci.*, **16**, 2485–94.

Horton, S. A., Dowson, D., Riley, F. L., Wallbridge, N. C., Broussaud, D. and Denape, J. (1986). The wear behaviour of sialon and silicon carbide ceramics in sliding contact. In *Proceedings of International Conference on Non-oxide Technical and Engineering Ceramics*, ed. S. Hampshire. London: Elsevier Applied Science Publishers, pp. 281–98.

Jepps, N. W. and Page, T. F. (1983). Polytypic transformations in silicon carbide – a review. *Prog. Cryst. Growth Charact.*, **7**, 259–307.

Jepps, N. W., Smith, D. J. and Page, T. F. (1979). Direct identification of stacking sequences in silicon carbide polytypes by high resolution electron microscopy. *Acta Crystall.*, **A35**, 916–23.

Johnson, C. A. and Prochazka, S. (1977). Microstructures of sintered SiC. In *Ceramic Microstructures '76*, eds. R. M. Fulrath and J. A. Pask. Boulder, CO: Westview Press, pp. 366–78.

Kackell, P., Wenzien, B. and Bechstedt, F. (1994). Electronic properties of cubic and hexagonal SiC polytypes from *ab initio* calculations. *Phys. Rev. B*, **50**, 10761–8.

Katz, R. N. (1980). High temperature structural ceramics. *Science*, **208**, 841–7.

Katz, R. N. (1983). US national programmes in ceramics for energy conversion. In *Progress in Nitrogen Ceramics*, ed. F. L. Riley, NATO ASI Series E: Applied Sciences, No. 65. Boston, MA: Martinus Nijhoff, pp. 727–35.

Katz, R. N. (1994). Ceramic materials for rolling element bearing applications. In *Friction and Wear of Ceramics*, ed. S. Jahanmir. New York: Marcel Dekker, pp. 313–38.

Kennedy, P., Shennan, J. V., Braiden, P., Mclaren, L. and Davidge, R. W. (1973). An assessment of the performance of Refel silicon carbide under conditions of thermal stress. In *Ceramics for High-Temperature Engineering*, ed. D. J. Godfrey, *Proc. Br. Ceram. Soc.*, **22**. Stoke-on-Trent: The British Ceramic Society, pp. 67–87.

Keppeler, M., Reichert, H. G., Broadly, J. M., Thurn, G., Wiedmann, I. and Aldinger, F. (1998). High temperature mechanical behaviour of liquid phase sintered silicon carbide. *J. Europ. Ceram. Soc.*, **18**, 521–6.

Kim, Y., Zangvil, A., Goela, J. S. and Taylor, R. L. (1995). Microstructure comparison of transparent and opaque CVD SiC. *J. Am. Ceram. Soc.*, **78**, 1571–9.

Kim, Y. W., Mitomo, M., Emoto, H. and Lee, J. G. (1998). Effect of initial alpha-phase content on microstructure and mechanical properties of sintered silicon carbide. *J. Am. Ceram. Soc.*, **81**, 3136–40.

Kingery, D. W. and Vasilos, T. (1954). Thermal conductivity: XI, conductivity of some refractory carbides and nitrides. *J. Am. Ceram. Soc.*, **37**, 409–14.

Kirschstein, G. and Koschel, D., Eds. (1984). *Gmelin Handbook of Inorganic Chemistry*, 8th edition. *Silicon, Supplement Volume B2, Silicon Carbide*. Berlin: Springer Verlag.

Kruse, B. D. and Hausner, H. (1981). Fracture toughness of sintered alpha silicon carbide. *Sci. Ceram.*, **11**, 453–8.

Lambrecht, W. R. L., Limpijumnong, S., Rashkeev, S. N. and Segall, B. (1997). Electronic band structure of SiC polytypes: a discussion of theory and experiment. *Phys. Stat. Solidi B*, **202**, 5–33.

Lamkin, M. A., Riley, F. L. and Fordham, R. J. (1992). Oxygen mobility in silicon dioxide and silicate glasses: a review. *J. Eur. Ceram. Soc.*, **10**, 347–67.

Lange, F. F. (1975). Hot-pressing behavior of silicon carbide powders with additions of aluminum oxide. *J. Mater. Sci.*, **10**, 314–20.

Lankford, J. (1983). Comparative study of the temperature dependence of hardness and compressive strength in ceramics. *J. Mater. Sci.*, **18**, 1666–74.

Lee, S. K. and Kim, C. H. (1994). Effects of alpha-SiC versus beta-SiC starting powders on microstructure and fracture toughnss of SiC sintered with Al_2O_3–Y_2O_3 additives. *J. Am. Ceram. Soc.*, **77**, 1655–8.

Li, Z. and Bradt, R. C. (1987). Thermal expansion and thermal expansion anisotropy of SiC polytypes. *J. Am. Ceram. Soc.*, **70**, 445–8.

Luthra, K. L. (1991). Some new perspectives on oxidation of silicon carbide and silicon nitride. *J. Am. Ceram. Soc.*, **74**, 1095–103.

McHenry, K. D. and Tressler, R. E. (1980). Fracture-toughness and high-temperature slow crack-growth in SiC. *J. Am. Ceram. Soc.*, **63**, 152–6.

McKee, D. W. and Chatterji, D. (1976). Corrosion of silicon carbide in gases and alkaline melts. *J. Am. Ceram. Soc.*, **59**, 441–4.

McLarnan, T. J. (1981). Mathematical tools for counting polytypes. *Z. Krist.*, **85**, 227–45.

Mendiratta, M. G. and Petrovic, J. J. (1976). Prediction of fracture-surface energy from microhardness indentation in structural ceramics. *J. Mater. Sci.*, **11**, 973–6.

Moissan, H. (1905). Etude du siliciure de carbone de la météorite de Cañon Diable. *Compt. Rend.*, **140**, 405–7.

Morgan Advanced Ceramics Manufacturers' data, <http://www.performancematerial.com>

Morrell, R. (1989). *Handbook of Properties of Technical & Engineering Ceramics: Part 1: An Introduction for the Engineer and Designer*. National Physical Laboratory, London: HMSO, pp. 77–84.

Mulla, M. A. and Krstic, V. D. (1991). Low-temperature pressureless sintering of a beta-silicon carbide with aluminium oxide and yttrium oxide additions. *Am. Ceram. Soc. Bull.*, **70**, 439.

Mulla, M. A. and Krstic, V. D. (1994). Mechanical properties of beta-silicon carbide pressureless sintered with Al_2O_3 additions. *Acta. Met. Mater.*, **42**, 303–8.

Munro, R. G. (1997). Material properties of a sintered alpha-SiC. *J. Phys. Chem. Ref. Data*, **26**, 1195–203.

Murata, Y. and Smoak, R. H. (1979). Densification of silicon carbide by the addition of BN, BP and B_4C and correlation to their solid solubilities. In *Proc. Int. Symp. of*

242 Silicon carbide

Factors in Densification and Sintering of Oxide and Non-Oxide Ceramics, Hakone
Japan, eds. S. Sōmiya and S. Saitō. Tokyo: Gakujutsu Bunken Fukyūkai, pp. 382–99.
Nakashima, S. and Harima, H. (1997). Raman investigation of SiC polytypes. *Phys.
Status Sol. A – Applied Research*, **162**, 39–64.
Nakashima, S., Matsunami, H., Yoshida, S. and Harima, H., Eds. (1996). *Silicon Carbide
and Related Materials 1995*, Proceedings of the 6th International Conference, 18–21
September, Kyoto Institute of Physics Conference, Series Number 142. Bristol:
Institute of Physics Publishing.
Ness, J. N. and Page, T. F. (1986). Microstructural evolution in reaction-bonded silicon
carbide. *J. Mater. Sci.*, **21**, 1377–97.
Niihara, K. (1984). Mechanical properties of chemically vapor deposited nonoxide
ceramics. *Am. Ceram. Soc. Bull.*, **63**, 1160–4.
North, B. and Gilchrist, K. E. (1981). Effect of impurity doping on a reaction-bonded
silicon carbide. *Am. Ceram. Soc. Bull.*, **60**, 549–54.
Norton Grinding Wheel Co. Ltd. (1958). *Process of Making Recrystallized Silicon
Carbide Articles*. British Patent No. 790,672, February 12th 1958.
Padture, N. P. (1994). In situ-toughened silicon carbide. *J. Am. Ceram. Soc.*, **77**, 519–23.
Padture, N. P. and Lawn, B. R. (1994). Toughness properties of a silicon carbide with an
in situ induced heterogeneous grain-structure. *J. Am. Ceram. Soc.*, **77**, 2518–22.
Page, T. F. and Sawyer, G. R. (1980). Comments on "Microstructural characterization of
REFEL (reaction-bonded) silicon carbide" – authors' reply. *J. Mater. Sci.*, **15**,
1850–6.
Palmour, J. W., Edmond, J. A., Kong, H. S. and Carter, C. H. (1993). 6H-Silicon carbide
devices and applications. *Physica B*, **185**, 461–5.
Pareek, V. and Shores, D. A. (1991). Oxidation of silicon carbide in environments
containing potassium salt vapour. *J. Am. Ceram. Soc.*, **74**, 556–63.
Patrick, L. and Choyke, W. J. (1969). Optical absorption in *n*-type cubic SiC. *Phys. Rev.*,
186, 775–7.
Pellisier, K., Chartier, T. and Laurent, J. M. (1998). Silicon carbide heating elements.
Ceram. Internat., **24**, 371–7.
Pensl, G. and Choyke, W. J. (1993). Electrical and optical characterization of SiC.
Physica B, **185**, 264–83.
Phillips, J. C. (1970). Ionicity of chemical bond in crystals. *Rev. Mod. Phys.*, **42**, 317–56.
Pierson, H. P. (1996). *Handbook of Refractory Carbides and Nitrides: Properties,
Characteristics, Processing and Applications*. Park Ridge, NJ: Noyes Publications.
Popper, P. (1960). The preparation of dense self-bonded silicon carbide. In *Special
Ceramics*, ed. P. Popper. London: Heywood, pp. 209–19.
Popper, P. and Mohyuddin, I. (1964). The preparation and properties of dense self-bonded
silicon carbide. In *Special Ceramics 1964*, ed. P. Popper. London: Heywood,
pp. 209–19.
Popper, P. and Riley, F. L. (1967). The texture of pyrolytic silicon carbide. *Proc. Brit.
Ceram. Soc.*, **7**, 99–109.
Powell, R. W. and Tye, R. P. (1969). In *Thermal Conductivity*, ed. C. Y. Ho. New York:
Plenum, pp. 575–83.
Price, R. J. (1969). Structure and properties of pyrolytic silicon carbide. *Bull. Am. Ceram.
Soc.*, **48**, 859.
Price, R. J. (1977). Properties of silicon carbide for nuclear-fuel particle coatings, *Nucl.
Tech.*, **35**, 320–6.
Price, R. J. and Hopkins, G. R. (1982). Flexural strength of proof-tested and neutron-
irradiated silicon carbide. *J. Nucl. Mater.*, **108**, 732–8.

Prochazka, S. (1975). The role of boron and carbon in sintering of silicon carbide. In *Special Ceramics 6*, ed. P. Popper. Stoke-on-Trent: The British Ceramic Research Association, pp. 171–81.

Prochazka, S. and Scanlan, R. M. (1975). Effect of boron and carbon on sintering of SiC. *J. Am. Ceram. Soc.*, **58**, 72.

Ramberg, C. E. and Worrell, W. L. (2001). Oxygen transport in silica at high temperatures: implication of oxidation kinetics. *J. Am. Ceram. Soc.*, **84**, 2607–16.

Ramberg, C. E., Cruciani, G., Spear, K. E., Tressler, R. E. and Ramberg, C. F. (1996). Passive-oxidation kinetics of high-purity silicon carbide from 800°C to 1100°C. *J. Am. Ceram. Soc.*, **79**, 2897–911.

Ramsdell, L. S. (1947). Studies on silicon carbide. *Am. Mineral.*, **32**, 64–82.

Rixecker, G., Wiedmann, I., Rosinus, A. and Aldinger, F. (2001). High temperature effects in the fracture mechanical behaviour of silicon carbide liquid-phase sintered with AlN-Y$_2$O$_3$ additives. *J. Europ. Ceram. Soc.*, **21**, 1013–19.

Roewer, G., Herzog, U., Trommer, K., Müller, E. and Frühauf, S. (2002). Silicon carbide – a survey of synthetic approaches, properties and applications. In *High Performance Non-Oxide Ceramics*, ed. M. Jansen, *Structure and Bonding*, Vol. 101. Berlin: Springer, pp. 59–135.

Rohm and Haas, Manufacturers' data, <http://www.cvdmaterials.com>.

Sawyer, G. R. and Page, T. F. (1978). Microstructural characterization of 'REFEL' reaction-bonded silicon carbide. *J. Mater. Sci.*, **13**, 885–904.

Sawyer, G. R., Sargent, P. M. and Page, T. F. (1980). Microhardness anisotropy of silicon carbide. *J. Mater. Sci.*, **15**, 1001–13.

Scace, R. I. and Slack, G. A. (1959). Solubility of carbon in silicon and germanium. *J. Chem. Phys.*, **30**, 1551–5.

Schlichting, J. and Riley, F. L. (1991). Silicon carbide. In *Concise Encyclopedia of Advanced Ceramic Material*, ed. R. J. Brook. Oxford: Pergamon Press, pp. 426–9.

Schmidt, W. R., Interrante, L. V., Doremus, R. H., Trout, T. K., Marchetti, P. S. and Maciel, G. E. (1991). Pyrolysis chemistry of an organometallic precursor to silicon carbide. *Chem. Mater.*, **3**, 257–67.

Schneider, B., Guette, A., Naslain, R., Catldi, M. and Costecalde, A. (1998). A theoretical and experimental approach to the active-to-passive transition in the oxidation of silicon carbide – experiments at high temperatures and low total pressures. *J. Mater. Sci.*, **33**, 535–47.

Schröder, F. A., Ed. (1986). *Gmelin Handbook of Inorganic Chemistry*, 8th edition. *Silicon, Supplement Volume B3, Silicon Carbide*. Berlin: Springer Verlag, pp. 53–62.

Schwetz, K. A. (2000). Silicon carbide based hard materials. In *Handbook of Ceramic Hard Materials*, ed. R. Riedel. Weinheim: Wiley-VCH, pp. 683–748.

Schwetz, K. A. and Lipp, A. (1980). Effect of boron and aluminiun sintering additions on the properties of dense alpha silicon carbide. In *Sci. Ceram. 10*, Proceedings of the tenth international conference 1–4th September, 1979, ed. H. Hausner. Bad Honnef: Deutsche Keramische Gesellschaft, pp. 149–58.

Shaffer, P.T.B. (1965). Effect of crystal orientation on hardness of beta silicon carbide. *J. Am. Ceram. Soc.*, **48**, 601–5.

Shaffer, P.T.B. (1971). Refractive index, dispersion, and birefringence of silicon carbide polytypes. *Appl. Opt.*, **10**, 1034–6.

Shaffer, P. T. B. and Jun, C. K. (1962). Elastic modulus of dense polycrystalline silicon carbide. *Mat. Res. Bull.*, **7**, 63–70.

She, J. H., Ohji, T. and Deng, Z. Y. (2002). Thermal shock behaviour of porous silicon carbide ceramics. *J. Am. Ceram. Soc.*, **85**, 2125–7.

Simon, G. and Bunsell, A. R. (1984). Mechanical and structural characterization of the Nicalon fibre. *J. Mater. Sci.*, **19**, 3649–57.

Singer, F. and Singer, S. S. (1963). *Industrial Ceramics*. London: Chapman and Hall.

Singhal, S. C. (1976). Thermodynamics and kinetics of oxidation of hot-pressed silicon nitride. *J. Mater. Sci.*, **11**, 500–9.

Slack, G. A. (1964). Thermal conductivity of pure and impure silicon carbide and diamond. *J. Appl. Phys.*, **35**, 3460–6.

Slack, G. A. (1973). Non-metallic crystals with high thermal conductivity. *J. Phys. Chem. Solids*, **34**, 321–35.

Slack, G. A. and Bartram, S. F. (1975). Thermal expansion of some diamond-like crystals. *J. Appl. Phys.*, **46**, 89–98.

Snead, L. L., Nozawa, T., Katoh, Y., Byun, T. S., Kondo, S. and Petti, D. A. (2007). Handbook of SiC properties for fuel performance modelling. *J. Nucl. Mater.*, **371**, 429–77.

Srinivasan, M. (1989). The silicon carbide family of structural ceramics. In *Structural Ceramics*, ed. J. B. Wachtman, *Treatise on Materials Science and Technology*, Vol. 29. Boston: Academic Press, pp. 99–159.

Stutz, D. H., Prochazka, S. and Lorenz, J. (1985). Sintering and microstructure formation of β-silicon carbide. *J. Am. Ceram. Soc.*, **68**, 479–82.

Su, S. J., Zhou, J. G., Yang, B. and Zhang, B. Z. (1995). Effect of deposition temperature on the properties of pyrolytic SiC. *J. Nucl. Mat.*, **224**, 12–16.

Suzuki, H. (1983). Recent trends and theoretical background in sintering of silicon carbide. *Seramikkusu*, **18**, 3–9.

Taylor, A. and Jones, R. M. (1960). Silicon carbide, a high temperature semiconductor. In *Proc. First Conf. on Silicon Carbide, Boston, 1959*, eds. J. R. O'Connor and J. Smiltens. Boston, MA: Pergamon, pp. 147–54.

Torti, M. L., Alliegro, R. A., Richerson, D. W., Washburn, M. E. and Weaver, G. Q. (1973). Silicon carbide and silicon nitride for high-temperature engineering applications. *Proc. Brit. Ceram. Soc.*, **22**, 129–46.

Touloukian, Y. S., Ed. (1977). *Thermophysical Properties of High-Temperature Solid Materials – Thermophysical Properties of Matter, Vol. 5*. New York: Macmillan, pp. 125–8.

Tressler, R. E. (1994). Theory and experiment in corrosion of advanced ceramics. In *Corrosion of Advanced Ceramic Materials*, ed. K. G. Nickel, NATO Applied Science Series E: Applied Science, Vol. 267. Dordrecht: Kluwer Academic Publishers, pp. 3–22.

Veldkamp, J. D. B. (1975). Short fibre reinforced materials. *Philips Research Reports, Suppl.*, **4**, 1–91.

Verma, A. R. and Krishna, P. (1996). *Polymorphism and Polytypism in Crystals*. New York: Wiley.

Voice, E. H. (1969). Silicon carbide as a fission product barrier in nuclear fuels. *Mat. Res. Bull.*, **4**, 331–45.

Voice, E. H. and Scott, V. C. (1972). The formation and structure of silicon chemically deposited in a fluidised bed of microspheres. In *Special Ceramics 5*, ed. P. Popper. Stoke-on-Trent: The British Ceramic Research Association, pp. 1–31.

Wallbridge, N. C., Dowson, D. and Roberts, E. W. (1983). The wear characteristics of sliding pairs of high density polycrystalline aluminium oxide under both dry and wet conditions. In *Wear of Materials 1983, Proc. Int. Conf. on the Wear of Materials*, ed. K. C. Ludema. Reston, VA: ASME, pp. 202–11.

Weaver, G. Q., Baumgartner, H. R. and Torti, M. L. (1975). Thermal shock behaviour of sintered silicon carbide and reaction-bonded silicon nitride. In *Special Ceramics 6*, ed. P. Popper. Manchester: The British Ceramic Research Association, pp. 261–81.

Wei, G. C. (1983). Beta SiC powders produced by carbothermic reaction of silica in a high-temperature rotary furnace. *Comm. Am. Ceram. Soc.*, C111–13.

Wiederhorn, S. M., Hockley, B. J. and French, J. D. (1999). Mechanisms of deformation of silicon nitride and silicon carbide at high temperatures. *J. Eur. Ceram. Soc.*, **19**, 2273–84.

Yean, D. H. and Riter, J. R. (1971). Estimates of isothermal bulk moduli for group (IVA) crystals with zinc blende structure. *J. Phys. Chem. Solids*, **32**, 653.

Yu, R., Zhang, X. F., De Jonghe, L. C. and Richie, R. O. (2007). Elastic constants and tensile properties of Al_2OC by density functional calculations. *Phys. Rev. B*, **75**, Art. No. 104114.

Zhang, X. F. and De Jonghe, L. C. (2003). Thermal modification of microstructures and grain boundaries in silicon carbide. *J. Mater. Res.*, **18**, 2897–913.

Zheng, Z., Tressler, R. E. and Spear, K. E. (1992). The effects of C_{12} on the oxidation of single-crystal silicon carbide. *Corr. Sci.*, **33**, 557–67.

5

Silicon nitride

5.1 Description and history

Silicon nitride provides the basis for the second group of non-oxide ceramics. In many ways silicon nitride is very similar to silicon carbide, but there are marked differences – it has no electrical conductivity, or polytypism, for example. There is one crystalline compound with the chemical formula Si_3N_4 (nitrogen is 3-valent and silicon 4-valent), which has two common structures, designated α and β (Hardie and Jack, 1957). The Si–N bonding is directed and appreciably covalent: a covalency of 70% is often quoted, though a smaller value – about 50% – has been calculated (Robertson, 1991). There are also less common crystal structures (the γ- and δ-phases), which are only formed under very high pressures, and are metastable under normal conditions (Zerr *et al.*, 1999; Zerr *et al.*, 2006). They are of some interest because of their hardness; ~47 GPa has been estimated (Zerr, 2001), but they have not yet been commercially developed. There is a non-crystalline (amorphous) form of silicon nitride, which is often non-stoichiometric, and denoted by the formula SiN_x. In the dense form it is important for its extensive use as a thin masking film in the production of silicon chips for microelectronic applications, and in microelectronics more generally, but discussion of this specialised form of the material is outside the scope of this book. Amorphous silicon nitride is also available as a very fine powder, and this has been of some interest for the small-scale production of ceramic materials. The *Gmelin Handbook of Inorganic and Organometallic Chemistry*, Silicon Supplement Volume B 5e (Kurtz and Schröder, 1994), provides a comprehensive review of the literature on SiN_x up to ~1993.

Silicon nitride does not melt (at least, under normal pressures). As the temperature is raised it gradually dissociates and loses nitrogen according to a (reversible) reaction which can most simply be expressed:

$$Si_3N_{4(s)} = 3Si_{(s,l)} + 2N_{2(g)}. \tag{5.1}$$

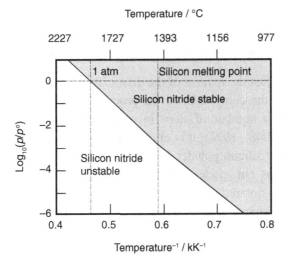

Figure 5.1 The equilibrium pressure of nitrogen over silicon nitride, resulting from the dissociation reaction (5.1). For externally applied nitrogen pressures greater than the equilibrium values, silicon nitride is stable (shaded area). (After Chase, 1998.)

At temperatures below ~1000 °C the dissociation is practically imperceptible, but at much higher temperatures it becomes very important (Chase, 1998). This is particularly the case in low-pressure systems, where the nitrogen can, in effect, be pumped off, leaving just the silicon. When silicon nitride is heated in a *closed system* (one in which the components are completely isolated from the external environment), the pressure of nitrogen reaches 1 atm at ~1884 °C. Above this temperature solid silicon nitride can only exist if a higher (external) nitrogen pressure is applied (Pehlke and Elliott, 1959). The shaded area in Fig. 5.1 indicates the conditions of temperature and external nitrogen pressure for which silicon nitride is stable. Although temperatures of 1800–1900 °C are often used in the production of silicon nitride ceramics, special precautions have to be taken. The normal application temperatures for silicon nitride components are not much above 1400 °C.

Silicon nitride does not occur naturally in bulk in the Earth's crust, for the same reasons that silicon carbide does not, but (like silicon carbide) traces have been found in meteoritic rocks (Lee *et al.*, 1995), and called *nierite*. Because there is no natural large-scale source of material, silicon nitride powder has to be manufactured. There are many methods available, and the literature has been compiled and reviewed in considerable detail (Sangster, 2005). The main powder production process (which tends to give the α-phase) is the direct reaction at high temperature between loosely compacted silicon powder and nitrogen (*nitridation*) at temperatures usually between ~1250 and ~1450 °C, formally expressed by:

$$3Si_{(s)} + 2N_{2(g)} = Si_3N_{4(s)}; \quad \Delta G_{1500K} = -248.9 \text{ kJ mol}^{-1}. \tag{5.2}$$

This, of course, is just the reverse of the dissociation reaction (5.1). However, the equation conceals the complexity of the process. In fact the nitridation reaction normally follows a number of steps involving both silicon vapour and liquid silicon (Atkinson *et al.*, 1974). It is of considerable interest that the nitridation of densely compacted silicon powder can, very conveniently, be used to produce a very useful, porous but moderately strong material, with strengths which can match those of a sintered alumina. This is the *reaction-bonded* (or *reaction-sintered*) form of silicon nitride (Popper, 1977), and one more similarity with silicon carbide will be recognised.

The reaction between silicon dioxide powder and nitrogen in the presence of carbon at temperatures of 1450–1500 °C – a type of reaction sometimes referred to as *carbothermal reduction* – can also be made to yield silicon nitride (Komeya and Inoue, 1975):

$$3SiO_{2(s)} + 6C_{(s)} + 2N_{2(g)} = Si_3N_{4(s)} + 6CO_{(g)}. \tag{5.3}$$

This reaction also involves several stages, including the production of the vapour phase silicon monoxide SiO (already met in the active oxidation of silicon carbide) and its reaction with nitrogen. Of these two production processes, the lower temperature, direct nitridation of silicon is the more commonly used. Both processes are very slow and normally require many hours with 1 atm pressure of nitrogen, depending on the particle size of the starting powders (which must be of micrometre dimension). The silicon nitride formed by nitriding loosely packed silicon powder is normally in the form of a lightly sintered, and thus agglomerated, coarse powder: milling is required to give powders that are fine enough to be sintered to a fine-grained ceramic (needing submicrometre silicon nitride particles). Because of the low cost and ready availability of the raw materials (nitrogen, and silicon or silicon dioxide), it has been suggested that silicon nitride ceramics should be very cheap to produce, but as with many ceramic materials it is the powder and component production stages that largely account for the cost (Davidge, 1977).

There are other methods for the production of very fine particle size silicon nitride powder, which use chemically more complex processes. These include the production from silicon tetrachloride, and decomposition, of silicon diimide $(Si(NH)_2)$ (Satoh, 1938; Yamada, 1993). Very fine (<100 nm) amorphous silicon nitride powder can also be formed from gaseous reactants at relatively low temperatures using laser light energy. Although the resulting powders are of very

Table 5.1 *Silicon nitride powders: examples of the properties of three common forms of powder. The impurities reflect the sources of the raw materials and the silicon nitride production routes. (After van Dijen et al., 1994.)*

Property / unit	Silicon nitridation	Carbothermal	Diimide
Particle size (d_{50} value) / μm	0.6	1.6	0.6
Specific surface area / $m^2 g^{-1}$	22	12	12
α-phase content / %	98	99	98
Oxygen / % by weight	1.9	2.2	1.7
Carbon / % by weight	0.2	0.8	0.1
Chlorine / ppm	40	<10	500
Silicon carbide / % by weight	0.5	2.5	<0.2

high purity, they are correspondingly more expensive, and their commercial use is more restricted. Table 5.1 shows three common types of silicon nitride powder produced by these methods, and representative typical properties. The actual properties of a powder will vary from manufacturer to manufacturer, and possibly from batch to batch (as is normal for any ceramic starting powder). One important feature, influencing sintering characteristics, is the oxygen content of the powder. Oxygen is difficult to avoid for a number of reasons, and this aspect of silicon nitride will be referred to again in the next section.

Closely similar chemical routes are used for the production of thin, dense, films of silicon nitride (and the amorphous powder which has already been referred to). These are generally known as chemical vapour deposition or CVD methods (Airey *et al.*, 1973; Kingon, 1983), similar in principle to those used for the production of CVD silicon carbide (Section 4.4). While there has been considerable interest in this technique for producing a bulk form of polycrystalline material (as it has for silicon carbide), and the method has been of use for growing small single crystals of silicon nitride, it has yet to find significant commercial application.

Man-made silicon nitride has been known as a chemical substance for almost 150 years; the first reports appear in the chemical literature in the middle of the nineteenth century. It was speculated that, during the formation of the Earth, the reaction of red-hot silicon nitride with water vapour was responsible for the production of the gaseous ammonia thought to be necessary for the start of life (Oparin, 1952; Miller, 1953). Interestingly this is a topical issue in the context of the use of silicon nitride components in hot gas-turbine environments where there is exposure to water vapour at high temperature (Smialek *et al.*, 1999), and it is leading to work on the development of protective coatings (Ueno *et al.*, 2006). The possibilities for the use of silicon nitride as a bulk refractory material were first explored in the 1950s. An early patent (Nicholson, 1952) describes the development of ". . . mechanically strong articles consisting of and/or bonded by silicon nitride . . .". By 1955 one form of silicon nitride was in use as a heat-resistant material with a range of applications, including that of thermocouple sheathing better able to resist thermal shock than those made from sintered alumina. In the 1970s silicon nitride was included, together with silicon carbide, in programmes of development work intended to demonstrate the practicality of ceramic components in the hot zones of new internal combustion (gas-turbine and diesel) engines. There were ambitious hopes that these materials would ultimately provide the basis for higher efficiency engines (Katz, 1980; Katz, 1983). Although engines were developed and tested, it became apparent that the costs of mass-production of components of adequate reliability at high temperature under load would be prohibitive. However, more modest applications were found in the engine area for small silicon nitride components working at much lower temperatures and loads, such as in turbocharger rotors, and a number of stationary components.

The literature on silicon nitride is extensive. Much of the earlier work up to about 1993 concerning these and other applications of silicon nitride has been recorded and reviewed in the Silicon Supplement Volume B 5e of the *Gmelin Handbook of Inorganic and Organometallic Chemistry* (Kurtz and Schröder, 1994), and briefly by Popper (1994). A more recent review selected chemical aspects of silicon nitride for treatment (Petzow and Herrmann, 2002). The silicon nitrides are notable in that they have probably been the most intensively studied of all the structural ceramics. Since 1955 more than 70 000 abstracts (for papers published in scientific journals) have been indexed by the Chemical Abstracts service of the American Chemical Society, and more single volumes of the *Gmelin Handbook of Inorganic and Organometallic Chemistry* have been devoted to silicon nitride than any other single compound (Sangster, 2005). While a lot of the work has undoubtedly been repetitive, it is a consequence of this effort that silicon nitride can be regarded as one of the most mature of the structural ceramics, in terms of the understanding of it as a material, and of the ability to control its microstructures and thus properties (Riley, 2000; Lange, 2006). It provides a good demonstration of what can be achieved when research and development effort (and thus funding) is directed towards a specific material. High-grade silicon nitride materials are now firmly established as structural ceramics, and they have many (though less dramatic) applications (Briggs, 2007). Current applications are based primarily on their strength, and their resistance to wear and thermal shock. The quantities of silicon nitride produced for these structural ceramic applications are very small on the overall materials scale (taking in iron and steel, plastics and concrete), perhaps only several hundred tonnes a year in Europe and the USA, most of which is the porous reaction-bonded form. Where it is used, significant improvements are obtained in life or reliability that justify the extra costs. It is also of interest that the uses of thin films of silicon nitride in the microelectronic areas, though also very small in terms of quantity, are very large in terms of the value of the components depending on them, though this is a specialised area which cannot be treated here.

As is the case with the silicon carbide group, there are several forms of ceramic material consisting entirely, or mainly, of silicon nitride, and there are significant differences between them. In discussions of microstructures and properties it will again be more convenient to treat each member of this group as a separate material; their applications will be treated by broad category, rather than the type of material used.

5.2 Basic aspects

The crystal structure of silicon nitride is more complex than those of aluminium oxide and silicon carbide, because it has to accommodate 4-coordinated silicon atoms (*tetrahedral coordination*) in SiN_4 units with N–Si–N bond angles of about 106°, and 3-coordinated nitrogen in NSi_3 units (*trigonal planar coordination*), with Si–N–Si bond angles of 120°. These basic units are illustrated in Fig. 5.2. As with silicon carbide, these bonding conditions preclude the development of any kind of close-packed structure, but in solving the problem of how to arrange the units, Nature provides not just one solution but several. The two phases of practical importance are referred to as α- and β-silicon nitride, of which the β-phase is considered to be the more stable (Ruddlesden and Popper, 1958). The structures of both forms are based on puckered 8- and 12-membered rings of silicon and nitrogen atoms, with strong, predominantly covalent bonds, linked to build what, for convenience, can be regarded as sheets (*ab* sheets) of atoms. Figure 5.3 shows schematically one sheet of this structure, with its rings of N and Si atoms. The corners indicated in the individual SiN_4 tetrahedra are the nitrogen atoms shared in forming Si–N–Si bridges between the sheets; the shadings indicate the orientation of the tetrahedra. In β-Si_3N_4 the sheets are stacked exactly on top of each other, and linked to give the simpler structure, which can be expressed as ... *abab* .... In α-Si_3N_4, every other sheet is first inverted (giving a mirror image, which in these terms is a *ba* sheet), and then repositioned (that is, sheared) slightly with respect to the underlying *ab* sheet, before completion of the bridges to give an ... *abba* ... structure. To achieve this arrangement and to get the *ab* and *ba* sheets of rings to fit, a small amount of bond angle distortion is required: the structure therefore has corresponding strain energy (in fact the β-phase also departs very slightly, but to a much lesser extent, from the idealised structure). Because the α-structure contains *ab* and *ba* layers, the unit cell *c*-axis dimension is twice that of

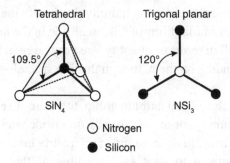

Figure 5.2 The basic structural units of silicon nitride: 4-coordinated silicon in SiN$_4$ (*tetrahedral coordination*) units with N–Si–N bond angles of ~109°; 3-coordinated nitrogen in NSi$_3$ units (*trigonal planar coordination*), with Si–N–Si bond angles of 120°.

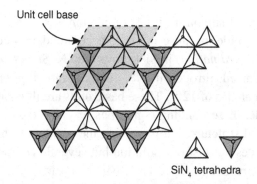

Figure 5.3 One sheet of the silicon nitride hexagonal symmetry crystal structure. This can be regarded as constructed from SiN$_4$ units, bonded by the nitrogen atoms. The sheets are linked by shared N atoms and are extensively buckled. (After Riley, 2000, reprinted with permission of Wiley-Blackwell Publishing.)

the β-phase, though the *a*-axis dimensions are closely similar. Lattice dimensions for the hexagonal cells of the two phases are given in Table 5.2. Further consequences of these two stacking arrangements are that the β-structure contains continuous channels along the *c*-axis, and the α-structure contains polyhedric interstices along the *c*-axis; these interstices can be occupied by large cations which appear to be able to further stabilise the α-structure. Perhaps surprisingly, it is the strained α-Si$_3$N$_4$ structure that is usually formed during silicon nitride production reactions. While the β-structure might thermodynamically be the most stable, it is nonetheless the α-structure which is the more readily constructed from reacting atomic or molecular species at temperatures in the region of 1300–1400 °C. It does tend, as might be expected, to convert (*reconstructively*) at higher temperatures to the (relatively) unstrained β-Si$_3$N$_4$ structure (Messier *et al.*, 1978).

Table 5.2 *The unit cell dimensions of the α- and β-phases of silicon nitride. (After Wang et al., 1996.)*

Phase	*a*-axis dimension / pm	*c*-axis dimension / pm	*c/a* ratio	Density / Mg m^{-3}
α-Si$_3$N$_4$	774.9–781.8	559.1–562.2	0.715–0.726*	3.167–3.19
β-Si$_3$N$_4$	760.8–759.5	291.1–290.2	0.383–0.392*	3.192

* The idealised ratios are α-Si$_3$N$_4$, 0.667, and β-Si$_3$N$_4$, 0.385.

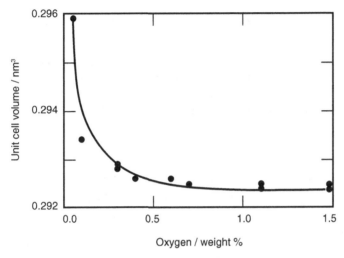

Figure 5.4 The variations in unit cell volume of α-silicon nitride with oxygen content. The dimensions of the β-silicon nitride unit cell are more consistent. (After Wang *et al.*, 1996.)

This involves the complete undoing, and then reassembly, of the α-Si$_3$N$_4$ atomic network; volatilisation, or dissolution in a solvent, is therefore required. The reverse β- to α-silicon nitride transformation has never been observed, probably because it is a lower-temperature process (because the α-phase seems to be stable only at lower temperatures), and therefore too slow to be detectable.

A second interesting (and at one time, controversial) feature of the α-structure is that it is able to accommodate small amounts of impurity oxygen (as high as 2%) located on nitrogen lattice sites, though it is clear that oxygen is not necessary for the stability of the structure (Kato *et al.*, 1975), and may in fact actually destabilise it (Liang *et al.*, 1959). The oxygen has a small but measurable effect on the unit cell volume (Jack, 1983). Figure 5.4 shows the slight variations in unit cell volume (about 1%) with oxygen content (Wang *et al.*, 1996). In contrast the unit cell

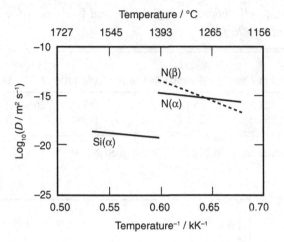

Figure 5.5 Diffusion coefficients (*D*) for Si in α-silicon nitride, and for N in α- and β-silicon nitride, shown in standard semi-logarithmic form. All the values are very low in this temperature range, though the nitrogen appears to be the more mobile. (After Riley, 2000, reprinted with permission of Wiley-Blackwell Publishing.)

dimensions of the β-phase are more consistent. The two forms of single crystal have similar densities (just under 3.2 Mg m^{-3}) and, as far as can be measured, practically identical physical and mechanical properties. While these crystallographic aspects may seem in some ways to be rather academic, they do in fact turn out to have an important influence on the development of the microstructure of sintered silicon nitride, and thus its mechanical properties. This aspect will be returned to in the sections dealing with the individual materials.

As is the case with silicon carbide, lattice diffusion coefficients are very low. Values for pure silicon nitride are experimentally difficult to measure, partly because of the problem of volatility at the high temperatures necessary to achieve measurable diffusion rates. Data plotted in the standard Arrhenius form (the logarithm of the diffusion coefficient as a function of reciprocal thermodynamic temperature) in Fig. 5.5 suggest that diffusion of the larger silicon atoms is the slower (as it is in the case of silicon carbide), and therefore rate-controlling, process (Riley, 1996). The values of the diffusion coefficient for silicon over this temperature range can be compared with those for the diffusion of silicon in silicon carbide: numerically they are similar, but the temperatures are very much lower. However, the importance of surface diffusion, and especially the volatility, has to be kept in mind when considering how material may be transported during the sintering process. In this respect silicon nitride and silicon carbide present similar problems for sintering to full density.

5.3 Intrinsic physical and mechanical properties

The property values reviewed here are those of the "normal" α- or β-phase materials. Most measurements of physical and mechanical properties have been made on polycrystalline materials containing either mainly β-Si_3N_4 or a mixture of α- and β-phases, that is, on sintered silicon nitride or reaction-bonded silicon nitride. This is because special methods are required to obtain large single crystals of pure silicon nitride. The largest single crystals of α-phase (which can be of mm dimension) can be grown by CVD techniques. Property measurements have been made on individual large β-phase crystals in polycrystalline materials using nanoindentation. Additional factors to be taken into account in using measurements made on bulk materials, are that the sintered forms usually contain fairly large amounts (up to around 10% by volume) of silicate sintering additive, and reaction-bonded silicon nitride, although it can be of high purity, is always very porous (usually around 25%). Extrapolations of values to pure silicon nitride and zero porosity are therefore required, and even then uncertainties exist with regard to the effects of crystal size and the grain boundaries on property value.

Most of the important publications up to about 1995 dealing with the properties of silicon nitride have been systematically summarised and reviewed in the *Gmelin Handbook of Inorganic and Organometallic Chemistry*, 8th Edition, Silicon Supplement Volume B 5b1, *Silicon Nitride: Mechanical and Thermal Properties* (Schröder, 1994). Much of the earlier physical and mechanical property data referred to in the following sections can be found in this volume. Early publications on the thermal decomposition of silicon nitride, and its chemical reactions with metals and metal oxides (including silicon dioxide and aluminium oxide), are dealt with in the *Gmelin* Silicon Supplement Volume B 5d1, which covers the literature up to about 1994 (Schröder, 1995). This volume also reviews the properties of silicon nitride powders, although a more recent comprehensive review of silicon nitride powder production has already been referred to (Sangster, 2005).

5.3.1 Colour

Like silicon carbide, the high-purity single-crystal form of silicon nitride is always colourless, but because of its wider impurity band gap of 3.4–5.3 eV (Ye *et al.*, 2000) even slightly impure silicon nitride normally has no distinctly coloured forms (Niihara and Hirai, 1979), and remains an electrical insulator. However, most commercially produced powders and nominally purer forms of silicon nitride ceramic are of varying shades of grey. This is commonly assumed to be the direct result of trace impurities, present as very small inclusions

(Jack, 1983), in particular transition metals capable of multiple oxidation states such as iron, perhaps at part per million levels. Dense sintered silicon nitride, containing a high level of sintering additive, is often very dark grey to almost black – again associated with either transition metal impurities, which are likely to be in a mixture of reduced oxidation states because of the low oxygen potentials present during sintering, or free silicon (Herrmann and Göb, 2001).

5.3.2 Density

The measured densities for the α-crystal phases are in the range 3.167–3.19 Mg m^{-3} (Wang *et al.*, 1996). The variations are associated with the presence of very small amounts of impurity oxygen within the α-structure. The density of the β-phase is more consistent at ~3.192 Mg m^{-3}. These low values are very close to that of silicon carbide, because carbon and nitrogen have similar atomic sizes, and the directed Si–C and Si–N bonds in the crystal structures are of similar lengths (~175 pm).

5.3.3 Young modulus and hardness

The mean elastic modulus (the Young modulus) measured on low additive content, fully dense, materials, is estimated to be in the region of 300–330 GPa (Himsolt *et al.*, 1979). Because of the anisotropy of the hexagonal symmetry crystal structure, properties can also be markedly anisotropic. As an extreme case (Hay *et al.*, 1998), large β-grains elongated along the *c*-axis have moduli parallel to the *c*-axis in the range 540–590 GPa, and perpendicular to this axis only 180–280 GPa. The bulk modulus is estimated to be 124 GPa, and the Poisson ratio is ~0.25–0.35.

Measurements of hardness using small, CVD, α-phase single crystals with standard loads have provided values in the range of 30–35 GPa (Reimanis *et al.*, 1996). All these values indicate that while the silicon nitride crystal structure is certainly very rigid and strong, and a little stiffer than α-aluminium oxide, it is deformed slightly more easily than silicon carbide. The difference is, however, probably only really significant for applications as ball and roller bearings, where Hookian contact loads are very high.

5.3.4 Fracture toughness

Fracture toughness (K_{Ic}) values obtained from indentation measurements on small crystals (Reimanis *et al.*, 1996) are low, at ~2 MPa $m^{1/2}$, and similar to those for single-crystal silicon carbide. However, for dense sintered polycrystalline materials discussed later, values are very much higher, and controlled by the composition and microstructure. This aspect of microstructure control has proved to be of great importance for the development of high strength in sintered silicon nitride, and resistance to crack growth; considerable effort has been directed towards understanding these relationships and developing the best microstructures. Manufacturers' values for commercial sintered materials are usually in the approximate range 5–8 MPa m$^{1/2}$. In contrast K_{Ic} values for porous reaction-bonded forms are much lower; commercial materials have values in the range 1–4 MPa m$^{1/2}$. Improving the fracture toughness of polycrystalline silicon nitride has been a major target, and this will be discussed in more detail in a later section.

5.3.5 Thermal expansivity

Like the Young modulus, the mean thermal expansion coefficient of silicon nitride also shows slight anisotropy between the *a*- and *c*-axes, with a mean value over the temperature range 25–1000 °C of about 3.6 MK^{-1}, which is slightly smaller than that of silicon carbide (Henderson and Taylor, 1975). For this reason silicon nitride materials might be expected to have a resistance to thermal shock as good as, if not slightly better than, those of materials based on silicon carbide. As was described above, the low thermal expansivity of silicon nitride was initially instrumental in attracting attention to the silicon nitride group of materials, in the search for ceramics able to withstand the violent changes of temperature experienced as hot-zone components in gas-turbine and diesel engines. An indication of the thermal shock resistance of silicon nitride relative to that of a sintered alumina is provided by the retained strength after a standard quench from high temperature into cold water. While the alumina would show a severe loss of strength after a quench of only 250 °C or so, good-quality sintered silicon nitride can withstand a quench of ~700 °C, before suffering the same extent of microstructural damage. The typical influence of quench temperature on the residual strength of a dense sintered silicon nitride is shown in Fig. 5.6 (Schneider and Petzow, 1994).

5.3.6 Thermal conductivity

The thermal conductivity of dense liquid phase sintered silicon nitride is more "normal" for a ceramic material. Reported values for dense polycrystalline materials are in the approximate range of 15–30 W m^{-1} K^{-1} at 25 °C, decreasing slowly as the temperature rises. This is about 10% of the value for silicon carbide at these temperatures. Part of the reason for the relatively low value must be the presence of the grain boundaries, and the mainly amorphous materials trapped

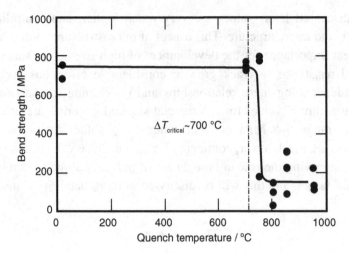

Figure 5.6 The influence of a quench into water on the residual strength of gas-pressure sintered silicon nitride. There is a catastrophic loss of strength for a quench ($\Delta T_{critical}$) of >700 °C. (After Schneider and Petzow, 1994. Reprinted by kind permission of the Max-Planck-Institute for Metals.)

there. The thermal conductivity of porous forms of material (such as reaction-bonded silicon nitride) is reduced further by the gas-filled porosity itself, and values can be as low ~ 3 W m^{-1} K^{-1} at room temperature. The much higher thermal conductivity of silicon carbide should in theory provide it with extra protection against thermal shock by smoothing severe surface temperature gradients, and silicon carbide would be expected to have the edge over silicon nitride. In practice, however, with very fast temperature changes the thermal conductivity does not seem to have time to operate, and silicon nitride and silicon carbide materials have similar responses to thermal shock. Calculated values for some of the commonly applied thermal shock (the Hasselman R) parameters are shown in Table 5.3.

5.3.7 Electrical conductivity

Silicon nitride is a good electrical insulator in all its common forms because of its large gap between valence and conduction bands of up to ~5.5 eV, and absence of impurity band conduction (Robertson, 1991; Ye *et al.*, 2000). The room-temperature volume conductivity value (manufacturers' data) is of the order of 10^{-14} S m^{-1} (10 fS m^{-1}). Reaction-bonded silicon nitride has for this reason been considered for small components where electrical insulation is required in combination with high temperatures and rapid temperature changes, such as heating element supports (traditionally made from oxide ceramics such as cordierite or

Table 5.3 *Selected thermal shock (R) parameters for silicon nitride, with the values of the components making up the parameters.*

Material property / unit	Bend strength / MPa	Young modulus / GPa	Poisson ratio	Expansivity / MK^{-1}	Thermal conductivity/ W m^{-1} K^{-1}	R / W m^{-1}	R' /kW m^{-1}	R''' / MPa^{-1}
Reaction-bonded	250	250	0.3	3.0	10	230	2.3	5.7
Sintered	800	320	0.3	3.6	20	490	9.7	0.7

$R = \sigma (1-v) / E\alpha$; $R' = \sigma (1-v) \lambda/E\alpha$; $R''' = E / [\sigma^2 (1-v)]$.

alumina, which are considerably cheaper to produce). The higher cost is one barrier to the more widespread use of silicon nitride, and which of course has to be justified in terms of increased performance or reliability.

5.3.8 Mechanical strength

Calculated single-crystal strengths are 51 GPa for α-phase and 57 GPa for the β-phase (Ogata *et al.*, 2004). These of course provide no indication of the actual strengths of the polycrystalline forms of material, which, like all ceramics, are strongly microstructure and flaw dependent. Strengths of individual ceramic forms of materials are reviewed under their sections; they have a very wide range of 100 to more than 1500 MPa, depending on the exact form of the material and its microstructure.

5.3.9 High-temperature behaviour – thermal stability

It might seem surprising that a hard, stiff, "refractory", material which is still a solid at 1850 °C should none the less, in effect, simply evaporate at a temperature marginally above this. Part of the reason that silicon nitride tends to lose nitrogen by reversal of the formation reaction (5.1) is the very high stability of the nitrogen molecule, N_2, with its six-electron triple bond. Silicon nitride, like silicon carbide (graphite is also a very stable species), is not particularly stable in a chemical sense, and like silicon carbide it reacts readily with many metals at relatively low temperatures (~1000 °C) to form metal silicides and release nitrogen. One immediate practical consequence is that silicon nitride based materials cannot be used for long periods of time at high temperature under atmospheres where equilibrium cannot be maintained, such as those deficient in nitrogen (for example a vacuum, or an inert gas such as argon), or in the presence of a chemical "sink" for silicon vapour, such as carbon or iron. Under most practical conditions, however, the dissociation of the nitride and loss of nitrogen is likely to be physically impeded by the presence of a surface film of liquid silicon, though in the longer term chemical stability must be compromised.

The silicon nitride dissociation reaction also becomes a problem of practical importance for the sliding wear behaviour of silicon nitride based materials in contact with ferrous alloys, where local frictional temperatures of well in excess of 1000 °C can be attained (Vleugels *et al.*, 1994). Under these conditions the formation of low melting-point (~1200 °C) iron silicides is possible, depending on the gaseous atmosphere and the silicon content of the alloy (Glemser *et al.*, 1957), leading to the rapid loss of silicon nitride (Schuster *et al.*, 1988; Silva *et al.*, 1998). Dissociation of the nitride is also aided by solution of nitrogen and silicon in the alloy (Bennett and Houlton, 1979). In contrast, the development of protective interfacial films leading to protection or non-wetting behaviour seems to provide resistance to attack by molten aluminium (Loehman, 1989; Mouradoff *et al.*, 1994; Yan and Fan, 2001), a feature that led to the development of silicon nitride components for use in the aluminium casting industry, and provided one of the first commercial markets for reaction-bonded silicon nitride structural ceramic.

5.3.10 *High-temperature behaviour – oxidation*

Oxidation can occur according to the reaction equations, depending on the partial pressure of oxygen (Singhal, 1997)

$$passive: Si_3N_{4(s)} + 3O_{2(g)} = 3SiO_{2(s)} + 2N_{2(g)};$$
$$\Delta G^o_{1500K} = -1685 \text{ kJ mol}^{-1} \tag{5.4}$$

$$active: Si_3N_{4(s)} + 3/2O_{2(g)} = 3SiO_{(g)} + 2N_{2(g)};$$
$$\Delta G^o_{1500K} = -434 \text{ kJ mol}^{-1}. \tag{5.5}$$

Silicon nitride is therefore very similar to silicon carbide in its oxidation behaviour, except that gaseous nitrogen is lost instead of carbon dioxide or carbon monoxide (Jacobson, 1993). Much of the background to the oxidation behaviour of silicon nitride is identical to that for silicon carbide described in Chapter 4, including the discussion of the oxidation parabolic rate constant.

During passive oxidation, the final product is primarily a protective surface film of silicon dioxide: during active oxidation no protection is possible and oxidation is much faster (though at a rate controlled by the actual availability of oxygen). The switch between active and passive modes depends on the oxygen pressure, and the temperature (Vaughn and Maars, 1990; Narushima *et al.*, 1994), and is of the order of a few Pa ($\sim 10^{-2}$–1 mbar) at temperatures in the region of 1300–1400 °C. However, during passive oxidation the picture is slightly more complex than that of silicon carbide, because silicon is also able to form a solid silicon oxynitride, of composition Si_2N_2O. The interface between silicon nitride and the outer surface film of silicon dioxide must now consist, in accord with the stability relationships expressed by the Si–N–O phase equilibrium diagram, of silicon oxynitride (or an oxynitride of composition approximating to this). This is shown schematically in Fig. 5.7. In fact the passive oxidation of pure silicon nitride occurs, under matching conditions, distinctly more slowly than that of silicon carbide (Riley, 1980; Tressler, 1994) as shown in Fig. 5.8, although it might have been expected that the two rates, if controlled simply by oxygen

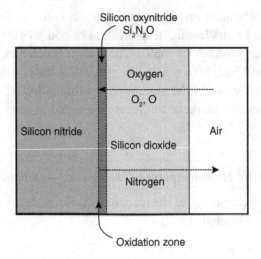

Figure 5.7 The oxidation of pure silicon nitride. This occurs by the inwards diffusion of oxygen through a surface silicon dioxide barrier layer. At the interface between silicon nitride and the silicon dioxide there is a thin region of silicon oxynitride, which forms an additional barrier layer.

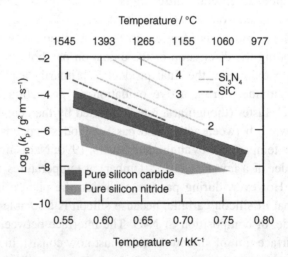

Figure 5.8 A comparison of the oxidation rates, using parabolic rate constants (k_p), of pure and liquid phase sintered silicon carbide and silicon nitride. The dotted lines represent liquid phase sintered materials. Pure silicon nitride oxidises appreciably more slowly, because of the development of the thin interfacial film of Si_2N_2O. The liquid phase sintered silicon nitride oxidises more quickly because there is a larger volume of intergranular silicate. Key. 1: 4% Al_2O_3; 2: 1% MgO; 3: 2% Y_2O_3; 4: 5% MgO.

transport through a silicon dioxide film, would be identical. The difference is attributed to a slower rate of transport of oxygen through the very thin interphase layer of an oxynitride of composition approximating to Si_2N_2O (Tresssler, 1994). As with silicon carbide, rates of oxidation start to become significant only at temperatures above 1000 °C, and the protection conferred by the silicon dioxide film allows the use of pure silicon nitride over many thousands of hours at temperatures of 1400 °C. Oxidation rate is linked to temperature by the standard relationship of the Arrhenius form, but other factors such as creep or slow crack growth may also play a part in determining the practical life of a load-bearing component (Billy, 1988). Sintered silicon nitride containing secondary, inter-granular, silicate phases has much faster oxidation rates, the reasons for which are explained in the section dealing with this material.

As an addendum to the subject of the oxidation, it should be noted that a silicon nitride particle surface becomes coated automatically with a thin protective silicon dioxide film the moment the particle (which has probably been produced under a nitrogen atmosphere) is exposed to air. These films are normally very thin (a few nm in thickness), but because the particles themselves may be of the order of only a few hundred nm in dimension, the total oxygen content of the powder can be significant, and equivalent to several per cent by weight (Szépvölgyi *et al.*, 1996). This becomes important for the sintering of silicon nitride, because it is this silicon dioxide that initially reacts with the additive oxides (the sintering aids) to form liquid silicates. It should also be remembered that α-phase silicon nitride may in addition to the oxide external film, contain up to 2% of oxygen *internally*, with the oxygen atoms occupying nitrogen lattice sites. This means that the total oxygen content of a silicon nitride powder may easily amount to several per cent by weight. All this oxygen becomes available (in the form of its silicon dioxide equivalent) when the silicon nitride particle dissolves in the liquid silicates during sintering, and will modify the liquid content and composition of the overall additive system. It is therefore important, if the sintering process is to be fully controlled, to have information on the total oxygen content of a silicon nitride powder; this information is usually part of the powder specification.

5.3.11 High-temperature behaviour – corrosion

Silicon nitride will react with most metal oxides (in the absence of gaseous oxygen), to release nitrogen, and form silicon dioxide and the metal. For example:

$$Si_3N_{4(s)} + 6Na_2O_{(s)} = 3SiO_{2(s)} + 12Na_{(l)} + 2N_{2(g)}. \tag{5.6}$$

This is because over temperature ranges of practical interest, silicon dioxide is generally thermodynamically more stable than the metal oxide (Loehman, 1983).

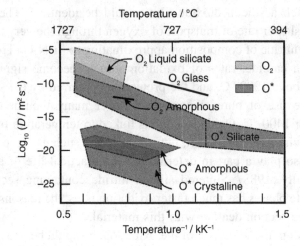

Figure 5.9 The mobility of oxygen, in atomic (O^*), and molecular (O_2) forms, in amorphous and crystalline silicon dioxide, and a selection of silicates. Logarithm of diffusion coefficient is shown as a function of reciprocal thermodynamic temperature. The mobility of both species is considerably greater in the silicates. (After Lamkin *et al.*, 1992.)

Two of the few oxides which are more stable are Al_2O_3 and ZrO_2. However, silicon nitride is normally used under oxidising atmospheres and such oxygen-free conditions are in practice not common. In this case there will be a surface film of silicon dioxide already in place on the nitride, and the metal oxide reacts with this, and not the nitride. The (potentially protective) silicon dioxide film is thereby converted to a liquid silicate. Because oxygen diffusion through the (liquid) silicate is (under comparable conditions) much faster than through pure silicon dioxide (Lamkin *et al.*, 1992), there is now much easier access of oxygen to the underlying nitride, and oxidation rates are accelerated. A comparison of the mobilities, expressed in terms of diffusion coefficients, of oxygen in atomic and molecular forms in silicon dioxide and a selection of Group IA metal (sodium and potassium) silicates, is shown in Fig. 5.9. During oxidation in the presence of metal oxides, bubbling and bloating of the liquid silicate layers, caused by the escaping nitrogen gas, usually occurs. The overall reaction can be regarded as a form of accelerated oxidation. Oxidative corrosion caused by metal oxide vapour can be an important factor determining the life at high temperature of silicon nitride components (Mayer and Riley, 1978). It is particularly important to prevent contamination by the volatile Group IA metal oxides, Na_2O and K_2O, and sources of these oxides such as the metal halides, which give very low melting-point silicates. The behaviour of silicon nitride and silicon carbide in this respect is essentially the same.

5.4 Introduction to the development of silicon nitride ceramics

The result of heating compacted pure silicon nitride powder to very high temperature is a slightly "recrystallised" and weakly bonded porous material of low strength (Greskovich and Rosolowski, 1976). In the case of silicon carbide the recrystallised form of material has important practical applications (Section 4.5). The same might have been true for silicon nitride, except that it turns out that there is an alternative (and much more convenient) lower temperature process available for producing a porous but reasonably strong silicon nitride; this is the reaction of silicon powder with nitrogen, reaction (5.2).

The use of materials formed from the nitridation of silicon powder was first reported in the context of a review of silicon nitride as a refractory (Collins and Gerby, 1955). It was then discovered that it was possible to form a reasonably strong, silicon nitride ceramic (usually termed *reaction-bonded* or *reaction-sintered* silicon nitride) by the nitridation of compacted and shaped silicon powder (Parr *et al.*, 1960; Popper and Ruddlesden, 1961). Because there is no overall dimensional change during nitridation, the product is an exact replica of the original shape (Popper, 1977). The material is porous, with the pore volume being a direct function of the packing density of the initial silicon powder. The strength, and very good thermal shock resistance, of this form of the material were instrumental in making it a candidate for use in high-temperature gas-turbine and diesel engines.

5.4.1 Densification by hot pressing

The sintering of pure silicon nitride powder presents problems similar to those found with silicon carbide, and for a long time it was categorised as "unsinterable" (Greskovich *et al.*, 1977). As was shown by Fig. 5.5, the mobilities of silicon and nitrogen atoms in the silicon nitride crystal lattice are extremely low, and attempts to accelerate the process by raising the temperature merely result in an increased evaporation rate, through the reversal of the formation reaction (5.2). Within the compacted powder, silicon nitride is evaporating (in the form of silicon or silicon monoxide vapour, and nitrogen gas), to condense on surfaces of reduced chemical potential. The dominant material transport process is therefore a form of evaporation and condensation (Ashby, 1974; Swinkels and Ashby, 1976). While this certainly allows the redistribution of silicon nitride and some degree of interparticle bonding (as in silicon carbide), there is still only very slow movement of material from grain boundaries by lattice or grain boundary diffusion. In principle, surface diffusion should also be occurring, but in any case the observed consequence is that sintering shrinkage is negligible.

Success in achieving the conversion of silicon nitride powder to a dense and strong solid was first achieved in 1960 (Deeley *et al.*, 1960). This approach followed the discovery that silicon carbide powders could be densified using aluminium oxide by pressure-assisted sintering – the technique conventionally known as *hot pressing* (Alliegro *et al.*, 1956). Densification of silicon nitride powder was obtained by blending it with small quantities of metal oxides able to form liquids at more easily attainable temperatures. The liquid phase provides a faster (and thus dominant) mass transport path. In practice hot pressing consists of applying uniaxial mechanical pressure to a fine powder (predominantly α-phase silicon nitride) contained within a cylindrical die (usually graphite), while simultaneously heating the die to a high temperature. A wide range of additive metal oxides and nitrides was explored, including magnesium oxide, magnesium nitride, aluminium oxide, and beryllium oxide, with the emphasis turning to magnesium oxide. Attention was then paid to the liquid-forming densification systems, and it was quickly established that the initial function of the magnesium oxide was to react with the natural oxide film coating the silicon nitride particles, to form (above the appropriate eutectic temperature) liquid silicates. As noted above, in effect, the equivalent of several weight per cent of silicon dioxide can be present as a natural contaminant of the silicon nitride powder. The first reaction sequence suggested (Wild *et al.*, 1972), assumed the availability of oxygen, at first in the form of surface silicon dioxide on the α-phase particles, then by solution of the α-silicon nitride particles in the liquid magnesium silicate to release lattice oxygen in the α-silicon nitride structure:

$$2MgO + SiO_2(\text{surface oxide}) = Mg_2SiO_4 \tag{5.7}$$

$$Si_3N_{4(\alpha)} + SiO_2(\text{internal oxygen}) + Mg_2SiO_4 = Si_3N_{4(\beta)} + 2MgSiO_3. \tag{5.8}$$

The silicon nitride crystallises, with *reconstructive transformation*, as the more stable β-phase. Densification is therefore occurring through the normal process of liquid phase sintering, with the solubility of the silicon nitride in the silicate liquid increased by the applied pressure, operating at particle contact points to increase the chemical potential (Bowen and Carruthers, 1978).

The dense fine-grain products first obtained by hot pressing contained a considerable proportion of intergranular silicate. Even so, with improved control of processing, room-temperature strengths of 900 MPa could be obtained (Lange, 1980), and high strength was retained to high temperatures; for example ~670 MPa at 900 °C, and >400 MPa at 1200 °C (Ashcroft, 1975). These are short-term strength values; for practical applications long-term strengths are important, as well as stress-induced slow crack growth, leading to failure under lower stresses (Ashcroft, 1973).

Densification of silicon nitride powder without the use of liquid-forming additives (liquid silicates) can be achieved, but very intense conditions are needed. It is possible to produce fully dense high-purity silicon nitride (Tanaka, 1989) on a very small scale by the direct application to powder contained in a small die of *ultra-high* pressures (above 1 kbar, or 100 MPa) at high temperature. An easier method is to use *hot-isostatic pressing*, when pressure is applied by nitrogen gas (that is, the pressure is applied *isostatically*) to a pre-compacted powder encapsulated in an inert material such as silica glass. Nitrogen pressures of 200 MPa and higher can be applied to the powder, using a "cold wall" pressure vessel, with powder temperatures above 1700 °C (Heinrich *et al.*, 1988). Because specialised equipment is needed, the production process becomes even more expensive than standard hot pressing.

A variation on the hot-isostatic pressing technique can be (and on a limited commercial scale is) used to complete the full densification of silicon nitride materials that have initially been pressureless sintered using normal conditions and quantities of liquid phase. Because the conditions are less extreme, far less expensive equipment is required. The first sintering stage is continued to the point where all the porosity is closed, and the subsequent application of pressure iso-statically using nitrogen gas leads to more rapid final densification. The gas pressures used are of the order of 10 MPa, at 1700–2000 °C (Giachello *et al.*, 1980). Better control over the microstructure (a more homogeneous micro-structure, with less grain orientation) can usually be achieved by isostatic pressing. This use of isostatic hot pressing also means that more complex shapes of fully dense material can be obtained relatively cheaply; the alternative is to hot press (uniaxially) a billet of material, which then needs to be diamond-machined. The increased microstructural uniformity obtained by isostatic pressing is an advantage when uniformity of response to contact load is required, such as in ball or roller bearings. In uniaxially hot-pressed materials the silicon nitride grains tend to be aligned normal to the pressing axis, and this microstructural anisotropy can lead to similar anisotropy in modulus and wear behaviour.

5.4.2 Liquid-forming additives

It is perhaps at first sight surprising that silicon nitride, a nitride, should be soluble in a liquid (oxide) silicate. But it has to be remembered that the nitrogen atom and the oxygen atom (we are dealing with materials which are to a large extent covalent) are similar in character and not too dissimilar in size, so that many crystalline silicates are able to replace a proportion of their oxygen by nitrogen, while retaining the oxide structure (Lang, 1977). The same is true for silicate glasses, and liquids; in many cases as much as 10% of the oxygen can be replaced by nitrogen, forming what are known as *nitrogen glasses*. The main, and from the materials angle, important difference is that the oxygen atoms are 2-coordinated (bonded to two silicon atoms) and the nitrogen 3-coordinated. This has implications for atom diffusion coefficients and high-temperature viscosities (Hampshire, 1993). Most of the silicate systems used for the liquid phase sintering of alumina have been tested on silicon nitride, with varying degrees of success.

Initially magnesium oxide, in quantities of a few per cent by weight, had been identified as being the most effective (Deeley *et al.*, 1960); yttrium oxide (Gazza, 1975), and aluminium oxide (Jack, 1973) later became the focus of attention. Practically full densification of μm dimension silicon nitride powders could be achieved in times of the order of a few minutes, under pressures of up to 30 MPa, at temperatures of around 1850 °C. The die system, typically, was fabricated from dense graphite (then often nuclear reactor grade), which, though easily able to withstand the high temperatures, was not strong enough to allow much higher pressures to be applied (a further advantage of graphite is that because of its electrical conductivity it allows the use of high-frequency AC induction heating). The densified silicon nitride product, usually in the form of a cylindrical rod or billet, requires diamond-tooled machining to final shapes, a difficult and slow process, which is as would be expected – silicon nitride is notable for its resistance to wear! The direct mechanical hot pressing of more complex shapes is possible, but requires elaborate die systems. The best of these hot-pressed materials had room-temperature strengths of the order of 600 MPa, and 400 MPa at 1200 °C, considerably better than liquid phase sintered alumina.

The ternary $MgO–SiO_2–Si_3N_4$ densification system was the first to be studied in detail (Brook *et al.*, 1977). This simply requires the addition of magnesium oxide, because, as has been noted, the silicon nitride powder generally contains sufficient "natural" silicon dioxide to produce an adequate volume of silicate (if not, then more can easily be added as very fine, amorphous, silicon dioxide powder). It was significant that densification appeared to start between 1500 and 1600 °C, which corresponds to the temperatures at which liquid appears in the $MgO–SiO_2$ binary system, 1543 and 1557 °C, depending on the MgO/SiO_2 ratio (Levin *et al.*, 1979, Fig. 266). This suggested that silicon nitride could be dissolving in a standard silicate liquid. Work on the magnesium silicate densification system and the properties of its products produced a tentative phase equilibrium diagram, containing a ternary nitrogen-containing eutectic at 1515 °C, and indicating that it is perfectly possible to describe these mixed nitride–oxide systems in terms of simple phase equilibrium diagrams (Jack, 1978; Lange, 1978). Figure 5.10 is an illustration of a very early attempt (Jack, 1978) to express the behaviour of this system, and to identify compositions which provide liquid at

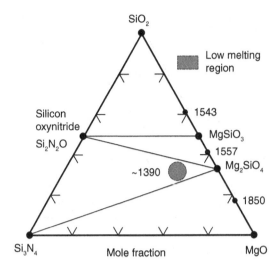

Figure 5.10 An early MgO–SiO_2–Si_3N_4 phase equilibrium diagram. It shows the postulated region in the Si_3N_4–Si_2N_2O–Mg_2SiO_4 compatibility triangle, forming liquid at ~1390 °C. (After Jack, 1978.)

~1600 °C. The area marked in the figure identifies the compositional region presumed to contain a ternary eutectic at ~1390 °C. The presence of nitrogen, resulting from the solution of silicon nitride in the liquid, thus appeared to lower the eutectic temperature by about 150 °C, explaining why full densification of fine silicon nitride powder could readily be achieved under pressure at temperatures in the region of 1600–1700 °C.

Other additives quickly attracted attention, and interest in the use of magnesium oxide alone as a sintering additive decreased. One reason is that magnesium silicates appeared to be slow to crystallise and were retained as intergranular glasses, that then soften at high temperature to reduce high-temperature strength. The similar ternary system, Y_2O_3–SiO_2–Si_3N_4 (outlined in Fig. 5.11) was also studied in detail. The yttrium-containing silicates are of higher viscosity (at the same temperatures) than silicates in the MgO system, and also more readily crystallise: yttrium oxide therefore appeared to be more useful as an additive because of its potential to give silicon nitride materials of better high-temperature strength (Gazza, 1973; Gazza, 1975; Hampshire, 1993). One drawback to the use of yttrium oxide is that sintering tends to be slower, and requires higher temperatures. Another recognised disadvantage of yttrium and the rare earth oxides was their reduced availability and much higher price. MgO and Y_2O_3 became a commonly used blend (Giachello *et al.*, 1980), often together with aluminium oxide. The inclusion of additional oxides as sintering systems such as the quaternary MgO–Al_2O_3–SiO_2–Si_3N_4, produces even more complex phase systems,

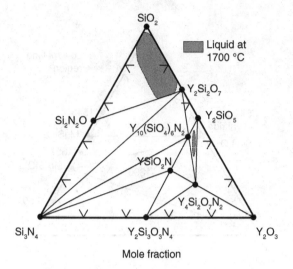

Figure 5.11 The Y_2O_3–SiO_2–Si_3N_4 phase equilibrium diagram, showing an extended liquid-forming region at 1700 °C. (After Jack, 1978.)

which require multi-dimensional phase equilibrium diagrams to represent adequately phases and phase stability fields (Gauckler and Petzow, 1977; Huang *et al.*, 1986; Yen and Sun, 1993).

The use of aluminium oxide in conjunction with yttrium oxide has been favoured as a densification aid for a number of reasons. One is that the ternary phase yttrium aluminium garnet ($Y_3Al_5O_{12}$, or $3Y_2O_3.5Al_2O_3$) can crystallise from the liquid on cooling, to reduce the volume of residual intergranular glass. This is also a desirable reaction product, because of its very high melting-point of ~1970 °C. In general it is found that post-densification annealing treatments do tend to improve strength, because they aid the crystallisation of intergranular glass.

5.4.3 *Pressureless sintering*

Early attempts to produce silicon nitride of significantly increased density or strength by standard pressureless liquid-phase sintering failed, and only porous materials were obtained. It was then found that silicon nitride powder could be densified without the application of pressure, provided it was in a totally enclosed system (such as the cavity of a graphite die), and the need for the full control of the evaporation of silicon nitride was realised (Terwilliger, 1974; Terwilliger and Lange, 1975). Suppression of the natural high-temperature volatility of silicon nitride can be achieved either by the application of nitrogen pressures of the order of 1 MPa (Mitomo, 1976), or, more easily, under nitrogen at lower pressure by immersing the article being sintered in a bed of loosely packed silicon nitride powder, blended with an inert powder such as boron nitride to prevent sintering (Giachello *et al.*, 1979). The function of the powder bed is to supply the equilibrium pressures of silicon vapour phase species ($Si_{(v)}$ and $SiO_{(v)}$) and nitrogen gas, reducing silicon nitride evaporation and migration from the sintering component. Carbon crucibles are generally used, which have been coated with silicon carbide to minimise the silicon "sink" action of the carbon.

Pressureless sintered silicon nitride is now produced on a commercial scale for a wide range of small components. The amounts of additive required are generally similar to those used in hot pressing, that is 3–15 mol% (~2–10 weight%), but the temperature needs to be 200–300 °C higher: 1800–1900 °C is not uncommon. Careful control of the conditions to minimise volatilisation of silicon nitride, and the oxide additives, is required, which entails using high nitrogen pressures, and powder beds containing the oxide additive. With submicrometre powder particle sizes, and very efficient additive blending, full densities can be obtained in 1–2 hours at peak temperature, with the full cycle time determined by furnace heating and cooling rates.

Much of the earlier work on the mechanical properties of dense silicon nitride, sintering aid systems, and the linking of mechanical property to microstructure, was carried out using a range of trial additive compositions. These materials were investigated in some detail for possible commercial applications, including high-temperature gas-turbine components. Major problems were found to be the cost of machining billets of material into the required refined shapes, or the production of more complex (*near net*) shapes directly by hot pressing. The hot pressing technique severely restricts the range of shapes that can be formed readily, and following the realisation that silicon nitride could be satisfactorily pressureless sintered, interest in hot-pressing diminished. However, much of the very large amount of information on additive systems developed for hot pressing was later successfully transferred to the production of sintered silicon nitride.

5.4.4 Sialons

Silicon nitride reacts at high temperatures (around 1770 °C) with aluminium oxide to form a range of crystalline phases that are usually referred to as *sialons*, coined from the elements, Si, Al, O, and N making up the simplest of these phases (Oyama and Kamigaito, 1971; Jack and Wilson, 1972; Jack, 1977). The liquid phase sintering of silicon nitride in the presence of aluminium oxide therefore involves more than just the physical solubility, and crystallisation, of silicon nitride. Complex chemical reactions also take place, with the partial or complete conversion of the binary silicon nitride to one or more sialon phases. This is more than just of academic interest: the use of aluminium oxide as a component of the sintering additive mixture has a number of strong points in its favour. The first is that liquids are likely to be formed at lower temperatures. The simple ternary oxide Al_2O_3–MgO–SiO_2 system (Fig. 3.6) contains two eutectics with compositions close to that of cordierite ($2MgO.2Al_2O_3.5SiO_2$), and melting at around 1360 °C, though when greater proportions of aluminium oxide are present the eutectics have much higher melting-points (Levin *et al.*, 1964, Fig. 712). In the quaternary (MgO–Al_2O_3–SiO_2–Si_3N_4) system formed with the quantities of MgO and Al_2O_3 normally used, the lowest eutectic melting-point is believed to be around 1470 °C, lowering the temperature required for sintering, or the time needed at higher temperatures. Aluminium oxide as a constituent of the intergranular liquid may also assist densification by lowering the liquid viscosity (see also Section 2.7), thereby increasing Si and N mobility (Urbain *et al.*, 1981; Hampshire, 1993). The second point is that the liquid may, at least in principle, be *transitory*. This means that it either decreases in volume, or converts to crystalline phases, while densification is occurring (or during a subsequent annealing stage). One of the reasons for this is that some of the aluminium oxide can be incorporated (effectively dissolved) in the silicon nitride crystal lattice with the formation of a sialon phase. A third factor may be the greater ease with which aluminates crystallise, compared with silicates (and ideally these processes should of course be delayed until densification is completed).

An important variation on sintering silicon nitride powder blended with the densification aid, consists of sintering reaction-bonded silicon nitride, containing the standard liquid phase sintering aids (Mangels and Tennenhouse, 1980). The additives can be incorporated into the silicon powder before nitridation, or later impregnated as solutions into the porous reaction-bonded material. The sintered reaction-bonded silicon nitride has properties very similar to materials obtained by the normal sintering process with silicon nitride powders (Govila *et al.*, 1985).

5.5 Ceramic materials

5.5.1 Reaction-bonded silicon nitride

Production

Reaction-bonded silicon nitride is produced by heating compacted fine silicon powder, in a nitrogen atmosphere, at temperatures in the region of 1250–1450 °C (Moulson, 1979). Several days' heating may be needed for full conversion to the nitride. To provide a large reaction interface area the silicon must be in the form of a fine (high surface area), powder, normally with particle sizes in the approximate range 1–10 μm. One simple reason for these long reaction times is that the reaction of nitrogen with solid silicon is intrinsically slow, because (as has been seen) the diffusion coefficients for silicon and nitrogen in silicon nitride are very small at these temperatures. This means that the product silicon nitride forms a thin protective film on the silicon particle surface, through which the reactants then have to diffuse. Although kinetically the silicon nitridation reaction normally tends to be slow, it is also strongly exothermic, with a reaction enthalpy of ~740 kJ mol^{-1} at 1400 °C (the silicon powder might be thought of as "burning" in nitrogen). A large mass of silicon powder can undergo uncontrollable self-heating, and its temperature can rapidly rise above the furnace set temperature. Silicon has a relatively low melting-point of ~1412 °C, and it is necessary to prevent premature melting and fusion of the silicon powder particles, which would make complete nitridation impossible. Close control of the reaction rate (and thus the rate of release of heat) is needed, and programming of the rate of rise of the furnace temperature is required, particularly in the early stages of the reaction. There are a number of ways in which this can be done, varying from simple step-wise increases in furnace power, to linking the power input, or nitrogen supply rate, directly to the rate of nitrogen consumption (Mangels, 1981).

In principle, reaction-bonded silicon nitride could also be made by nitriding a mixture of silicon nitride powder and silicon, the method parallel to that for reaction-bonded silicon carbide. However, the resulting materials do not have as good mechanical properties as reaction-bonded silicon nitride prepared from silicon alone, possibly because the oxygen inevitably introduced into the system with the silicon nitride powder interferes with the development of microstructure.

The nitride formation, and "reaction bonding", process involves an interlinked set of steps. In the initial stages of the reaction, at temperatures below about 1250 °C, silicon nitride is nucleated and the nuclei grow, initially to form a thin, intermittent, film of α-phase on the surfaces of the silicon particles. The nitride film forms a barrier layer (as silicon dioxide does in oxidation) to further reaction. Film development stops, but silicon nitride growth can still continue outwards, by reaction of silicon vapour escaping from still open areas of the silicon surface (Atkinson *et al.*, 1976). This vapour phase reaction produces silicon nitride crystals with a whiskery (high aspect ratio) morphology. Figure 5.12 shows this process schematically. A linked network of fibrous silicon nitride (sometimes referred to as a matte) now forms, to isolate and retain unreacted silicon particles. At this stage strong interparticle bonding is initiated. To complete nitridation it is normally necessary to raise temperatures up to ~1450 °C, to melt the remaining silicon particles. There are now two consequences: the silicon vapour pressure increases, exponentially, in accord with the Arrhenius relationship, facilitating the vapour phase reaction; but more importantly silicon nitride growth now also takes places inwards into each liquid droplet, giving larger β-phase crystals of a more equiaxed morphology (Fig. 5.13). The final product therefore consists of a mixture of

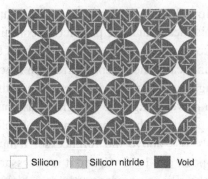

Silicon ▨ Silicon nitride ▨ Void ■

Figure 5.12 The nitridation of silicon powder, shown schematically: the first stage, with the initial development of silicon nitride whiskers. Compare this figure with that for a general reaction bonding process, Fig. 1.22.

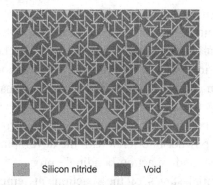

Silicon nitride ▨ Void ■

Figure 5.13 The completion of silicon nitridation, shown schematically: full nitridation of the silicon particles, and unavoidable residual porosity.

interlinked α- and β-phase crystals. It has been considered that the α/β ratio of the reaction-bonded silicon nitride, which can be controlled by the heating schedule, has an influence on the material's mechanical properties (Ziegler *et al.*, 1987). This is likely to be a consequence of the proportions of the whisker, and equiaxed, silicon nitride crystals in the microstructure. Figure 5.14 is a light micrograph of the polished surface of compacted powder which has been partially nitrided; the loss of silicon by evaporation from exposed surfaces is shown by the development of surface pits. Figure 5.15 shows the typical, very fine, whiskers of α-phase silicon nitride (the *matte*) formed within the voids between the silicon particles. An important feature of the reaction-bonding process is that, although bonding of the growing silicon nitride grains is certainly occurring, there is no sintering shrinkage of the component (another similarity with reaction-bonded silicon carbide will be

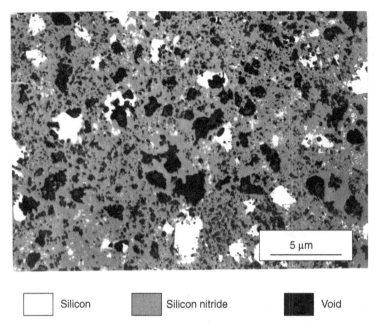

| | Silicon | | Silicon nitride | | Void |

Figure 5.14 A light micrograph of the polished surface of almost completely reacted silicon: the unreacted silicon is highly reflecting, silicon nitride is the grey phase, and the void space is black. The microstructural detail is otherwise on too fine a scale to be resolved at this magnification.

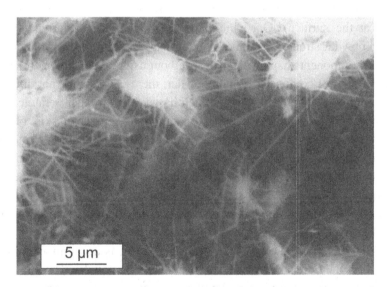

Figure 5.15 A scanning electron micrograph of α-silicon nitride whiskers, formed in the voids surrounding the silicon particles.

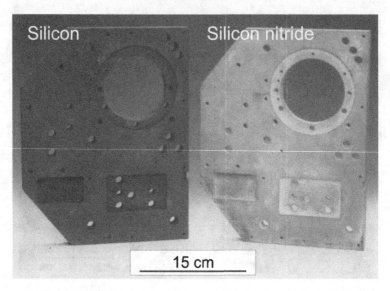

Figure 5.16 An illustration of the absence of shrinkage during nitridation, and the ability to retain precisely a pre-machined form. (After Riley, 2000, reprinted with permission of Wiley-Blackwell Publishing.)

recognised). The reason for the absence of shrinkage is that much of the initial development of silicon nitride is by a form of the *evaporation and condensation* process, in which silicon vapour evaporates from silicon particles, to condense as silicon nitride at other points on particle surfaces (on silicon nitride already nucleated at the surface). Once interparticle bridging has occurred, the whole compacted powder structure is dimensionally fixed, and no further volume change is possible. Component shaping can be carried out using a relatively soft and easily machinable billet of compacted silicon powder; there is no significant dimensional change during conversion to the very much harder and stronger ceramic. All that may be required is a final machining operation to remove a skin of slightly more porous material, to leave a surface with the required degree of smoothness. Figure 5.16 shows a complex shaped component in the compacted (and lightly sintered) silicon powder, and in the fully nitrided state: dimensional changes are not detectable.

A second very useful feature of the silicon nitridation reaction is that the volume of silicon nitride is greater than the volume of the silicon from which it was formed. This volume expansion, of around 21%, is accommodated entirely within the original pore space of the compacted silicon powder. There is therefore a bonus, in that during the reaction-bonding process (although there is no overall shrinkage) densification has occurred. In theory, silicon powder compacted to a solid volume fraction of 0.823 (17.7% porosity), which is certainly achievable, should yield a pure, 100% dense, silicon nitride component. However, there is a catch: compacted silicon powder of such a high density would be impossible to nitride fully, because the nitrogen would not be able to reach the interior of the compact once the outer pores had become completely filled by silicon nitride. The maximum density of silicon powder that can be fully reacted (and even then only with components of thin cross-section), is about 75% of theoretical, which gives a reaction-bonded silicon nitride of bulk density corresponding to ~91% of the theoretical. Most commercial materials have porosities much higher than this, in the region of 15–30% (and densities of about 2.2–2.7 Mg m^{-3}), which represents a compromise between property value such as strength, and ease of attaining full nitridation of the silicon powder.

General properties

Reaction-bonded silicon nitride was studied in some detail for possible development as a refractory material with low electrical conductivity, but also with lower thermal expansion coefficient, and therefore better resistance to thermal shock, than sintered alumina ceramics (Popper, 1977). It also received considerable attention because of its resistance to attack by molten aluminium, which is a very corrosive material for oxide ceramics because of its ability to abstract oxygen from the oxide (Yan and Fan, 2001).

Density and porosity

Material formed from micrometre dimension silicon powders using the standard powder shaping processes of uni- or multi-axial die-pressing, or slip casting, has densities in the region of 80% of the true solid density. This corresponds to a bulk density of around 2.5 Mg m^{-3}, though the spread of values reported for commercial materials is wide. The microstructures of these materials consist of very fine, micrometre dimension, silicon nitride crystals. Because much of the microstructure is of a very fine scale, and also fibrous, it can be difficult to obtain good micrographs, except by using high-resolution scanning electron microscopy. A distinction can be made between "bonding nitride" and "non-bonding" nitride. The first corresponds to crystals which have grown to directly link other, neighbouring, crystals. Some of this material will have grown in the liquid silicon, in much the same way that the secondary silicon carbide does through the carbon–silicon reaction. The second is the whiskery or fibrous (mainly α-phase) material which has simply grown from a surface into a void (corresponding to the spaces between the original silicon particles), and therefore not making any contribution to bulk strength or stiffness (although its presence will be registered in measurements of bulk density). It is also difficult to speak of pore size in the normal sense with this material: all space not occupied by (mostly) β-silicon nitride formed by the nitridation of liquid silicon is to a greater or lesser extent filled with the whisker form of the α-phase nitride. The pore size can be equated to the space between individual α-silicon nitride whiskers, which is certainly on an extremely fine scale. However, this is not very helpful in discussions of mechanical properties, and it is better, from the point of view of identifying critical (Griffith) flaws, to think in terms of regions of "low whisker density", as corresponding to the "pores" in a normal sintered ceramic.

The quality of a microstructure of any sintered ceramic is dependent on the homogeneity of powder particle distribution in the compacted starting powder. The relationship is especially marked with reaction-bonded silicon nitride, which can be said to be a particularly "unforgiving" material, because the microstructure depends very strongly on the silicon powder microstructure. Although the microstructure of the compacted silicon powder has to some extent been modified during nitridation (because the silicon particles were partly vaporised during their conversion to silicon nitride) the resulting reaction-bonded microstructure retains a clear "memory" of this structure, in the form of the size and distribution of the whisker filled void spaces between the silicon particles. Moreover, because the liquid silicon is not able to assist particle movement, there is very little scope for large-scale rearrangement or redistribution of material, and smoothing of microstructural inhomogeneities. The void sizes and size distributions in the silicon powder compact (or their equivalents) are retained largely unchanged in the nitride. A fine, homogeneously packed, silicon powder will tend to produce reaction-bonded silicon nitride containing a homogeneous distribution of small voids, and coarse-grained or badly packed silicon powders will generate an irregular distribution of large voids. There are important implications for the mechanical strength of reaction-bonded silicon nitride, in which the strength-controlling critical defect is likely to be a low-density region of dimension much larger than the average (Stephen and Riley, 1993). These larger "porous" regions can often be of multi-micrometre dimension and easily identifiable using optical microscopy.

Young modulus and hardness

The Young modulus values depend on the total pore volume (that is, bulk density), and for commercial materials are generally in the region of 80–250 GPa. Data fitted to the standard semi-logarithmic relationship for modulus as a function of pore fraction have extrapolated values for (though unattainable by the reaction-bonding route) fully dense material, in the region of 320 GPa. This is close to values obtained for dense sintered silicon nitride. Figure 5.17 shows data for a range of commercial reaction-bonded silicon nitrides, taken from manufacturers' data sheets. Because there is no intergranular amorphous

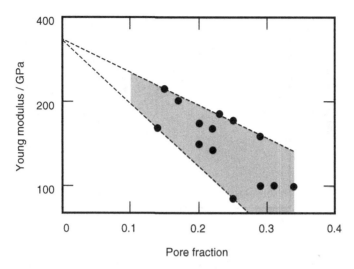

Figure 5.17 The influence of porosity on the Young modulus of reaction-bonded silicon nitride: manufacturers' data, as a function of pore fraction in semi-logarithmic form.

material of the type found for example with alumina ceramics modulus values do not change significantly with temperature until very high temperatures (~1400 °C) are reached. With long-term use at high temperature there will be deterioration in stiffness, but this is more likely to be the result of oxidation. The Vickers indentation hardness values in the region of 10 GPa are not particularly high, because of the porosity of the material: there can be a wide scatter of measured values because of the varying densities from point to point in the silicon nitride.

Mechanical strength

Widely different strength values can be obtained for reaction-bonded silicon nitrides of the same bulk density, and of apparently similar silicon nitride α-phase to β-phase ratios, and crystallite sizes. This should not be surprising because the strength is a function of two, partially but not completely related, factors: the total pore fraction (related to the bulk density), and the size of the critical Griffith defect, which will most probably be a pore much larger than the average (Eddington *et al.*, 1975). The size of the largest "pore" (or low-density region) in the nitride can in turn be traced back to the silicon particle size, and homogeneity of packing, of the powder. With reaction-bonded silicon nitride prepared from finer particle sized silicon, bend strengths can exceed 300 MPa at

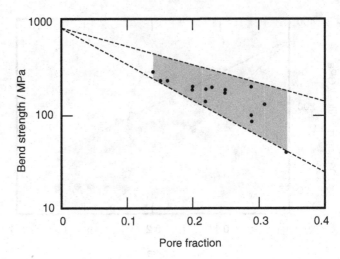

Figure 5.18 The influence of porosity on the strength of reaction-bonded silicon nitride: manufacturers' data for bend strength, as a function of pore fraction, in semi-logarithmic form.

room temperature. Materials of lower density or with poorer microstructural homogeneity prepared from coarser, or badly mixed silicon powders, have room-temperature strength values as low as 80 MPa. Most commercial materials have bend strengths in the region of 200–250 MPa. Figure 5.18 indicates the broad spread of manufacturers' data for bend strength, treated as a function of pore fraction in standard semi-logarithmic form. Extrapolations of room-temperature strength values to zero porosity using standard empirical expressions indicate that strengths approaching 900 MPa should be possible. It has been of interest that this extrapolated value is of the same order as that for the strength of dense, fine-grain, sintered or hot-pressed silicon nitride – as would be expected if grain size were the dominating factor. This observation served to counter earlier suggestions that the bonding in reaction-bonded silicon nitride was in some way different from normal intergranular bonding in a sintered material. Defect size is also a factor influencing strength, as is shown by the Griffith equation. In Fig. 5.19 strength is shown as a function of the largest apparent pore size (measured in random cross-sections). This shows the importance of the larger pores in a (necessarily) porous material, for strength.

The strength of fully nitrided materials is maintained up to temperatures of 1400 °C (Fig. 5.20), because there is no intergranular amorphous or low melting-point phase (Noakes and Pratt, 1972). Above this temperature oxidation becomes increasingly important, and leads to gradual loss of strength and creep, although the overall behaviour is complex, and can include some improvement

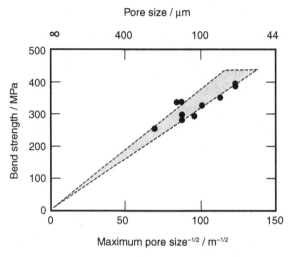

Figure 5.19 Strength of reaction-bonded silicon nitrides of constant pore fraction, as a function of the apparent *maximum* pore size. (After Stephen and Riley, 1993.)

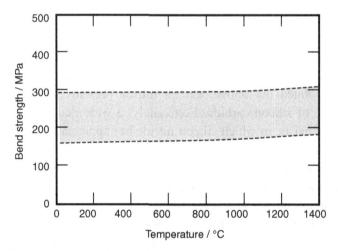

Figure 5.20 Manufacturers' data for the strength of reaction-bonded silicon nitride, as a function of temperature. The absence of low melting grain boundary phases means that strength is maintained unchanged to a very high temperature.

in room-temperature strength through surface pore or damage healing (Davidge *et al.*, 1972). As with the Young modulus, the reduction in strength over long times at high temperature can be attributed to oxidation, and the loss of bonding nitride.

Fracture energy and fracture toughness

Measured fracture surface energies are in the region of 30 J m^{-2}. The fracture toughness of reaction-bonded silicon nitride is low, as would be expected for a material with a low Young modulus resulting from its high levels of porosity. Measured K_{Ic} values are of the order of 2–3 MPa m$^{1/2}$ at room-temperature, increasing with increasing density. The extrapolated value for fully dense material is about 5 MPa m$^{1/2}$ (Barnby and Taylor, 1972), which matches well the measured values for dense sintered silicon nitride with the more equiaxed β-phase grains.

Thermal properties

The mean thermal expansion coefficient, as measured for a range of commercial reaction-bonded materials over the temperature range of 0–1000 °C, is 3.0 MK^{-1}. Thermal conductivity values show a wide scatter, centred around 10 W m^{-1} K^{-1} at room temperature. The resistance to thermal shock, as estimated by a standard quench test, is very good, with critical quench temperatures in the region of 400 °C (Weaver *et al.*, 1975). This puts reaction-bonded silicon nitride, on a level with reaction-bonded silicon carbide.

5.5.2 Sintered silicon nitride

Production

As has been pointed out above, silicon nitride presents sintering difficulties similar to those of silicon carbide. Fortunately, a wide range of liquid-forming silicate systems exists, in which silicon nitride has appreciable solubility at high temperature; conventional liquid phase sintering is therefore readily possible. When the silicon nitride dissolves in the liquid it loses its identity as the species Si_3N_4 and dissociates into silicon and nitrogen atoms, which are then accommodated within the silicate Si–O network. It is therefore more accurate to say that the sintering system is a silicon nitride oxide (*oxynitride*) system, which is able to contain most (if not all) of the metal atoms of the silicate. Another requirement for successful sintering densification (as it is with silicon carbide) is an extremely fine powder, which means a median particle size usually in the range 100–200 nm, corresponding to a specific surface area of the order of 20–10 m^2 g^{-1}.

During sintering the (initially generally equiaxed) α-phase silicon nitride particles slowly dissolve, and then recrystallise from the liquid as the stable β-phase silicon nitride (or as β′-phase sialon when aluminium oxide is present – see the following section). This process tends to give the microstructure a high proportion of high aspect ratio (1:10 and higher) "needle-like" or fibrous grains (Kramer

Figure 5.21 A scanning electron micrograph of fully dense sintered silicon nitride, which has been chemically etched to dissolve the intergranular glass, and reveal the high aspect ratio morphology of the remaining β-silicon nitride grains. (Courtesy of Paul Andrews.)

et al., 1993), elongated along the crystal *c*-axis. The scanning electron micrograph, Fig. 5.21, shows a sample of sintered silicon nitride which has been chemically treated to remove the intergranular glass: the elongated silicon nitride grains are revealed clearly. Median grain dimensions (perhaps more meaningfully, lengths) are normally in the range 1–10 μm so that these are fine-grain materials. Figure 5.22 is a scanning electron micrograph of a polished and then lightly etched surface of sintered silicon nitride, showing the extensive volume of intergranular glass; this feature is seen more clearly in the transmission electron micrograph in Fig. 5.23. Crack bridging by the elongated grains also contributes to high fracture toughness (*R*-curve behaviour) and very high strengths (~1000 MPa) can be obtained (Lange, 1980; Becher, 1991). Materials have therefore been referred to as "*in situ* reinforced ceramics", and the sintering process requires not only densification, but control of grain development (Mitomo *et al.*, 2006).

For a typical amount of additive of the order of 5 mol%, and allowing for the incorporation of silicon dioxide, the total volume of secondary phase will be in the range of 5–10%. The exact amounts present vary from material to material, and at present there are no standard compositions. The additives remain in the microstructure for the most part as amorphous, or crystalline, intergranular phases, but there is commonly a thin (1–2 nm) amorphous film at the silicon nitride grain boundaries (Kleeb *et al.*, 1993), which will be responsible to some

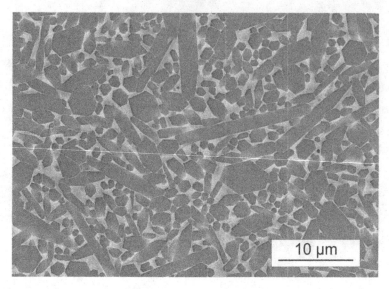

Figure 5.22 A scanning electron micrograph of sintered silicon nitride. The polished surface has been etched chemically to show more clearly the high aspect ratio silicon nitride grains responsible for crack diversion and increased toughness. (Courtesy of Mamoru Mitomo.)

Figure 5.23 A transmission electron micrograph of sintered silicon nitride, showing the continuous intergranular glass (light phase). (Courtesy of Chong-Min Wang.)

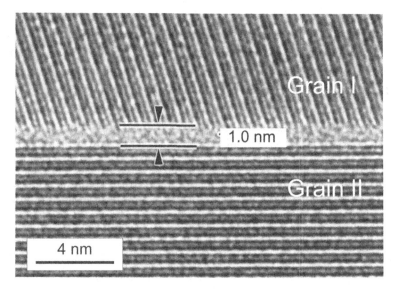

Figure 5.24 High-resolution transmission electron micrograph, showing the typical ~1 nm grain boundary amorphous film. (Courtesy of Chong-Min Wang.)

extent for the loss of strength and rigidity at high temperature. An example of this type of film is shown in Fig. 5.24.

Physical and mechanical properties

The mechanical properties of sintered silicon nitride are determined by the intrinsic properties of silicon nitride itself, modified by the microstructure, and the quantity and nature of the residual intergranular phase material (Hoffmann and Petzow, 1993). Above about 1000 °C the intergranular, amorphous, phase controls the mechanical properties. Creep sets in and strength is dependent on the time under load because of subcritical crack growth. The larger the volume of secondary amorphous phase, the greater is the susceptibility to cavitational creep and strength degradation (Luecke *et al.*, 1995; Wiederhorn *et al.*, 2005). Important microstructural features are the proportion of interlocked high aspect ratio silicon nitride grains, and the degree of crystallisation of the intergranular liquid generated by the sintering additive. Because these two aspects of the microstructure can vary widely between materials, depending on system composition and sintering conditions (and are difficult to quantify precisely), it would also be expected that there would be corresponding wide variations in mechanical property value (in particular fracture toughness, and high-temperature strength and creep rate). Silicon nitride powder purity is also an important factor, in that traces of impurity oxides such as calcium and iron can lower glass viscosity at high temperatures, and inhibit its crystallisation (Iskoe *et al.*, 1976). Factors such as these are responsible for the wide variations in property seen.

Commercial materials have densities in the range 3.1–3.3 Mg m^{-3}, depending on the composition of the sintering additive system and the amount used, and there is very little residual porosity. The Young modulus typically is ~300 GPa, so that these materials have a similar stiffness to a fully dense liquid phase sintered alumina. Fracture toughness is normally in the region of 5–6 MPa m$^{1/2}$, but with care much higher values (>10 MPa m$^{1/2}$) can be obtained. This can be achieved through controlling the proportion of β-phase silicon nitride in the starting powder, in order to minimise the concentration of nuclei for the growth of new β-phase during sintering (Lange, 1973, 1979). While small β-silicon nitride grains are required in the final microstructure (needing a high nuclei density), each developing β-silicon nitride grain requires adequate space if it is to develop to the optimum size and aspect ratio (needing a low nuclei density). The final microstructure is therefore determined by the fraction, mean size, and size distribution of the β-phase nuclei (Kramer *et al.*, 1993). Ideally the initial β-silicon nitride content should be no more than 30% (which is not difficult to obtain, since most silicon nitride powders consist predominantly of the α-phase). Figure 5.25 shows one relationship established between fracture toughness and starting powder composition (Lange, 1979). Fracture toughness is also influenced by thermal expansion coefficient differences between silicon nitride and the grain boundary phases, the residual stresses from which promote intergrain debonding and microcrack development, leading to improved fracture toughness (Sun *et al.*, 1998).

Bend strengths can be very high: short-term, room-temperature, values can be higher than 1000 MPa, though 600–800 MPa is more normal. Silicon nitride is therefore a much stronger material than liquid phase sintered alumina, or solid state sintered silicon carbide. This is a consequence of the much higher fracture toughness achieved through careful control of the microstructure, coupled with the finer grain sizes attainable from the use of ultra-fine starting powders. Even higher strengths can be obtained by reducing the volume of intergranular amorphous material. Post-densification annealing can give materials with bend strengths as high as 1500 MPa (Tsuge *et al.*, 1975). However, to obtain strengths as high as these, careful attention to test sample preparation is required, particularly with regard to the quality of surface finish.

Short-term strength, and creep resistance, are usually maintained up to 1000 °C; above this temperature values start to decrease, because of the softening of amorphous intergranular and secondary phases (Wiederhorn *et al.*, 1994). By 1400 °C over 80% of strength will have been lost. The way in which strength decreases with rising temperature is determined mainly by the composition and volume of the intergranular glass: manufacturers' data are summarised in Fig. 5.26.

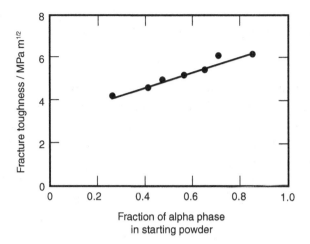

Figure 5.25 The relationship between fracture toughness and proportion of high aspect ratio β-silicon nitride grains in the material influenced by the phase composition of the starting silicon nitride powder. (After Riley, 2000, reprinted with permission of Wiley-Blackwell Publishing.)

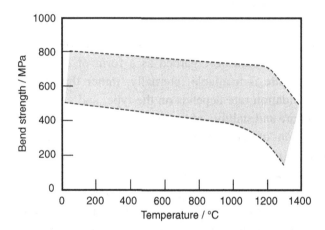

Figure 5.26 The influence of temperature on the strength of fully dense silicon nitride: manufacturers' data, showing the decreasing strengths at temperatures above ~1000 °C, depending on the nature and volume of intergranular second phase.

Mean values of thermal expansivity over the range 0–1000 °C for commercial materials vary with composition, but are in the range of 3.0–3.5 MK^{-1}. Thermal conductivity values likewise vary considerably with composition, but with a mean at 25 °C of around 20 $W\ m^{-1}\ K^{-1}$. The critical quench temperatures required to cause significant loss of strength also show some variation with microstructure. Most reported values are in the range 600–700 °C, which is considerably better than sintered alumina – the result of the higher strengths achievable without matching increases in Young modulus (Schneider and Petzow, 1994).

Oxidation and corrosion

The oxidation behaviour of liquid phase sintered silicon nitride is controlled to a large extent by the nature of the intergranular material, in particular by the continuous, amorphous silicate phase (see Figs. 5.22–5.24). Oxidation at temperatures above 1000 °C initially leads to the formation of a potentially protective surface film of (pure) silicon dioxide on the silicon nitride. However, this phase is not in equilibrium with the intergranular silicates formed by the liquid-forming metal oxides, and interdiffusion occurs in an attempt to bring the system to equilibrium. The metal ion concentration gradient between the interior of the material and the external silicon dioxide is responsible for the outwards leakage, by diffusion through the intergranular silicate, of cations (with their balancing oxide ions) towards the silicon dioxide. This process leads to the production of surface silicates of greater permeability towards oxygen (Clarke and Lange, 1980) and is shown schematically in Fig. 5.27 (see also Fig. 4.10). For this reason the oxidation rates of sintered (and hot-pressed) silicon nitride materials can, under the same conditions, be several orders of magnitude faster than those of pure silicon nitride (Fig. 5.28).

The oxidation of liquid phase sintered silicon nitride materials containing intergranular silicates could be regarded as a form of corrosion, in which the supply of metal oxide is available internally, rather than being external. The acceleration of oxidation rate depends on the type, and amount, of intergranular phase, and the nature and stability of the product silicate. It would be expected that if the concentration of subsurface intergranular cation could be reduced, a decreased rate of oxidation would be obtained. This is what is seen, together with an improvement in high-temperature strength, after a pre-oxidation treatment, following which the oxidised surface film is removed by grinding (Lange, 1984).

The oxidation of many nitrides leads to a destructive volume expansion, because of the relatively low densities of silicon dioxide, and silicon oxynitride (Table 5.4). Where this expansion takes place within a (new) surface film, the associated stresses can be accommodated. Expansions taking place within the subsurface grain boundaries, caused by the internal oxidation of the boundary oxynitride glass, or secondary crystalline phases, cannot. The resulting strains can lead to very high surface stresses, and the catastrophic failure of the material (Gazza, 1975). This effect was first demonstrated for silicon nitride densified with yttrium oxide, where the overall composition was such that yttrium silicon oxynitride ($Y_2Si_3O_3N_4$ – *melilite*) crystallised within the boundaries. To avoid this type of phenomenon, it is necessary to ensure that the densification additive

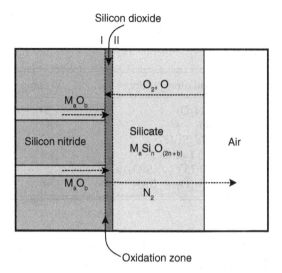

Figure 5.27 The oxidation of silicon nitride containing intergranular silicate. The metal oxide concentration gradient from the interior to the surface causes the outwards diffusion of metal ions down the grain boundaries, and the breakdown of the silicon dioxide oxidation barrier with the formation of silicates.

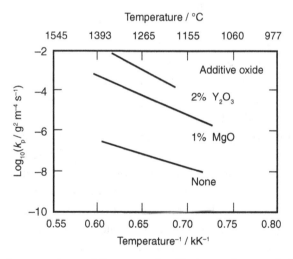

Figure 5.28 A comparison of the ease of oxidation, using parabolic rate constants, of pure silicon nitride, and silicon nitride containing liquid phase sintering additives: expanded detail of Fig. 5.8. The grain boundary silicate facilitates the inwards diffusion of oxygen to the nitride.

Table 5.4 *Volume expansions on oxidation of common components of sintered silicon nitride systems, showing the potential for disruption of the material. (After Lange et al., 1977.)*

Phase	Oxidation product	Volume change / %
Si_3N_4	Si_2N_2O	-20
Si_3N_4	SiO_2	$+87$
$YSiO_2N$	$0.5Y_2Si_2O_7$	$+12$
$Y_5(SiO_4)_3N$	$0.75Y_2SiO_5 + 0.75Y_{4.67}(SiO_4)_3O$	$+4$
$Y_2Si_3O_3N_4$	$Y_2Si_2O_7 + SiO_2$	$+30$

has a sufficiently high SiO_2/Y_2O_3 ratio (Lange *et al.*, 1977). This provides a good illustration of the significance of small variations in the silicon dioxide content of a silicon nitride powder for the final microstructure and properties of the material, and the importance of relatively small amounts of secondary phase material for a material's high-temperature properties.

Wear

Dense, fine-grain, sintered silicon nitride also has very good resistance to wear of all types, and in sliding mode, low friction coefficients. The reasons for this are not entirely clear, but are certainly related to the intrinsic hardness of silicon nitride, and to the small grain size, and high fracture toughness. However, because temperatures can readily exceed 1000 °C during sliding wear, combined with the natural chemical reactivity of silicon nitride, interactions with the counter-material may occur. Surface oxidation is therefore often an important component of the wear process, leading to the production of silicon dioxide or silicate wear films and debris, and under wet conditions slow solution of the silicon nitride may occur. These processes, involving chemical reactions, are usually referred to as *tribochemical* wear. The normal microfracture processes lead to grain detachment and the physical loss of material. In these respects the behaviour of silicon nitride is similar to that of silicon carbide, and comparisons were made in Chapter 4. Reactions may also occur between the relatively unstable silicon nitride and ferrous alloys, which tend to react with solution of nitrogen into the alloy, and the formation of low melting ferrosilicon phases. This leads to the rapid loss of material, and faster wear rates (Fischer and Tomizawa, 1985; Horton *et al.*, 1986; Gee and Butterfield, 1993; Dong and Jahnmir, 1999). A comparison of the rates of wear of dense silicon nitride with silicon carbide and alumina was shown in Fig. 4.28.

5.5.3 Hot-pressed silicon nitride

This material is included here briefly, simply because historically it represented an important stage in the development of dense silicon nitride ceramics. However, with the subsequent appearance of pressureless sintered silicon nitride of high quality, the initial advantages of the hot-pressed form have been submerged, and its use for the production of silicon nitride ceramics is largely (though not entirely) superseded. The properties of uniaxially hot-pressed silicon nitride can be taken, on an equal density basis, to be virtually identical to those of pressureless sintered materials. The materials are usually fully dense, and their microstructures reflect to some extent the uniaxial nature of the applied stress, with its influence on grain orientation: grain growth tends to take place along axes normal to the hot-pressing axis, because of the pressure gradients within individual grains. This means that there will be corresponding degrees of anisotropy in mechanical properties of strength and toughness, and in related properties such as wear resistance (Weston, 1980; Lee and Bowman, 1992). Where highly isotropic dense materials are required, in the production of ball bearings for example, described in the section dealing with applications, hot isostatic pressing may be preferred. In this case the component may be first sintered to the stage where all the porosity is closed (corresponding to around 90% of theoretical density) and then hot pressed under high gas pressures to full density. Where hot pressing still retains advantages is in the production of silicon nitride matrix composite materials, such as those containing silicon carbide particles, for which the application of pressure is useful in accelerating densification.

5.6 Sialons

The sialons, introduced above, require more explanation and a separate section. *Sialon* is the term often used for a very large class of materials, related to silicon nitride in that all (or almost all – the meaning is stretched occasionally) contain silicon and nitrogen (Jack, 1976, 1993). More formally, sialons could be termed *aluminium silicon nitride oxides* (or *oxynitrides*), and can be considered to bridge the oxide and non-oxide ceramics. An alternative view is that these phases should be regarded simply as β-silicon nitride solid solutions, or α-silicon nitride solid solutions (Petzow and Herrmann, 2002).

The range of known sialon phases is very large, and attention here is focused mainly on the β'-sialon phase because it can be the major phase in a liquid phase sintered silicon nitride. Details of others can be found in literature reviews (Ekstrøm and Nygren, 1992). The β'-sialon phase was first identified in 1971, as a product of the reaction at 1750–1800 °C between silicon nitride and aluminium oxide. β'-Sialon has variable composition (as do many sialons), and was at first thought to be a solid solution of aluminium oxide in silicon nitride and for a time referred to as an *alloy*, borrowing the metallurgical term (Jack, 1973), although after a more full investigation this description turned out not to be quite accurate. The maximum aluminium oxide content of the β'-sialon phase corresponds to about 70% by weight. A simple solid solution of Al_2O_3 in Si_3N_4, with Al atoms on Si sites, and O atoms on N sites, would require the compositions to lie on the Si_3N_4–Al_2O_3 join, but this was found not exactly to be the case. Compositions are more accurately expressed by $Si_{(3-x)}Al_xO_xN_{(4-x)}$, where x takes values between 0 and about 2.1. This means that small amounts of Si and O must be lost at high temperature during the preparation process (which is not unlikely, given the volatility of silicon monoxide, SiO). The metal/non-metal ratio is maintained constant at 3:4, with the other end member at $x = 3$, and in principle aluminium oxynitride, Al_3O_3N (though the series is not continuous, and there is a break of compositions for x values between 2.1 and 3).

β'-Sialon has the same basic crystal structure as β-Si_3N_4, in which some of the silicon atoms (up to about half) can be replaced by aluminium atoms, and nitrogen by oxygen (to maintain, at least formally, charge neutrality – species are usually treated as though they have their normal ionic charges). This compositional range exists because the chemistries of aluminium and silicon, and the Si–N and Al–O bond lengths, are similar (~175 pm). The Al–N (187 pm) and Si–O (162 pm) bond lengths are not too different either; an Al–O unit therefore seems able to replace an Si–N unit in the β-nitride crystal structure without introducing too much lattice strain, and raising the energy of the structure. Because the hexagonal crystallographic unit cell of β-silicon nitride contains two Si_3N_4 units, that is, it can also be represented as "Si_6N_8", the β'-sialon phase is more often written as $Si_{(6-z)}Al_zO_zN_{(8-z)}$, where the maximum value of z is about 4.2.

One way of representing the simple β'-sialon system, in a type of phase-equilibrium diagram, is shown in Fig. 5.29 (Gauckler, 1975). The diagram corresponds to a section through a tetrahedron, the corners of which are the four component elements, Al, Si, N, and O. The axes of this diagram are unusual in that they use equivalents (%): this means that the compositional points represented by the axes are of constant *equivalent*, that is, constant formal positive or negative ionic charge (rather than of constant weight normally used in phase diagrams). The charge on each component is (12+) or (12−), and this accounts for the rather odd way of expressing the basic components and phases. Each primary component has 4Al (charge 3^+) or 3Si (charge 4^+) atoms, and 4N (3^-) or 6O (2^-) atoms. For example, in one molecular unit of mullite ($3Al_2O_3.2SiO_2$ or $Al_6Si_2O_{13}$) there are 26 units of charge on the metallic or oxygen atoms, so that to obtain 12 units of

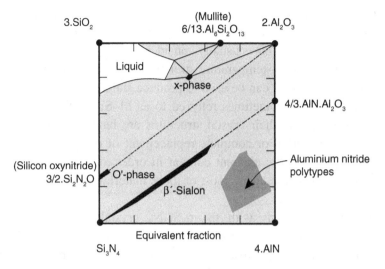

Figure 5.29 The Si_3N_4–SiO_2–Al_2O_3–AlN phase equilibrium system: the axes are expressed as *equivalent fractions*. This means that all compositions are adjusted so that each contains, notionally, 12 positive units of charge, and 12 negative units. The liquid region is at 1700 °C. (After Jack, 1983.)

charge 6/13 of the molecular unit is needed. Calculations of composition on the basis of mass then use quantities of material calculated from these numbers of moles. This simplified diagram shows the main phases, and the existence of a region which is liquid at 1700 °C, extending from the SiO_2 corner (the simple Al_2O_3–SiO_2 diagram in Fig. 2.5 shows the binary eutectic temperature at 1587 °C). This liquid would certainly assist the sintering densification of a mixture of silicon nitride and aluminium oxide powders, but in practice densification is still very slow unless other additive metal oxides (MgO for example) are also incorporated, to increase the liquid volume and decrease its viscosity.

The *AlN polytypes* shown in this diagram near the aluminium nitride corner are a number of structurally similar phases with constant metal/non-metal ratios. They are not polytypes in the strictest sense, because they have variable compositions. The term *polytypoid* has been suggested by the International Union of Crystallography.

The practical consequence of sintering silicon nitride powder with a liquid forming additive system (MgO for example) which also contains aluminium oxide, is that the product "β-Si_3N_4" phase will inevitably contain some lattice Al and O, and thus formally be a (normally "low z-value") β'-sialon. This situation can be described as the *partition* of the Al and O atoms between the intergranular phases and the silicon nitride phase, where the proportion transferred is determined in part by the initial chemical composition of the intergranular liquid phase.

A further consequence is that, through this transfer during or at the completion of sintering, the volume of liquid phase is decreased by an amount corresponding to the loss of the Al and O. This is usually considered to be desirable from the point of view of improving high-temperature strength and creep resistance.

Other metallic elements can be used to produce sialons. Where metals besides Al are present they are sometimes referred to as M–Si–Al–O–N phases. Many sialon compositions and their crystal structures are based on those of the aluminosilicates, with partial or complete replacement of oxygen by nitrogen, and adjustments to the metallic element content in order to preserve formal charge neutrality (*formal*, because again, most of these materials are probably to a large extent covalent).

Corresponding α'-sialons with the α-Si_3N_4 crystal structure can also be obtained, but only when silicon nitride powder is sintered (and chemically reacted) with aluminium oxide in the presence of an oxide of a second metal (M), such as lithium, calcium, yttrium, or a lanthanide (and a wide range of lanthanide elements has been investigated). The α-Si_3N_4 crystal structure is able to accommodate not only small amounts of oxygen (though probably not more than one O per unit cell of $Si_{12}N_{16}$), but also a range of metallic elements, provided their atomic radii are within specific limits. These metal atoms occupy the large cellular interstices of the (Si,Al)–(N,O) network (Hampshire *et al.*, 1978). It is therefore possible to think of the binary nitride, Si_3N_4, as a special case of the general class of "sialon", in that it is one limit to a range of compositions, in which the M, Al, and O contents happen to be zero. Some types of liquid phase sintered silicon nitride are referred to as "sialons", for the reason that the aluminium oxide often used as a sintering additive introduces aluminium and oxygen into the phases present in the sintered material. Indeed, because it is difficult to exclude traces of aluminium as an impurity in silicon nitride powders (or accidental introduction during powder preparation and sintering), the dividing line between a liquid phase sintered silicon nitride and a liquid phase sintered sialon becomes rather blurred. It is therefore not particularly helpful to try to distinguish between sintered *silicon nitride* and *sialon*, used simply as materials' descriptions. From the point of view of microstructure and related properties, however, the presence of aluminium oxide in the system can make a very important difference.

Solid solution phases with the α-Si_3N_4 structure, the α'-sialons, have more complex compositions and contain, typically, aluminium, oxygen, and a metal (M) of larger atomic size, such as calcium or yttrium. There is, in effect, a partial replacement of Si by Al in the α'-Si_3N_4 lattice, with charge balance (formally) being achieved by a combination of the replacement of N^{3-} by O^{2-}, and the insertion of the modifier metallic ions M. Because these phases now contain another metal atom, three-dimensional diagrams (known as Janecke prisms) must

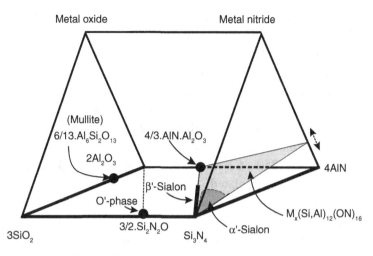

Figure 5.30 The α'- and β'-sialon behaviour diagram. This is the three-dimensional diagram, conventionally drawn as the pyramid, with six primary components, four of the Si_3N_4–SiO_2–Al_2O_3–AlN system, and a new metal oxide and its nitride. The axes expressed as equivalent fractions again correspond to 12 units of notional charge. The α'-sialon solid solutions lie on a small plane in this diagram at the Si_3N_4 corner. (After Jack, 1983.)

be used to show phase relationships. The α'-sialon compositions lie on a small two-dimensional triangular plane lifted off the β'-sialon base plane, with corners at Si_3N_4, Al_3NO_3 and (Al, M)N (Fig. 5.30). Compositions can be described by the rather complicated general formula $M_x(Si_{12-(m+n)}Al_{(m+n)}O_nN_{(16-n)})$, where x is determined both by the valence of M, and by m, in order to maintain charge neutrality. Unlike the β'-sialons, there appears to be a small compositional gap between α-silicon nitride and α'-sialon, with the smallest value of m of ~1. Some α'-sialon compositions are stable only at higher temperatures (typically >1600 °C), and revert on annealing at lower temperatures to a β'-sialon with the release of excess metal nitride to the intergranular liquid. The α'–β'-sialon transformation is reversible (unlike that of silicon nitride), and like the α-Si_3N_4 to β-Si_3N_4 transformation it is associated with important microstructural changes, in that the α'-sialons tend to be of equiaxed morphology, while the β'-sialon grains are of higher aspect ratio. However, α'-sialon grains can also be produced with high aspect ratio morphology, so that high toughness and strength can be obtained (Chen and Rosenflanz, 1997; Kim *et al.*, 2000). In principle therefore it should be possible to use the transformations, and the conversion of a predominantly α-silicon nitride starting powder to α'- and β'-sialons, to develop and to optimise the composition and morphology of the dense product (Mandal *et al.*, 1993; Mandal *et al.*, 1999).

Fully dense α'-sialons are readily obtained by pressureless sintering appropriate mixtures of oxide and nitride powders under a controlled atmosphere. The physical properties of the most commonly produced lanthanide oxide α'-sialons seem to be very similar to those of materials consisting predominantly of α-silicon nitride, and many materials appear to be thermally stable under inert atmospheres up to 1750 °C. The Vickers indentation hardness is reported to be in the region of 22 GPa, and they have the mechanical properties and thermal shock resistance expected from a fine-grain, low expansion coefficient ($\sim$3.4 MK^{-1} over 0–1250 °C) material. Mechanical properties of fracture toughness and strength are related closely to the microstructure, and in particular to the proportion of high aspect ratio grains (Cao and Metselaar, 1991; Mandal, 1999).

It is clear that the α'- and β'-sialon systems containing Al_2O_3 and other metal oxides, in contrast with simpler silicon nitride materials, have a wider and more complex range of phase relationships. These offer considerable scope for developing specific microstructures, which can be used to improve mechanical properties. The transfer of information from the laboratory to the larger-scale, less well-controlled, commercial production of materials presents a considerable challenge.

5.7 Applications

All three of the major forms of silicon nitride ceramic (reaction bonded, pressureless sintered, hot pressed) have current commercial applications; noncrystalline, thin films of silicon nitride, though of considerable importance in their own right, are not discussed here. As was the case for silicon carbide, this section, for convenience, will be arranged by application, rather than silicon nitride type. In practice no real distinction is made between dense silicon nitrides sintered with aluminium oxide as one of the liquid-forming additives (and therefore formally *sialons*) and those not; both types will be regarded as silicon nitride.

Silicon nitride materials have a range of small but important niche markets. Although the total quantity sold accounts for less than 1% by weight of the European and USA markets for four of the major structural ceramics considered in this volume (porcelain is excluded in this assessment because of its large domestic market), the products are generally of high added value, and the proportion is estimated to be around 8% by value (Briggs, 2007). Several reviews and surveys outline the general application areas for these materials (Lange *et al.*, 1991; Katz, 1993; Popper, 1994).

5.7.1 General industrial

The earliest reported applications for silicon nitride ceramics were in what can be regarded as the metal processing fields. The silicon nitride then available was the porous reaction-bonded type, with its advantage of the relatively easy forming (for a ceramic) of complex shapes; in fact with complexity matching those of typical metallic articles. Silicon nitride was regarded as a versatile refractory material with good resistance to thermal shock. It had the additional very useful ability to resist attack by molten aluminium, which is corrosive towards oxide refractories because of its ability to abstract their oxygen (as noted, aluminium oxide is thermodynamically one of the most stable oxides). Early applications providing an incentive to develop reaction-bonded silicon nitride were in the aluminium casting industry, where there was a need to pump and channel high-purity molten aluminium. Metals released from oxide refractories can lead to undesirable contamination of the aluminium. Tubes and other simple pump components are readily manufactured in reaction-bonded silicon nitride, by shaping and then nitriding compacted silicon powder. Accurate machining (using conventional metal machining techniques) of a compacted silicon powder billet is possible, and because of the virtual absence of dimensional change during nitridation, the amount of final diamond machining required is minimal. The strength of reaction-bonded silicon nitride, though much less than that of the fully dense forms, is quite adequate for these applications (it is on a level with that of normal liquid phase sintered alumina). Figures 5.31 and 5.32 show the elements making up the cost of producing in small numbers a simple (tubular) component in reaction-bonded, and sintered, silicon nitride.

Of course, the actual cost of a component depends considerably on other factors, such as its complexity, and the scale of production envisaged, but a significant part of the production cost in all cases is in the final precision diamond machining. Reducing the requirement for dimensional tolerance decreases the cost considerably. Another successful early application for reaction-bonded silicon nitride was in tubular sheaths for thermocouples inserted into molten metals where protection of the metallic thermocouple is required. These applications depended to a large extent on the ability of silicon nitride to withstand violent change of temperature (up to 1500 °C), and to resist attack (at least in the short-term) by the alloy. Again, very high mechanical strength was not usually necessary. Figure 5.33 shows a small selection of components used in these types of application, fabricated from reaction-bonded silicon nitride.

Reaction-bonded silicon nitride has also been considered for many other small-scale applications, where a reasonably strong and high melting-point material is

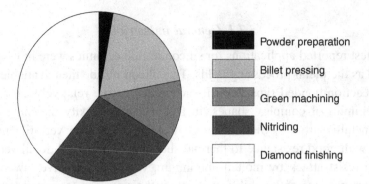

Figure 5.31 An approximate indication of the breakdown of the costs of producing a simple shape in reaction-bonded silicon nitride. (Courtesy of Robert Wordsworth.)

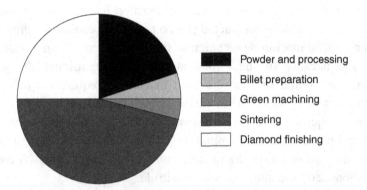

Figure 5.32 An approximate indication of the breakdown of the costs of producing a simple shape in sintered silicon nitride. (Courtesy of Robert Wordsworth.)

required with better thermal shock resistance than alumina. It has, for example, been successfully marketed as the small conical, threaded, end-pieces of inert gas welding torches, which experience sudden changes of temperature as the torch is turned on and off. It has also been used as an electrical heating element support material where exceptionally high reliability is needed. In these cases it is electrical resistivity, and ability to withstand high temperatures, which are important. The mechanical load is almost negligible, though the thermal shocks can be considerable. All these application requirements are easily met by the porous reaction-bonded form of material, and there is no need for the much more

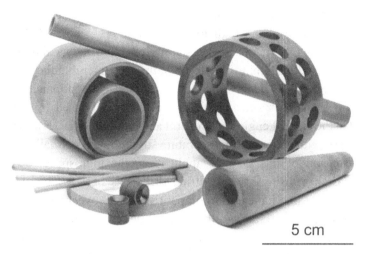

5 cm

Figure 5.33 An illustration of the range of reaction-bonded silicon nitride components, used in metal forming operations. (After Riley, 2000. Reprinted with permission of Wiley-Blackwell Publishing.)

expensive fully dense sintered, or hot-pressed, forms. The porosity in itself is not a problem, because impermeability towards gases, or very low viscosity fluids, is not required. Some internal oxidation, within the pores, may occur at high temperatures under oxidising conditions, but again this is not normally a serious problem.

5.7.2 Metal working

The dense, sintered or hot-pressed, forms of silicon nitride have a very fine-grain microstructure, which gives very high strength and good resistance to wear (Katz, 1994, 1999). Careful refinement of the microstructure, to ensure a high proportion of high aspect ratio grains, provides very good toughness. The ability of silicon nitride (as a sialon) to cut metals at high speeds was shown in the 1970s (Lumby *et al.*, 1978; Lumby, 1991), and metal machining now provides an important market for the fully dense materials. The silicon nitride piece is usually in the form of a small (1–2 cm dimension) tablet, with varying numbers of corners, which is clamped in a holder. It is disposed of when wear or damage becomes excessive. Chemical reactions with ferrous alloys at high temperature (generated by friction) can cause loss of cutting tool material, so that the composition of the alloy, and the temperatures reached during cutting, must be taken into account in the choice of materials, and conditions of use. Silicon nitride based tools are routinely used for the very high-speed turning and milling of brittle, low-to-medium strength, cast iron (though "low" strength, here, means ~200 MPa ultimate tensile strength).

However, the stronger and tougher steels generally prove to be too reactive towards silicon nitride at the high local temperatures generated. Some silicon nitride compositions, particularly the sialons prepared using aluminium oxide densification aid, with aluminosilicate or aluminate intergranular phases, show definite advantages over dense silicon nitride (and conventional metal-cutting materials), for the high-speed machining of the nickel-based (nimonic) alloys widely used in the aircraft engine, and the oil and gas, industries. For these types of alloy, sialon tips with better high-temperature strength, will normally be preferred.

As an extension of this type of application, the high-speed, and accurate, drilling of cast iron using silicon nitride drills has been explored with some success. There is a fine balance in competitiveness between silicon nitride and conventional tungsten carbide, but the possibilities for the uses of silicon nitride in areas previously dominated by metals and hard metals (such as tungsten carbide) are clearly illustrated.

5.7.3 Bearings

Fully dense silicon nitride has considerable potential advantages over bearing steel (Katz, 1994). The much greater hardness, lower density and higher stiffness of silicon nitride compared to steel can result in lower wear rates, and less vibration during high-speed running. A lower ball density means lower centrifugal loading, and reduced stresses between the ball and the outer race, which in turn reduces friction and wear. The strength and fracture toughness of high-quality silicon nitride are easily able to support the contact stresses developed in the bearing, particularly at high speeds. Silicon nitride also tolerates limited periods of dry running (resulting from lubrication failure) better than steel. Although it is possible to produce all-ceramic ball (or roller) bearing systems, these are expensive, and normal practice has been to use silicon nitride balls or rollers in a steel support ring or cage – the *hybrid* bearing (Horton, 1991). Interestingly, the use of mixed balls – a small number (as low as one or two) of silicon nitride, with the majority steel – can also give significant improvements compared with the corresponding pure steel ball system. This is possibly because the hard silicon nitride ball flattens the normal wear debris on to the race surface, to minimise further abrasive wear.

It is not difficult to produce highly polished silicon nitride balls and rollers of very good dimensional tolerances (in fact they can be made smoother than steel balls), but it is generally found that an initial hot-isostatic pressing (though relatively expensive) provides dense blanks with highly isotropic microstructures, which assist the accuracy of the subsequent grinding processes, and thus the roundness of the ball. The use of a pressure-assisted densification process has the further advantage that lower levels of liquid-forming sintering aids are required, giving smaller volumes of intergranular phase and better high-temperature strengths. While silicon nitride balls and rollers are considerably more expensive (of the order of 5–10 times) than standard steel components, their advantages in terms of bearing running characteristics and life can make the increased overall cost of the hybrid bearing acceptable.

Silicon nitride hybrid bearings are used where maximum stiffness and minimum vibration are necessary. An important example of this kind of application is in machine tools for the high-speed precision machining of metals. Silicon nitride bearing systems are also used in a wide range of smaller scale, specialised, applications, such as turbomolecular pumps operating in corrosive environments, and in precision instruments. The electrical insulation characteristics of silicon nitride bearings provide a possible advantage in electric motors, as a means of reducing internal arcing. Ceramic bearings are being considered by the aviation industry for use in flap control motors, because of their longer life, and reliability.

Figure 5.34 shows a selection of small components produced in dense sintered silicon nitride, for a wide range of applications. Easily identifiable are small cutting tool inserts, ball bearings, and pump sealing rings. Other components are intended for use in automobile engines in valve mechanisms, as parts of the valve mechanism where resistance to sliding wear is needed.

5.7.4 Transport

In the early 1970s silicon nitride was singled out as a promising light-weight material, for use as hot-zone components in new ranges of higher-temperature and (therefore) higher-efficiency gas-turbine and diesel engines. Much of the research activity on silicon nitride carried out in the 1970s and 1980s was directed towards these ambitious applications. The very attractive features of silicon nitride, apart from its low density, were its high mechanical strength, a good proportion of which could be maintained to 1300 °C, and its good resistance to thermal shock. Potentially, it had much better high-temperature properties than the nimonic alloys (*super-alloys*) based on nickel and cobalt, and 2500 °F (~1371 °C) was set in the USA as a target (gas inlet) temperature for the gas-turbine engine. Test engines using silicon nitride hot-zone components were developed and satisfactorily evaluated, but interest began to decline in the more ambitious types of application when it was realised that major problems would be the high cost of the mass production of components of adequate strength and long-term reliability (Komeya, 1993; Heinrich and Krüner, 1994). Figure 5.35 shows an example of an early small nozzle guide vane assembly for a ceramic gas turbine (intended to direct hot gases on to the rotor blades) made in reaction-bonded silicon nitride, illustrating the versatility of the slip casting process. Difficulty in

Figure 5.34 An illustration of the range of components in dense, sintered silicon nitride. (Reproduced by kind permission of Ceradyne Inc.)

obtaining consistency and reliability is not a problem unique to silicon nitride, but because of the inherent difficulty of liquid phase sintering silicon nitride powder to full density, and the tendency for the microstructure of the compacted silicon powder to be retained in reaction-bonded silicon nitride, both forms of material are particularly unforgiving of defective powder processing. However, development work has continued over a number of years on a range of less severely stressed components in reciprocating and gas-turbine engines. Examples of reciprocating engine components, which could potentially be partially or entirely replaced by silicon nitride, include valves, and associated operating components (Woetting *et al.*, 1996), turbocharger rotors, and diesel glow plugs and swirl chambers. Of the materials' advantages, in addition to low density and resistance to high temperature, low thermal conductivity to reduce heat losses, and wear resistance, were seen as potentially important. These possibilities were all explored with varying degrees of success, leading in some cases to moderate commercial use. However, the high costs of silicon nitride powder, and of

Figure 5.35 A demonstration turbine nozzle guide vane ring, in reaction-bonded silicon nitride.

machining sintered material, have provided barriers to the commercial production of engine valves. On the other hand, smaller and simpler valve lifting components such as rocker arm tips, and cam rollers (which have used the sintered reaction-bonded form), are currently used in selected ranges of diesel engine. Additional reasons for considering silicon nitride for these components are its lower friction, and better wear resistance, than the standard metallic component.

5.8 Summary

Silicon nitride has provided the second example of a non-oxide structural ceramic, and makes a good subject for study (Lange, 2006). In common with alumina, a liquid phase (a silicate) is normally used to assist sintering. A combination of the flexibility of the liquid phase composition, and the silicon nitride crystal chemistry, allows the development of β-phase grains of high aspect ratio, which give strength and toughness values approaching those of the transformation toughened zirconia ceramics (Chapter 6). Because silicon nitride is also very hard, it has very good wear resistance and, like sintered alumina, sintered silicon nitride has important uses as metal cutting inserts and bearings (though the tougher silicon nitride is used more widely in large ball and roller bearings).

The lower density reaction-bonded form (like that of silicon carbide), with the absence of sintering shrinkage during silicon nitridation, allows detailed shaping to be carried out at the powder stage. This similarly makes the material well-suited for the production of large complex-shape components. Also in common with silicon carbide, silicon nitride has very good resistance to thermal shock, with the additional useful feature of being an electrical insulator. A complete new large field of related silicon oxynitride (or sialon) materials was opened up by the incorporation of aluminium oxide into the silicon nitride system, and the extra opportunities for phase and microstructural control provided by the α'- and β'- sialons give these systems a considerable potential advantage over the simpler silicon nitrides.

A large amount of time, energy and funding has been spent on silicon nitride over the last 40 years, partly because of the attractive vision that ceramics might be used, for the first time, as load-bearing hot-zone components in new higher efficiency engines, and the prospect of establishing the position of ceramics as "serious" engineering materials. There was also the intellectual challenge of fabricating components in a nitride (as opposed to the usual oxide) ceramic, the behaviour of which was not properly understood. Particular issues included the reaction-bonding process, the difficulty of sintering silicon nitride powder to full density, and in producing very high-strength materials. Failure to achieve (so far) the commercial realisation of a ceramic gas-turbine engine is more a consequence of the economics of producing silicon nitride (and silicon carbide) components of the required degree of reliability at the right price, than with any fundamental material's property. The position may change, however, with the now urgent needs to conserve carbon-based fuels, and to improve the efficiency of their use. While silicon nitride was first seen primarily as a thermal shock-resistant material for applications at very high temperature, many of the current applications rely on the wear resistance of the dense forms of the material at room temperature; it is the reaction-bonded form which has the higher temperature applications. Like silicon carbide, the silicon nitride systems have shown the importance of the non-oxides as structural ceramics, and the possibilities for developing quite sophisticated microstructures in order to achieve their best mechanical properties.

Questions

5.1. What is the volume expansion when a crystal of silicon reacts with nitrogen to form silicon nitride? Assume that the density of silicon is 2.33 Mg m^{-3}.

5.2. Silicon powder is compacted to a density of 60% of theoretical, and then fully nitrided. What is the bulk density of the reaction-bonded silicon nitride product?

5.3. How does the microstructure of sintered silicon nitride (Fig. 5.22) account for the higher fracture toughness values obtainable with this type of material, compared with a liquid phase sintered alumina?

5.4. Show that the volume change on oxidising silicon nitride to amorphous silicon dioxide (density ~2.2 Mg m^{-3}) is about +87%. How would you expect this increase to be accommodated during the oxidation of a liquid phase silicon nitride?

5.5. Give reasons for the differences in strength which might be expected between well-made, pore-free, bars of solid state sintered silicon carbide and sintered silicon nitride (a) at room temperature, (b) at 1400 °C.

5.6. Figure 5.9 shows that the larger oxygen (O_2) molecule appears to be more mobile in a silicate glass, than a small oxygen (O) atom. Why is this?

5.7. Suggest why pure silicon carbide is better at resisting oxidation than pure silicon nitride.

5.8. On the basis of the thermal shock R parameters, which might be better at maintaining strength after a sudden severe change in temperature, sintered silicon nitride or sintered silicon carbide?

5.9. Suggest reasons why the silicon nitride structure is able to accommodate aluminium and oxygen to form the silicons. What difference between the α-silicon nitride and β-silicon nitride crystal structures allows the α'-sialons containing large metal cations, to be formed?

5.10. Liquid phase sintered silicon nitride and aluminous porcelain can both be regarded as constructed from crystalline phases of high aspect-ratio morphologies, together with a bonding glass phase. Explain why they have quite different mechanical properties.

Selected reading

Lange, F. F. (2006). The sophistication of ceramic science through silicon nitride studies. *J. Ceram. Soc. Japan*, **114**, 873–9.

Petzow, G. and Herrmann, M. (2002). Silicon nitride ceramics. In *High Performance Non-oxide Ceramics II*, eds. M. A. Jansen and R. Haubner, *Structure and Bonding*, Vol. 102. Berlin: Springer Verlag, pp. 47–166.

Riley, F. L. (2000). Silicon nitride and related materials. *J. Amer. Ceram. Soc. Centennial Review*, **83**, 245–65.

Torti, M. L. (1989). The silicon nitride and sialon families of structural ceramics. In *Structural Ceramics*, ed. J. B. Wachtman Jr., *Treatise on Materials Science and Technology*, Vol. **29**. Boston, MA: Academic Press Inc., pp. 161–94.

References

Airey, A. C., Clarke, S. and Popper, P. (1973). Pyrolytic silicon nitride coatings. *Proc. Brit. Ceram. Soc.*, **22**, 305–20.

Alliegro, R. A., Coffin, L. B. and Tinkelpaugh, J. R. (1956). Pressure sintered silicon carbide. *J. Am. Ceram. Soc.*, **39**, 386–9.

Ashby, M. F. (1974). First report on sintering diagrams. *Acta Metall.*, **22**, 275–89.

Ashcroft, W. (1973). Mechanical properties of silicon nitride at elevated temperatures. *Proc. Br. Ceram. Soc.*, **22**, 169–79.

Ashcroft, W. (1975). The tensile and bend strengths of silicon nitride and hot-pressed silicon carbide. In *Special Ceramics 6*, ed. P. Popper. Manchester: The British Ceramic Research Association, pp. 245–60.

Atkinson, A., Leatt, P. J. and Moulson, A. J. (1974). Mechanism for nitridation of silicon powder compacts. *J. Mater. Sci.*, **9**, 981–4.

Atkinson, A., Moulson, A. J. and Roberts, E. W. (1976). Nitridation of high-purity silicon. *J. Amer. Ceram. Soc.*, **59**, 285–9.

Barnby, J. T. and Taylor, R. A. (1972). The fracture resistance of reaction-sintered silicon nitride. In *Special Ceramics 5*, ed. P. Popper. Manchester: The British Ceramic Society, pp. 311–28.

Bauer, R. A., Brecht, J. G. M. and Kruis, F. E. (1991). Laser synthesis of low-agglomerated submicrometre silicon nitride powders from chlorinated silanes. *J. Am. Ceram. Soc.*, **74**, 2759–68.

Becher, P. F. (1991). Microstructural design of toughened ceramics. *J. Am. Ceram. Soc.*, **74**, 255–69.

Bennett, M. J. and Houlton, M. R. (1979). Interaction between silicon nitride and several iron, nickel and molybdenum-based alloys. *J. Mater. Sci.*, **14**, 184–96.

Billy, M. (1988). Reactivity in nitrogen ceramics. *Sci. Ceram.*, **14**, ed. D. Taylor. Stoke-on-Trent: Institute of Ceramics, pp. 45–60.

Bowen, L. J. and Carruthers, T. G. (1978). Development of strength in hot-pressed silicon nitride. *J. Mater. Sci.*, **13**, 684–7.

Briggs, J. (2007). *Engineering Ceramics in Europe and the USA*. Worcester, UK: Menith Wood.

Brook, R. J., Carruthers, T. G., Bowen, L. J. and Weston, R. J. (1977). Mass transport in the hot-pressing of α-silicon nitride. In *Nitrogen Ceramics*, ed. F. L. Riley, NATO ASI Series E: Applied Science No. 23. Leyden: Noordhoff, pp. 383–92.

Cao, G. Z. and Metselaar, R. (1991). α′-Sialon ceramics – a review. *Chem. Mater.*, **3**, 242–52.

Chase, M. W., Ed. (1998). *NIST-JANAF Thermochemical Tables. Journal of Physical and Chemical Reference Data – Monograph – No. 9*. Washington, DC: American Chemical Society and the American Institute of Physics.

Chen, I. W. and Rosenflanz, A. (1997). A tough SiAlON ceramic based on alpha-silicon nitride with a whisker-like morphology. *Nature*, **389**, 701–4.

Clarke, D. R. and Lange, F. F. (1980). Oxidation of Si_3N_4 alloy: relationships and phase equilibria in the Si_3N_4–SiO_2–MgO system. *J. Am. Ceram. Soc.*, **63**, 586–93.

Collins, J. F. and Gerby, R. W. (1955). New refractory uses for silicon nitride reported. *J. Met.*, **7**, 612–15.

Cubicciotti, D. and Lau, K. H. (1978). Kinetics of oxidation of hot-pressed silicon nitride containing magnesia. *J. Am. Ceram. Soc.*, **63**, 512–17.

Davidge, R. W. (1977). Economic and energetic considerations for nitrogen ceramics. In *Nitrogen Ceramics*, ed. F. L. Riley, ASI Series E: Applied Science No. 23. Leyden: Noordhoff, pp. 653–7.

Davidge, R. W., Evans, A. G., Gilling, D. and Wilyman, P. R. (1972). Oxidation of reaction-sintered silicon nitride and effects on strength. In *Special Ceramics 5*, ed. P. Popper. Manchester: The British Ceramic Society, pp. 329–43.

Deeley, G. G., Herbert, J. M. and Moore, N. C. (1960). Dense silicon nitride. *Powder Metall.*, **8**, 145–56.

Dong, X. and Jahanmir, S. (1999). Wear transition diagram for silicon nitride. *Wear*, **165**, 1579–97.

Eddington, J. W., Rowcliffe, D. J. and Henshall, J. (1975). The Mechanical properties of silicon nitride and silicon carbide: Part I Materials and Strength. *Powder Metall. Int.*, **7**, 82–96. See also: Eddington, J. W., Roweliffe, D. J. and Henshall, J. (1975). The Mechanical properties of silicon nitride and silicon carbide: Part II Enginering properties. *Powder Metall. Int.*, **7**, 136–47.

Ekstrøm, T. and Nygren, M. (1992). Sialon ceramics. *J. Am. Ceram. Soc.*, **75**, 259–76.

Fischer, T. E. and Tomizawa, H. (1985). Interaction of tribochemistry and microfracture in the friction and wear of silicon nitride. *Wear*, **105**, 29–45.

Gauckler, L. J. (1975). Contribution to phase diagram Si_3N_4–AlN–Al_2O_3–SiO_2. *J. Amer. Ceram. Soc.*, **58**, 346–7.

Gauckler, L. J. and Petzow, G. (1977). Representation of multicomponent silicon nitride base systems. In *Nitrogen Ceramics*, ed. F. L. Riley, NATO ASI Series E: Applied Science No. 23. Leyden: Noordhoff, pp. 41–62.

Gazza, G. E. (1973). Hot-pressed Si_3N_4. *J. Amer. Ceram. Soc.*, **56**, 662.

Gazza, G. E. (1975). Effect of yttria additions on hot-pressed Si_3N_4. *Am. Ceram. Soc. Bull.*, **54**, 778–81.

Gee, M. G. and Butterfield, D. (1993). The combined effect of speed and humidity on the wear and friction of silicon nitride. *Wear*, **162**A, 234–45.

Giachello, A., Martinengo, P. C., Tommasini, G. and Popper, P. (1980). Sintering and properties of silicon nitride containing Y_2O_3 and MgO. *Am. Ceram. Soc. Bull.*, **59**, 1212–15.

Giachello, A., Martinengo, P. C., Tommasini, G. and Popper, P. (1979). Sintering of silicon nitride in a powder bed. *J. Mater. Sci.*, **14**, 2825–30.

Glemser, O., Beltz, K. and Naumann, P. (1957). Towards an understanding of the silicon-nitrogen system. *Z. Anorg. Allg. Chem.*, **291**, 51–66.

Govila, R. K., Mangels, J. A. and Baer, J. R. (1985). Fracture of yttria-doped, sintered reaction bonded silicon nitride. *J. Am. Ceram. Soc.*, **68**, 413–18.

Greskovich, C. and Rosolowski, J. H. (1976). Sintering of covalent solids. *J. Am. Ceram. Soc.*, **59**:7–8, 336–43.

Greskovich, C., Prochazka, S. and Rosolowski, J. H. (1977). The sintering behaviour of covalently-bonded materials. In *Nitrogen Ceramics*, ed. F. L. Riley, NATO ASI Series E: Applied Science No. 23. Leyden: Noordhoff, pp. 351–7.

Hampshire, S. (1993). Oxynitride glasses and glass ceramics. In *Silicon Nitride Ceramics: Scientific and Technological Advances*, eds. I.-W. Chen, P. F. Becher, M. Mitomo, G. Petzow and T.-S. Yen, Materials Research Society Symposium Proceedings, Vol. **287**. Pittsburgh, PA: Materials Research Society, pp. 93–104.

Hampshire, S., Park, H. K., Thompson, D. P. and Jack, K. H. (1978). $α'$-Sialon ceramics. *Nature*, **274**, 880–2.

Hardie, D. and Jack, K. H. (1957). Crystal structures of silicon nitride. *Nature*, **180**, 332–3.

Hay, J. C., Sun, E. Y., Pharr, G. M., Becher, P. F. and Akexander, K. B. (1998). Elastic anisotropy of beta-silicon nitride whiskers. *J. Am. Ceram. Soc.*, **81**, 2661–9.

Heinrich, J., Backer, E. and Böhmer, M. (1988). Hot isostatic pressing of Si_3N_4 powder compacts and reaction bonded Si_3N_4. *J. Amer. Ceram. Soc.*, **71**, C-28–31.

Heinrich, J. G. and Krüner, H. (1994). Silicon nitride materials for engine applications. In *Tailoring of Mechanical Properties of Si_3N_4 Ceramics*, eds. M. J. Hoffmann and G. Petzow, NATO ASI Series E: Applied Sciences, Vol. **276**. Dordrecht: Kluwer Academic Publishers, pp. 19–41.

Henderson, C. M. B. and Taylor, D. (1975). Thermal expansion of the nitrides and oxynitrides of silicon in relation to their structures. *Trans. J. Br. Ceram. Soc.*, **74**, 49–53.

Herrmann, M. and Göb, O. (2001). Colour of gas-pressure-sintered silicon nitride ceramics Part II. Thermodynamic considerations. *J. Eur. Ceram. Soc.*, **21**, 461–9.

Himsolt, G., Knoch, H., Hübner, H. and Kleinlein, F. W. (1979). Mechanical properties of hot-pressed silicon nitride with different grain structures. *J. Am. Ceram. Soc.*, **62**, 29–32.

Hoffmann, M. J. and Petzow, G. (1993). Microstructural design of Si_3N_4 based ceramics. In *Silicon Nitride Ceramics: Scientific and Technological Advances*, eds. I.-W. Chen, P. F. Becher, M. Mitomo, G. Petzow and T.-S. Yen, Materials Research Society Symposium Proceedings, Vol. **287**. Pittsburgh, PA: Materials Research Society, pp. 15–27.

Horton, S. A. (1991). Hybrid silicon nitride bearings. In *3rd European Symposium on Engineering Ceramics*, ed. F. L. Riley. London: Elsevier Applied Science, pp. 35–50.

Horton, S. A., Dowson, D., Riley, F. L., Wallbridge, N. C., Broussaud, D. and Denape, J. (1986). The wear behaviour of sialon and silicon carbide ceramics in sliding contact. In *Proceedings of International Conference on Non-oxide Technical and Engineering Ceramics*, ed. S. Hampshire. London: Elsevier Applied Science Publishers, pp. 281–98.

Huang, Z. K., Tien, T.-Y. and Yen, T.-S. (1986). Subsolidus phase relationships in Si_3N_4–AlN–rare-earth oxide systems. *J. Am. Ceram. Soc.*, **69**, C241–2.

Iskoe, J. L., Lange, F. F. and Dias, E. S. (1976). Effect of selected impurities on high-temperature strength of mechanical properties of hot-pressed silicon nitride. *J. Mater. Sci.*, **11**, 908–12.

Jack, K. H. (1973). Nitrogen ceramics. *Trans. J. Br. Ceram. Soc.*, **72**, 376–84.

Jack, K. H. (1976). Sialons and related nitrogen ceramics. *J. Mater. Sci.*, **11**, 1135–58.

Jack, K. H. (1977). The crystal chemistry of the sialons and related nitrogen ceramics. In *Nitrogen Ceramics*, ed. F. L. Riley, NATO ASI Series E: Applied Science, No. 23. Leyden: Noordhoff, pp. 109–28.

Jack, K. H. (1978). The relationship of phase diagrams to research and development of sialons. In *Phase Diagrams: Materials Science and Technology*, Vol. **5**, ed. A. M. Alper. London: Academic Press, pp. 241–85.

Jack, K. H. (1983). The characterization of α'-sialons and the α–β relationships in sialons and silicon nitrides. In *Progress in Nitrogen Ceramics*, ed. F. L. Riley, NATO ASI Series E: Applied Sciences, No. 65. Boston, MA: Martinus Nijhoff, pp. 45–60.

Jack, K. H. (1993). Sialon ceramics: retrospect and prospect. In *Silicon Nitride Ceramics: Scientific and Technological Advances*, eds. I.-W. Chen, P. F. Becher, M. Mitomo, G. Petzow and T.-S. Yen, Materials Research Society Symposium Proceedings, Vol. **287**. Pittsburgh, PA: Materials Research Society, pp. 15–27.

Jack, K. H. and Wilson, W. I. (1972). Ceramics based on Si-Al-O-N and related systems. *Nature Phys. Sci.*, **238**, 28–9.

Jacobson, N. S. (1993). Corrosion of silicon-based ceramics in combustion environments. *J. Am. Ceram. Soc.*, **76**, 3–28.

Kato, K., Inoue, Z., Kijima, K., Kawada, J. and Tanaka, H. (1975). Structural approach to the problem of oxygen content in alpha silicon-nitride. *J. Am. Ceram. Soc.*, **58**, 90–1.

Katz, R. N. (1980). High temperature structural ceramics. *Science*, **208**, 841–7.

Katz, R. N. (1983). US national programmes in ceramics for energy conversion. In *Progress in Nitrogen Ceramics*, ed. F. L. Riley, NATO ASI Series E: Applied Sciences, No. 65. Boston, MA: Martinus Nijhoff, pp. 727–35.

Katz, R. N. (1985). Applications of high-performance ceramics in heat engine design. *Mat. Sci. Eng.*, **7**, 227–49.

Katz, R. N. (1993). Applications for silicon nitride based ceramics in the U.S. In *Silicon Nitride Ceramics: Scientific and Technological Advances*, eds. I.-W. Chen, P. F. Becher, M. Mitomo, G. Petzow and T.-S. Yen, Materials Research Society Symposium Proceedings, Vol. **287**. Pittsburgh, PA: Materials Research Society, pp. 197–208.

Katz, R. N. (1994). Ceramic materials for rolling element bearing applications. In *Friction and Wear of Ceramics*, ed. S. Jahanmir. New York: Marcel Dekker, pp. 313–38.

Katz, R. N. (1997). Applications of silicon nitride based ceramics. *Ind. Ceram.*, **17**, 158–64.

Katz, R. N. (1999). Wear applications of silicon carbide and nitride ceramics. In *Innovative Materials in Advanced Energy Technologies*, ed. P. Vincenzini, Advances in Science and Technology 24. Faenza: Techna Publishing.

Kim, J., Rosenflanz, A. and Chen, I. W. (2000). Microstructure control of in-situ toughened alpha-SiAlON ceramics. *J. Am. Ceram. Soc.*, **83**, 1819–21.

Kingon, A. I., Lutz, L. J. and Davis, R. F. (1983). Thermodynamic calculations for the chemical vapour deposition of silicon nitride. *J. Am. Ceram. Soc.*, **66**, 551–8.

Kleeb, H.-J., Cinibulk, M. K., Tanaka, I., Bruley, Cannon, R. M., Clarke, D. R., Hoffmann, M. J. and Rühle, M. (1993). High-resolution electron microscopy observations of grain-boundary films in silicon nitride ceramics. In *Silicon Nitride Ceramics: Scientific and Technological Advances*, eds. I.-W. Chen, P. F. Becher, M. Mitomo, G. Petzow and T.-S. Yen, Materials Research Society Symposium Proceedings, Vol. **287**. Pittsburgh, PA: Materials Research Society, pp. 65–78.

Komeya, K. (1993). Progress in silicon nitride ceramics in Japan. In *Silicon Nitride Ceramics: Scientific and Technological Advances*, eds. I.-W. Chen, P. F. Becher, M. Mitomo, G. Petzow and T.-S. Yen, Materials Research Society Symposium Proceedings, Vol. **287**. Pittsburgh, PA: Materials Research Society, pp. 29–38.

Komeya, K. and Inoue, H. (1975). Synthesis of alpha-form of silicon-nitride from silica. *J. Mater. Sci.*, **10**, 1243–6.

Kramer, M., Hoffmann, M. J. and Petzow, G. (1993). Grain-growth kinetics of Si3N4 during alpha-beta transformation. *Acta Metall. Mater.*, **41**, 2939–47.

Kurtz, W. and Schröder, F., Eds. (1994). *Gmelin Handbook of Inorganic and Organometallic Chemistry,* 8th edition. Silicon Supplement Volume B 5e, *Non-Electronic Applications of Silicon Nitride. SiNx.SiNxH.* Berlin: Springer-Verlag.

Lamkin, M. A., Riley, F. L. and Fordham, R. J. (1992). Oxygen mobility in silicon dioxide and silicate glasses: a review. *J. Eur. Ceram. Soc.*, **10**, 347–67.

Lang, J. (1977). Some aspects of the structure and crystal chemistry of the nitrides. In *Nitrogen Ceramics*, ed. F. L. Riley, NATO ASI Series E: Applied Science No. 23. Leyden: Noordhoff, pp. 89–105.

Lange, F. F. (1973). Relation between strength, fracture energy, and microstructure of hot-pressed Si_3N_4. *J. Am. Ceram. Soc.*, **56**, 518–22.

Lange, F. F. (1978). Phase relations in system Si_3N_4–SiO_2–MgO and their interrelation with strength and oxidation. *J. Amer. Ceram. Soc.*, **61**, 53–6.

Lange, F. F. (1979). Fracture-toughness of Si_3N_4 as a function of the initial alpha-phase content. *J. Am. Ceram. Soc.*, **62**, 428–30.

Lange, F. F. (1980). Silicon nitride polyphase systems: fabrication, microstructure and properties. *Int. Met. Rev.*, **25**, 1–20.

Lange, F. F. (1984). Method of strengthening silicon nitride ceramics. US Patent 4,457,958, July 3rd 1984.

Lange, F. F. (2006). The sophistication of ceramic science through silicon nitride studies. *J. Ceram. Soc. Japan*, **114**, 873–9.

Lange, F. F., Singhal, S. C. and Kuznicki, R. C. (1977). Phase relations and stability studies in the Si_3N_4-SiO_2-Y_2O_3 pseudo ternary system. *J. Amer. Ceram. Soc.*, **60**, 249–52.

Lange, H., Wötting, G. and Winter, G. (1991). Silicon nitride – from powder synthesis to ceramic materials. *Angew. Chem. Int.*, **30**, 1579–97.

Lee, F. J. and Bowman, K. J. (1992). Texture and anisotropy in silicon nitride. *J. Am. Ceram. Soc.*, **75**, 1748–55.

Lee, M. R., Russell, S. S., Arden, J. W. and Pillinger, C. T. (1995). Nierite (Si_3N_4), a new material from ordinary and enstatite chondrites. *Meteoritics*, **30**, 387–98.

Levin, E. M., Robbins, C. R. and McMurdie, H. F. (1979). *Phase Diagrams for Ceramists*. Columbus, OH: The American Ceramic Society, fourth printing, Fig. 266.

Liang, J. J., Topor, L. and Navrotsky, A. (1999). Silicon nitride: enthalpy of formation of the alpha- and beta-polymorphs and the effect of C and O impurities. *J. Mat. Res.*, **14**, 1959–68.

Loehman, R. E. (1983). Preparation and properties of oxynitride glasses. *J. Non-cryst. Solids*, **56**, 123–34.

Loehman, R. E. (1989). Interfacial reactions in ceramic metal systems. *Am. Ceram. Soc. Bull.*, **68**, 891–96.

Luecke, W. E., Wiederhorn, S. M., Hockey, B. J., Krause, R. F. and Long, G. G. (1995). Cavitation contributes substantially to tensile creep in silicon nitride. *J. Am. Ceram. Soc.*, **78**, 2085–96.

Lumby, R. J. (1991). The commercial development of an engineering ceramic. In *Engineering Ceramics*, ed. F. L. Riley. London: Elevier Applied Science, pp. 27–33.

Lumby, R. J., North, B. and Taylor, A. J. (1978). Properties of sintered sialons and some applications in metal handling and cutting. In *Ceramics for High Performance Applications II*. Proc. 5th Army Materials Technology Conf. 1977, eds. J. J. Burke, E. N. Lenoe and R. N. Katz. Chestnut Hill, MA: Brook Hill, pp. 893–906.

Mandal, H. (1999). New developments in alpha-SiAlON ceramics. *J. Eur. Ceram. Soc.*, **19**, 2349–57.

Mandal, H., Thompson, D. P. and Ekstrøm, T. (1993). Reversible $\alpha' \leftrightarrow \beta'$-sialon transformation in heat-treated sialon ceramics. *J. Eur. Ceram. Soc.*, **12**, 421–9.

Mandal, H., Thompson, D. P. and Jack, K. H. (1999). The $\alpha \leftrightarrow \beta$-transformation in silicon nitride and sialons. In *Novel Synthesis and Processing of Ceramics*, eds. H. Suzuki, K. Komeya and K. Uemtasu, *Key Engineering Materials*, **159–160**. TransTech Publications, pp. 1–9.

Mangels, J. A. (1981). Effect of rate-controlled nitriding and nitriding atmospheres on the formation of reaction-bonded Si_3N_4. *Am. Ceram. Soc. Bull.*, **60**, 613–17.

Mangels, J. A. and Tennenhouse, G. F. (1980). Densification of reaction bonded silicon nitride. *Am. Ceram. Soc. Bull.*, **59**, 1219–22.

Mayer, M. I. and Riley, F. L. (1978). Sodium assisted oxidation of reaction bonded silicon nitride. *J. Mater. Sci.*, **13**, 1319–28.

Messier, D. R., Riley, F. L. and Brook, R. J. (1978). Apha-beta-silicon nitride phase-transformation. *J. Mater. Sci.*, **13**, 1199–205.

Miller, S. L. (1953). A production of amino acids under possible primitive earth conditions. *Science*, **117**, 528–9.

Mitomo, M. (1976). Pressure sintering of silicon nitride. *J. Mater. Sci.*, **11**, 1103–7.

Mitomo, M., Hirosaki, N. and Nishimura, T. (2006). Microstructure control in silicon nitride ceramics – a review. *J. Ceram. Soc. Japan*, **114**, 867–72.

Moulson, A. J. (1979). Reaction bonded silicon nitride: its formation and properties. *J. Mater. Sci.*, **14**, 1017–51.

Mouradoff, L., Lachaudurand, A., Desmaison, J., Labbe, J. C., Grisot, O. and Rezkhanlou, R. (1994). Study of the interaction between liquid aluminium and silicon nitride. *J. Eur. Ceram. Soc.*, **13**, 323–8.

Narushima, T., Goto, T., Yokoyama, Y., Hagiwara, J., Iguchi, Y. and Hirai, T. (1994). High-temperature active oxidation and active-to-passive transition of chemically vapour deposited silicon nitride in N_2, O_2 and Ar-O_2 atmospheres. *J. Am. Ceram. Soc.*, **77**, 2369–75.

Nicholson, K. C. (1952). *Manufacture of silicon nitride-bonded articles*. U.S. Patent No. 2618,565, November 18th, 1952.

Niihara, K. and Hirai, T. (1979). Growth, morphology and slip system of alpha-Si3N4 single crystal. *J. Mater. Sci.*, **8**, 1952–79.

Noakes, P. B. and Pratt, P. L. (1972). High temperature mechanical properties of reaction-sintered silicon nitride. In *Special Ceramics 5*, ed. P. Popper. Manchester: The British Ceramic Society, pp. 299–310.

Ogata, S., Hirosaki, N., Kocer, C. and Shibutani, Y. (2004). A comparative ab initio study of the 'ideal' strength of alpha- and beta-Si3N4. *Acta. Mater.*, **52**, 233–8.

Oparin, A. I. (1952). *The Origin of Life*. New York: Dover.

Oyama, Y. and Kamigaito, O. (1971). Solid solubility of some oxides in Si_3N_4. *Japan J. Appl. Phys.*, **10**, 1637.

Parr, N. L., Martin, G. F. and May, E. R. W. (1960). Preparation, microstructure and mechanical properties of silicon nitride. In *Special Ceramics*, ed. P. Popper. London: Heywood, pp. 102–35.

Pehlke, R. D. and Elliott, J. F. (1959). High temperature thermodynamics of the silicon, nitrogen and silicon-nitride system. *Trans. AIME*, **215**, 781–5.

Petzow, G. and Herrmann, M. (2002). Silicon nitride ceramics. In *High Performance Non-oxide Ceramics II*, eds. M. A. Jansen and R. Haubner, *Structure and Bonding*, Vol. **102**. Berlin: Springer Verlag, pp. 47–166.

Popper, P. (1977). Introductory lecture. In *Nitrogen Ceramics*, ed. F. L. Riley, NATO ASI Series E: Applied Science No. 23. Leyden: Noordhoff, pp. 3–19.

Popper, P. (1994). Applications of silicon nitride. In *Engineering Materials*, Vol. 89–91, Proceedings of the International Conference on Silicon Nitride-Based Ceramics. Switzerland: TransTech Publications, pp. 719–24.

Popper, P. and Ruddlesden, S. N. (1961). The preparation, properties and structure of silicon nitride. *Trans. Br. Ceram. Soc.*, **60**, 603–26.

Reimanis, I. E., Suematsu, H., Petrovic, J. J. and Mitchell, T. E. (1996). Mechanical properties of single crystal alpha-Si3N4. *J. Am. Ceram. Soc.*, **79**, 2065–73.

Riley, F. L. (1980). High temperature oxidation of gas turbine ceramics. In *Environmental Degradation of High Temperature Materials*, Proc. Inst. Metallurgists Spring Residential Conference, Series 3, No. 13, **1**. London: Chameleon Press, pp. 3/1–3/11.

Riley, F. L. (1996). Diffusion in silicon nitride. In *Gmelin Handbook of Inorganic and Organometallic Chemistry,* 8th edition, *Silicon Supplement Volume B 5b1: Silicon*

Nitride: Mechanical and Thermal Properties, ed. F. Schröder. Berlin: Springer Verlag, pp. 397–412.

Riley, F. L. (2000). Silicon nitride and related materials. *J. Amer. Ceram. Soc.*, **83**, 245–65.

Robertson, J. (1991). Electronic structure of silicon nitride. *Phil. Mag. B*, **63**, 47–77.

Ruddlesden, S. N. and Popper, P. (1958). On the crystal structures of the nitrides of silicon and germanium. *Acta Crystallogr.*, **11**, 465–68.

Sangster, R. C. (2005). *Formation of Silicon Nitride: From the 19th to the 21st Century*. Switzerland: Trans Tech Publications, Materials Science Foundations, pp. 22–4.

Satoh, S. (1938). Silicon amide and silicon nitride production. *Inst. Phys. Chem. Res.*, **34**, 144–9.

Schneider, G. A. and Petzow, G. (1994). Thermal shock behaviour of Si_3N_4. In *Key Engineering Materials*, Vol. **89–91**, Proceedings of the International Conference on Silicon Nitride-Based Ceramics, eds. M. J. Hoffmann, P. F. Becher and G. Petzow. Switzerland: TransTech Publications, pp. 563–68.

Schröder, F., Ed. (1994). *Gmelin Handbook of Inorganic and Organometallic Chemistry*, 8th edition, *Silicon Supplement Volume B 5b1, Silicon Nitride: Mechanical and Thermal Properties*. Berlin: Springer Verlag, pp. 397–412.

Schröder, F., Ed. (1995). *Gmelin Handbook of Inorganic and Organometallic Chemistry*, 8th edition, *Silicon Supplement Volume B 5d1. Silicon Nitride: Electrochemical Behaviour, Colloidal Chemistry and Chemical Reactions*. Berlin: Springer-Verlag.

Schuster, J. C., Weitzer, F., Bauer, J. and Nowotny, H. (1988). Joining of silicon nitride ceramics to metals – the phase diagram basis. *Mater. Sci. Eng.*, **A 105/106**, 201–6.

Silva, R. F., Oliveira, F. J., Castro, F. P. and Vieira, F. M. (1998). Modelling of chemical wear in ferrous alloys silicon nitride contacts during high speed cutting. *Acta Mater.*, **46**, 2501–7.

Singhal, S. C. (1977). Oxidation of silicon nitride and related materials. In *Nitrogen Ceramics*, ed. F. L. Riley, Proc. NATO Advanced Study Institute, Applied Science No. 23. Leyden: Noordhoff, pp. 607–26.

Smialek, J. L., Robinson, R. C., Opila, E. J., Fox, D. S. and Jacobson, N. S. (1999). SiC and Si_3N_4 recession due to SiO_2 scale volatility under combustor conditions. *Advanced Composite Mat.*, **8**, 33–45.

Smith, J. T. and Quackenbush, C. L. (1978). In *Proc. Symp. on Factors in Densification and Sintering of Oxide and Non-oxide Ceramics*, eds. S. Somiya and S. Saito. Tokyo: Gakujutsu Bunken Fukyukai, pp. 126–40.

Stephen, R. G. and Riley, F. L. (1993). Oxidation of silicon powder: its significance for the strength of reaction bonded silicon nitride. In *Brit. Ceram. Proc.*, **50**, *Engineering Ceramics: Fabrication Science and Technology*, ed. D. P. Thompson. London: The Institute of Materials, pp. 201–11.

Sun, E. Y., Becher, P. F., Plucknett, K. P., Hsueh, C. H., Alexander, K. B., Waters, S. B., Hirao, K. and Brito, M. E. (1998). Microstructural design of silicon nitride with improved fracture toughness: II Effects of yttria and alumina. *J. Am. Ceram. Soc.*, **81**, 2831–40.

Swinkels, F. B. and Ashby, M. F. (1976). A second report on sintering diagrams. *Acta Metall.*, **29**, 259–81.

Szépvölgyi, J., Riley, F. L., Mohai, I., Bertoti, I. and Gilbart, E. (1996). Composition and microstructure of nanosized, amorphous and crystalline silicon nitride powders before, during and after densification. *J. Mater. Chem.*, **6**, 1175–86.

Tanaka, I., Pezzotti, G. and Okamoto, T. (1989). Hot isostatic press sintering and properties of silicon nitride without additives. *J. Am. Ceram. Soc.*, **72**, 1656–60.

Terwilliger, G. F. (1974). Properties of sintered silicon nitride. *J. Am. Ceram. Soc.*, **57**, 48–9.

Terwilliger, G. F. and Lange, F. F. (1975). Pressureless sintering of silicon nitride. *J. Mater. Sci.*, **10**, 1169–74.

Tressler, R. E. (1994). Theory and experiment in corrosion of advanced ceramics. In *Corrosion of Advanced Ceramics: Measurement and Modelling*, ed. K. G. Nickel, NATO ASI Series E: Applied Sciences 267. Dordrecht: Kluwer Academic, pp. 3–22.

Tsuge, A., Nishida, K. and Komatsu, M. (1975). Effect of crystallising grain-boundary glass phase on high-temperature strength of hot-pressed Si_3N_4 containing Y_2O_3. *J. Am. Ceram. Soc.*, **58**, 323–36.

Ueno, S., Ohji, T. and Lin, H. T. (2006). Designing lutetium silicate environmental barrier coatings for silicon nitride and its recession behavior in steam jets. *J. Ceram. Processing Research*, **7**, 20–3.

Urbain, G., Cambier, F., Deletter, M. and Anseau, M. R. (1981). Viscosity of silicate melts. *Trans. J. Br. Ceram. Soc.*, **80**, 139–41.

van Dijen, F. K., Kerber, A., Vogt, W., Pfeiffer, W. and Schultze, M. (1994). A comparative study of three silicon nitride powders, obtained by three different syntheses. In *Key Engineering Materials*, **89–91**, *Proceedings of the International Conference on Silicon Nitride-Based Ceramics*, eds. M. J. Hoffmann, P. F. Becher and G. Petzow. Switzerland: Trans Tech Publications, pp. 19–28.

Vaughn, W. L. and Maars, H. G. (1990). Active-to-passive transition in the oxidation of silicon carbide and silicon nitride in air. *J. Am. Ceram. Soc.*, **73**, 1540–3.

Vleugels, J., Laoui, T., Vercammen, K., Celis, J. P. and van der Biest, O. (1994). Chemical interaction between a sialon cutting-tool and iron-based alloys. *Mater. Sci. Engng A – Structural Materials Properties Microstructure and Processing*, **187**, 177–82.

Wang, C. M., Pan, X., Rühle, M., Riley, F. L. and Mitomo, M. (1996). Review: silicon nitride crystal structure and observations of lattice defects. *J. Mater. Sci.*, **31**, 5281–98.

Weaver, G. Q., Baumgartner, H. R. and Torti, M. L. (1975). Thermal shock behaviour of sintered silicon carbide and reaction-bonded silicon nitride. In *Special Ceramics 6*, ed. P. Popper. Manchester: The British Ceramic Society, pp. 261–310.

Weston, J. E. (1980). Origin of strength and isotropy in hot-pressed silicon nitride. *J. Mater. Sci.*, **15**, 1568–76.

Wiederhorn, S. M., Krause, R. F., Lofaj, F. and Taffner, U. (2005). Creep behavior of improved high temperature silicon nitride. In *Advanced Si-Based Ceramics and Composites*, eds. H.-D. Kim, H.-T. Lin and M. J. Hoffmann, *Key Engineering Materials* **287**. Switzerland: TransTech Publications, pp. 381–92.

Wiederhorn, S. M., Quinn, G. D. and Krause, R. (1994). High temperature structural reliability of silicon nitride. In *Key Engineering Materials*, **89–91**, Proceedings of the International Conference on Silicon Nitride-Based Ceramics. Switzerland: TransTech Publications, pp. 719–24.

Wild, S., Grieveson, P. and Jack, K. H. (1972). The rôle of magnesia in hot-pressed silicon nitride. In *Special Ceramics 5*, ed. P. Popper. Stoke-on-Trent: British Ceramics Research Association, pp. 377–84.

Wills, R. R. and Brockway, M. C. (1981). Hot isostatic pressing of ceramics. In *Special Ceramics 7, Proc. Brit. Ceram. Soc.*, **31**, ed. D. Taylor. Stoke-on-Trent: British Ceramic Society, pp. 233–47.

Woetting, G., Lindner, H. A. and Gugel, E. (1996). Silicon nitride values for automotive engines. In *Advanced Ceramic Materials, Key Engineering* Materials, vols. **122–124**, ed. H. Mostaghaci. Switzerland: TransTech Publications, pp. 283–92.

Yamada, T. (1993). Preparation and evaluation of sinterable silicon nitride powder by imide decomposition method. *Am. Ceram. Soc. Bull.*, **77**, 99–100.

Yan, M. and Fan, Z. (2001). Durability of materials in molten aluminium alloys. *J. Mater. Sci.*, **36**, 285–95.

Ye, J., Furuya, K., Mitomo, Y., Matuo, K., Munakata, F., Ishikawa, I. and Akimune, Y. (2000). Introduction of color center into beta-Si_3N_4 single crystal. *J. Luminescence*, **87**, 574–6.

Yen, T. S. and Sun, W. (1993). Phase relationship studies of silicon nitride system – a key to materials design. In *Silicon Nitride Ceramics: Scientific and Technological Advances*, eds. I.-W. Chen, P. F. Becher, M. Mitomo, G. Petzow and T.-S. Yen, Materials Research Society Symposium Proceedings, Vol. 287. Pittsburgh, PA: Materials Research Society, pp. 39–50.

Zerr, A. (2001). A new high-pressure delta-phase of Si_3N_4. *Phys. Stat. Solidi B*, **227**, R4–6.

Zerr, A., Miehe, G., Serghiou, G., Schwarz, M., Kroke, E., Riedel, R., Fuess, F., Kroll, P. and Boehler, R. (1999). Synthesis of cubic silicon nitride. *Nature*, **400**, 340–2.

Zerr, A., Riedel, R., Sekine, T., Lowther, F. R., Ching, W.-Y. and Tanaka, I. (2006). Recent advances in new hard high-pressure nitrides. *Adv. Mater.*, **18**, 2933–48.

Ziegler, G., Heinrich, J. and Wötting, G. (1987). Review: relationships between processing, microstructure and properties of dense and reaction bonded silicon nitride. *J. Mater. Sci.*, **22**, 3041–86.

6

Zirconia

6.1 Description and history

Zirconium is a second row (Group IVA), transition metal, and has several oxidation states. Normally the (IV) state is the most stable, giving the nominally Zr^{4+} cation, although it never takes this purely ionic form. Zirconia ceramics are based on the principal oxide of zirconium, zirconium dioxide (ZrO_2), with a melting-point in the region of 2710 °C (Kisi and Howard, 1998). Traditionally zirconium dioxide is referred to as *zirconia*, and, as is the case with the other ceramics reviewed here, the single term represents a large family of materials. Zirconium dioxide is one of the most stable oxides with respect to reduction, or dissociation back to the elements, because it has a large negative Gibbs function of formation:

$$Zr_{(s)} + O_{2(g)} = ZrO_{2(m)}; \quad \Delta G^{\circ}_{298\,K} = -1040\,kJ\,mol^{-1}. \tag{6.1}$$

The vapour pressure of oxygen over zirconium dioxide at 2500 K is still only $\sim 10^{-14}$ atm and for this reason metallic zirconium is a very effective scavenger for oxygen (Barin, 1995). However, under very low oxygen pressures it is possible for the high-temperature cubic phase of ZrO_2 to become non-stoichiometric without change of crystal structure. These compositions can be expressed as $ZrO_{(2-x)}$, where x can take values up to 0.5, which corresponds to the zirconium[III] oxide, Zr_2O_3 (Kisi and Howard, 1998).

Zirconium is not a particularly rare element. It is estimated that the Earth's crust contains between 0.02 and 0.03% of zirconium dioxide, making it more abundant than many other metallic ores (Youngman, 1931). Large natural deposits of impure zirconium dioxide – the mineral known as baddeleyite – were discovered in the São Paulo region of Brazil in 1892. During the following 20 years the properties of zirconia and its possible applications were explored, and many efforts were made to use this chemically inert, high-melting-point ("non-fusible") material as an industrial refractory (Ruff and Ebert, 1929). It was found that useful refractory materials could indeed be made from *impure* zirconium dioxide minerals (which contained up to 20% of impurity oxides, primarily silicon dioxide, titanium dioxide and iron oxide). However, these had low strength (typically <75 MPa) and, because of the high thermal expansivity of zirconium dioxide (~10 MK^{-1}), poor thermal shock resistance (Ryshkewitch and Richerson, 1985c). Attempts to prepare articles from *pure* sintered zirconium dioxide invariably and completely failed. It was eventually recognised that this was a consequence of its "volume instability", the very large contractions and expansions, of the zirconium dioxide crystal that occurred during heating or cooling. Coherent, strong, pure sintered zirconia could not be obtained because of the severe cracking and crumbling which took place at around 1000 °C during cooling from production temperatures (Cohn and Tolksdorf, 1930; Ryshkewitch and Richerson, 1985a). These damaging volume changes were not the normal smooth changes with change of temperature, which are shown by practically all materials. Instead, they appeared to be the result of an abrupt change in the crystal volume, with an overall volume expansion (on cooling) of about 5% (Wolten, 1963).

The unusual electrical conductivity of zirconium dioxide sintered with a substantial proportion of yttrium oxide had been known since 1897 (Nernst, 1897), but the first attempt at the production for larger-scale structural applications of sintered zirconia containing deliberately added magnesium and calcium oxides (which were actually used to try to lower the sintering temperature) took place in 1928 (D'Ans, 1928, 1932). Commercial production of zirconium dioxide intentionally "stabilised" with magnesium oxide against the crystal volume change was initiated the same year (Ryshkewitch, 1928a, 1928b). This material contained about 3% by weight (equivalent to ~9 mole%) of magnesium oxide, and could be sintered at 1600 °C. Subsequent improvement of the process by sintering at 1800 °C gave a fairly dense (~5 Mg m^{-3}, or ~90% of theoretical density) material. The following 30 years saw the development of coarse-grained zirconia materials containing magnesium oxide of reasonably high strength (~145 MPa), which were used where resistance to corrosion and wear was needed (Duwez *et al.*, 1952; Ryshkewitch, 1953). One structural application of this material was for wear-resistant linings in dies used for metal extrusion. It was also discovered that zirconia refractories could be produced with microstructures giving good thermal shock resistance (Curtis, 1947); refractory applications of a specialised nature, such as pouring spouts for liquid metal, then represented the tonnage use of zirconia.

This early work on the behaviour of materials based on zirconium dioxide laid the foundations for the development of the very high-strength zirconia structural ceramics, through identification of the polymorphism of zirconium dioxide, and its three important crystal phases, *monoclinic*, *tetragonal*, and *cubic* (Ruff and

Ebert, 1929; Teufer, 1962; Smith and Newkirk, 1965). Work carried out in the 1960s and early 1970s in particular led to an appreciation of the significance of the tetragonal-to-monoclinic phase transition for zirconia ceramics. In a major advance, a bend strength of 650 MPa at room temperature was obtained in zirconium dioxide sintered with calcium oxide, and containing the tetragonal phase; this was comparable with the highest strengths obtainable with fine-grain sintered alumina. In contrast, the best strength for purely monoclinic zirconium dioxide was ~250 MPa (Garvie *et al.*, 1975).

The development of the present commercial range of very high-strength fine-grain zirconia ceramics has been the outcome of extensive research during the period ~1970 to ~1990, on the stability and transformability of the zirconium dioxide phases, the nature of the tetragonal-to-monoclinic phase transformation, and microstructure–property relationships in polycrystalline sintered zirconium dioxide. This led to greatly improved understanding of the tetragonal-to-monoclinic transformation, and of how to obtain even better mechanical properties in zirconia ceramics (Subbarao, 1981; Green *et al.*, 1989). The background literature is considerable, and there are many reviews, and conference proceedings (Heuer and Hobbs, 1981; Claussen *et al.*, 1984; Sōmiya *et al.*, 1988; Kisi, 1998; Hannink *et al.*, 2000).

Zirconium dioxide with its all-important phase transformation is now amongst the best understood of the ceramic oxides. Zirconia materials can be produced with strengths in three-point bend of more than 2 GPa (Tsukuma *et al.*, 1985), and fracture toughness values up to ~30 MPa m$^{1/2}$. Many commercial materials have bend strengths at room temperature of ~1.4 GPa, and fracture toughness of 10 MPa m$^{1/2}$, which are more than twice the values of a very good sintered alumina. However, there tends to be an inverse relation between strength and fracture toughness (Swain, 1985), and very high strength and very high fracture toughness cannot be achieved simultaneously in the same material (Readey *et al.*, 1987). To obtain fracture toughness values of >8 MPa m$^{1/2}$ some strength must be sacrificed. The drawback to these materials as a class is that the application temperature must generally be restricted to <~1000 °C, and even lower in the presence of water vapour. Nonetheless, overall, zirconium dioxide provides the strongest, and toughest single-phase oxide ceramics known, outside the groups of fibres and fibre composites, and zirconia ceramics have important industrial applications as structural materials (Briggs, 2007).

6.2 Powder production

Several methods are available for the production of zirconium dioxide. It is extracted from baddeleyite, which can contain up to around 95% by weight of zirconium dioxide, using a variety of chemical methods (Ryshkewitch and Richerson, 1985c). One process uses a mixture of carbon (as coke), and chlorine gas at high temperature to convert most of the metallic elements to their volatile chlorides; silicon dioxide remains behind:

$$M_2O_{3(s)}+3C_{(s)}+3Cl_{2(g)}= 2MCl_{3(v)}+3CO_{(g)} \tag{6.2}$$

$$MO_{2(s)}+2C_{(s)}+2Cl_{2(g)}= MCl_{4(v)}+2CO_{(g)}. \tag{6.3}$$

The chlorides are trapped and dissolved in water, from which insoluble zirconium oxychloride ($ZrO.Cl_2$) is precipitated. After washing, this is converted to the hydroxide, and then dehydrated to zirconium dioxide in the form of a very fine powder,

$$ZrO.Cl_{2(s)}+3H_2O_{(1)}= \text{``}Zr(OH)_{4(s)}\text{''}+2HCl_{(aq.)} \tag{6.4}$$

$$\text{``}Zr(OH)_{4(s)}\text{''}= ZrO_{2(s)}+2H_2O_{(v)}. \tag{6.5}$$

An alternative, and currently important, source is the mineral *zircon*, which is zirconium silicate, with the composition $ZrSiO_4$, containing about 67% by weight of zirconium dioxide. This mineral conveniently occurs in the form of a sand (*zircon sand*) in many parts of the world, notably Australia (New South Wales), India (Kerala) and the USA (Florida) (Farnworth *et al.*, 1981). The most common method for extracting zirconium dioxide from zircon is by fusion with sodium hydroxide at around 600 °C to form sodium zirconate (Na_2ZrO_3) and sodium silicate (a process similar in principle to that used for the extraction of aluminium oxide from bauxite). Treatment with water hydrolyses the sodium zirconate to insoluble hydrated zirconium hydroxide, and the water-soluble sodium silicate is removed by washing:

$$ZiSiO_{4(s)}+4NaOH_{(1)}= Na_2ZrO_{3(s)} + Na_2SiO_{3(soln)} + 2H_2O_{(1)} \tag{6.6}$$

$$Na_2ZrO_{3(s)}+3H_2O_{(1)}= \text{``}Zr(OH)_{4(s)}\text{''} + 2NaOH_{(soln)}. \tag{6.7}$$

The precipitated hydrated zirconium dioxide is washed, spray-dried, and heated to convert it to zirconium dioxide. Chemical treatments to remove the unwanted oxides from these minerals yield very high purity, and very fine (often submicrometre) particle size, powders.

Most zirconium dioxide minerals contain a small amount (up to ~2%) of hafnium dioxide (HfO_2), which because of its close chemical similarity to zirconium dioxide is difficult to remove during processing. Hafnium is also a Group IVA transition metal, in the third row of the series. The hafnium atoms can be considered to occupy random zirconium sites in the zirconium dioxide crystal lattice. However, from the structural ceramic point of view the presence of the hafnium can safely be disregarded. It is of interest that hafnium dioxide shows many of the crystallographic features of zirconium dioxide, except that the temperatures at which phase changes occur are higher (Wang *et al.*, 1992). But because hafnium is a much less common element, it seems unlikely that use of the pure oxide as a ceramic will become widespread.

6.3 Crystal chemistry

Zirconium dioxide is usually regarded as an ionic oxide consisting of Zr^{4+} and O^{2-} ions, but analysis of the electronic structure shows that there is a significant (~20%) covalent contribution to the Zr–O bonding (Orlando *et al.*, 1992). There are six known crystal phases, or *polymorphs*, at least four of which can exist at room temperature (Garvie, 1970). This aspect of its crystal chemistry sharply distinguishes zirconium dioxide (and the zirconia ceramics) from aluminium dioxide (and alumina ceramics). Because of their importance for the properties of the zirconia ceramics it is necessary to introduce at this stage the three most common polymorphs. The other intrinsic physical properties of zirconium dioxide will be examined later.

Under the normal 1 atm pressure, three crystal structures, with *cubic*, *tetragonal*, and *monoclinic* symmetry, can be stable, each over its specific temperature range (Garvie, 1970), as shown in Fig. 6.1. The simplest form, the **cubic**, can be regarded as a face-centred cubic array of zirconium ions, with large oxide ions occupying the tetrahedrally coordinated sites, and four ZrO_2 units in the unit cell (the radius of Zr^{4+} is ~80 pm, O^{2-} ~140 pm). This is the basic CaF_2 *fluorite* structure, with twice as many tetrahedral (4-coordinate) sites as there are defining atoms: the coordination number of each zirconium ion is 8, and each oxide ion 4. The **tetragonal** structure is formed from the cubic by a very small (~2%) elongation along the *c*-axis, with a smaller contraction on the *a*-axis, and the Zr and O coordination numbers remain unchanged. The primitive tetragonal cell contains two ZrO_2 units and is half the volume of the cubic and monoclinic unit cells (which both contain four ZrO_2 units). To allow readier comparisons with the other two crystal structures the tetragonal structure is usually described in terms of a *c-centred* (or face-centred) unit cell, which has twice the volume of the primitive cell, and contains four ZrO_2 units. The third common structure, and the form stable at room temperature, is the **monoclinic**. This can be regarded as being formed from the cubic structure by a small overall expansion, coupled with a shear distortion. The coordination number of each zirconium ion is now 7. There are two distinct Zr–O bond lengths of ~207 and 221 pm, and the oxide ions can be

Table 6.1 *Zirconium dioxide: unit cell dimensions of the three main phases.*

Structure	Monoclinic	Tetragonal	Cubic
Lattice dimensions / pm*	$a = 515$	$a = 507$	$a = 512$
	$b = 520$	$c = 519$	
	$c = 532$		
	$\beta = 81°$		
		$c/a = {\sim}1.02$	
Density / Mg m^{-3}	5.83	~6.1*	6.09
Unit cell volume / nm^3	0.142	0.137	0.134

* Dependent on composition.

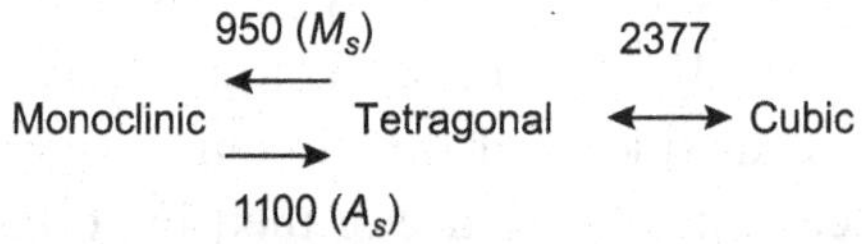

$$\text{Monoclinic} \underset{1100\ (A_s)}{\overset{950\ (M_s)}{\rightleftarrows}} \text{Tetragonal} \longleftrightarrow \text{Cubic} \quad 2377$$

Figure 6.1 Stability ranges and transformation temperatures (in °C), for the main phases of zirconium dioxide.

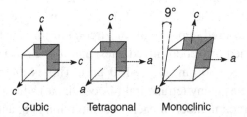

Cubic   Tetragonal   Monoclinic

Figure 6.2 A simplified picture of the spatial relationships between the cubic, tetragonal, and monoclinic unit cells of zirconium dioxide (the differences in dimension are exaggerated!).

considered to lie in buckled planes. These three unit cells, and the spatial relations between them, are illustrated schematically in Fig. 6.2, and their lattice dimensions and densities are listed in Table 6.1. The dimensional changes between the three structures are exaggerated for clarity. The other known crystal phases of zirconium dioxide are stable only under very high pressure, though one orthorhombic form can be obtained as a metastable phase at 1 atm pressure: they are therefore of less practical interest (Howard *et al.*, 1991; Dewhurst, 1998).

6.3.1 Phase stability

This section considers the three main crystalline phases of importance for the mechanical properties of zirconia ceramics, in order of decreasing high-temperature stability. Aspects of the transformation between the tetragonal and monoclinic phase are then examined in more detail, because they are central to the attainment of high fracture toughness, and strength.

Cubic

The existence of a phase with the simple cubic structure had long been known in impure zirconium dioxide materials. It was only established as a true high-temperature phase of pure zirconium dioxide in 1962 through the use of high-temperature X-ray diffraction techniques (Smith and Cline, 1962). The cubic phase melts at ~2710 °C. In the pure state it transforms on cooling to the tetragonal phase at ~2377 °C. However, the natural tendency of the cubic phase to transform can easily be prevented by the incorporation of small proportions of other metal oxides as solid solutions in the crystal lattice, usually in amounts of a few per cent by weight. The first oxides to be investigated in this context were magnesium oxide and calcium oxide; subsequently the less common yttrium oxide, and the rare earth oxides were studied in detail. This process of oxide incorporation is often referred to as *alloying* – borrowing the standard metallurgical term. Metal atoms of the added oxide occupy zirconium sites (M_{Zr}), and the oxygen atoms take up normal oxide positions in the zirconium dioxide lattice (represented as O_O). When di- or tri-valent metal oxides are used oxide site vacancies (V_O) are created, to maintain the metal:non-metal site ratio of 1:2:

$$MO = \{M_{Zr} + O_O + V_O\} \quad \text{(equivalent to one unit of } ZrO_2) \tag{6.8}$$

$$M_2O_3 = \{2M_{Zr} + 3O_O + V_O\} \quad \text{(equivalent to two units of } ZrO_2). \tag{6.9}$$

It has been suggested that the stabilising action of the additive oxides depends on the influence on lattice energies of the large concentrations of oxide ion vacancies generated by the aliovalent cations (Dwivedi and Cormack, 1990), but the detailed reasons may be more complex (French *et al.*, 1994; Stefanovich *et al.*, 1994). With large additions of suitable oxides, cubic zirconium dioxide can be sufficiently stabilised to exist at room temperature, and the material is then termed *fully stabilised* zirconia (though in some cases it is really only *metastable*). With lesser amounts of additive oxides (usually, in commercial materials, between 3 and 10 weight%), partial transformation may occur during cooling from production temperature, and the final material at room temperature can be a mixture of all three phases – the monoclinic, tetragonal, and cubic.

The cubic, fully stabilised, form of zirconium dioxide is of considerable importance in its own right because of its electrical properties (Subbarao and Maiti, 1988; Moulson and Herbert, 2002; Yokokawa *et al.*, 2005), but these will not be discussed in detail here.

Tetragonal

The tetragonal phase should be the stable phase over an intermediate temperature range, between the cubic and monoclinic phases. Pure tetragonal zirconium dioxide transforms spontaneously on cooling to the monoclinic phase at ~950 °C. This is not a sharply defined temperature: transformation starts at this temperature (sometimes referred to as the M_s temperature) and is completed over a range of ~200 °C. The exact temperature is also influenced by the form of the material – in particular its crystal size, an aspect returned to later. Transformation between the monoclinic and tetragonal structures is reversed on heating. However, it is possible to retain the tetragonal crystal structure at lower temperatures (as it is the cubic phase), by incorporating smaller amounts (usually up to ~5% by weight) of other oxides (Duwez *et al.*, 1952). Such materials are termed *partially stabilised* zirconias (PSZ). Transformation to the stable monoclinic phase (with changes in unit cell dimension) at lower temperatures can then be triggered by the application of mechanical stress, a feature with extremely important consequences for the mechanical properties of fracture toughness and strength.

Controlled cooling of the cubic phase containing oxides from its production temperature can also produce what is termed "non-transformable tetragonal" phase (*t'-phase*) identifiable, because it does not transform to the monoclinic phase at low temperature, even under the influence of mechanical stress (Miller *et al.*, 1981; Jue and Virkar, 1990; Mercer *et al.*, 2007). This is because each single crystal (which can be quite large) now consists of tetragonal phase variants, or domain orientations, each of size of the order of 100 nm, which collectively impede rearrangement of the structure to the monoclinic form.

The tetragonal-to-monoclinic transformation is usually described as *diffusionless,* or **martensitic**, because only very small movements in position of the zirconium and oxygen atoms are required, and atoms retain their nearest neighbours (Wolten, 1964). The martensitic transformation in steel was already well-known at the time development work on zirconia ceramics was being initiated, and the transfer of concepts from the metallurgical field was invaluable in advancing understanding of the transformation in zirconium dioxide materials (Antolovich and Singh, 1971; Antolovich and Fahr, 1972). The shifts in position of the atoms in the crystal lattice are of less than one interatomic distance, and mainly of the oxygen atoms, with minor shifts of the zirconium atoms (unlike the completely *reconstructive* transformation required for the interconversion of α- and β-silicon nitride). These movements, once initiated, are very fast (theoretically with lattice vibrational, or sonic, velocity), and because the phase changes normally occur only at high temperature, they are quite difficult to study (Fehrenbacher and Jacobson, 1965). While the transformation is normally initiated at one critical temperature, it only proceeds while the temperature is changing. At any one temperature within this process, the proportion of tetragonal to monoclinic phase is constant: the term *athermal* is therefore applied (Wayman, 1981). Orientational relationships between the lattice planes of the parent and product phases are of considerable importance for the interpretation of fine details of the changes in crystal morphology occurring during the transformation (Patil and Subbarao, 1970; Heuer and Rühle, 1984). The principal axes of the two phases remain approximately parallel, so that the (100) plane of a product monoclinic cell is parallel to the (110) plane of the parent tetragonal cell (Bansal and Heuer, 1974; Muddle and Hannink, 1986).

Monoclinic

The monoclinic, lower symmetry, polymorph (and met as the mineral baddeleyite) is the form stable at all temperatures below ~1170 °C. On heating, transformation to the tetragonal phase starts at ~1050 °C (the A_s temperature), and is completed over the range ~1050 to ~1200 °C (Garvie, 1970). There is therefore a large thermal hysteresis of ~200 °C, with the transformation temperature depending on the direction of travel over the temperature cycle (Curtis, 1947; Wolten, 1963). This behaviour is illustrated in Fig. 6.3. Although the three ZrO_2 crystal structures are closely related, and the differences between them are small, the consequences of these differences are considerable. The martensitic transformation between the tetragonal and the monoclinic polymorphs plays a very large part in controlling the mechanical properties of polycrystalline zirconia materials, and in the destructive volume changes first observed during the early development of zirconia refractories (Heuer and Rühle, 1985; Kisi and Howard, 1998). There is a slight similarity with silicon nitride, in that the formation of β-phase silicon nitride from the α-phase during sintering (and associated changes in microstructure) is also important for the development of maximum toughness and strength (though for quite different reasons). However, the effects of the phase changes in zirconium dioxide are far more dramatic, and the ways in which these changes affect the mechanical properties of fracture toughness and strength, are very much more complex.

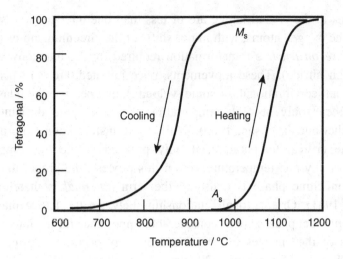

Figure 6.3 The hysteresis in the tetragonal-to-monoclinic phase transformation. (After Wolten, 1963. Reproduced with kind permission of Wiley-Blackwell Publishing.)

6.3.2 Unit cell dimensions

The phase transformations of zirconium dioxide would not be so significant, if it were not for the accompanying unit cell dimensional changes. The unit cells differ slightly in symmetry and volume, with the largest differences between the mono-clinic and tetragonal cells, as was shown in Table 6.1. The axis dimensions (and therefore unit cell volumes) of the c-tetragonal, and cubic, structures are closely similar, and the theoretical single-crystal densities are both ~6.1 Mg m^{-3} (varying slightly with composition). Because of its slightly larger b- and c-axis dimensions the monoclinic phase has the largest unit cell volume (and lowest density of ~5.8 Mg m^{-3}). It is also slightly skewed, with a β angle of ~81° (instead of the 90° of the tetragonal and cubic structures). Transition between the tetragonal and monoclinic crystal structure therefore has two consequences: there is a frac-tional increase in the unit cell volume of about 0.045, and a simultaneous distortion of the cell, corresponding to a shear of about 9° parallel to the tetragonal basal plane. This generates a shear strain of ~0.16 (Hannink *et al.*, 2000). The unit cell changes then attempt to cause equivalent changes in the dimensions of a zirconium dioxide crystal, or a grain in a polycrystalline material. At normal temperatures the com-bined expansion (*dilational*) and shear (*deviatoric*) strains cannot be relieved by atomic diffusion (that is, *plasticity*); instead, they have to be accommodated by elastic deformations (strains) in the crystal and its surrounding matrix. Through a number of mechanisms, these strains are able to increase significantly the fracture toughness of the material, which the following sections will outline.

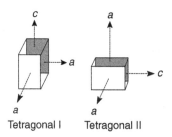

Tetragonal I   Tetragonal II

Figure 6.4 One flip between the *c*- and *a*-axes of the tetragonal unit cell, providing the possibility of ferroelastic toughening (note that the dimensional change is also exaggerated!).

The tetragonal unit cell itself has important properties. Because its *c*-axis to *a*-axis ratio is only ~1.02, the symmetry is not really far removed from that of cubic, and application of sufficient appropriately aligned stress can cause the cell *c*- and *a*-axes to flip, through concerted small movements of the zirconium and oxygen ions. This is the property of *ferroelasticity* (Fig. 6.4), and is also considered to make a significant contribution to the toughness of materials containing the transformable tetragonal phase (Virkar, 1998). It is also able to contribute towards the high fracture toughness of zirconia ceramics containing the *t'*-phase introduced above (Evans and Cannon, 1986; Virkar and Matsumoto, 1988). The ferroelastic transition has a potential further advantage: it is less temperature-sensitive than the martensitic tetragonal–monoclinic transition, and it therefore offers a possible high-temperature toughening mechanism (Virkar, 1998). Zirconium dioxide is not unique in having important phase transformations. The main crystalline phases of silicon dioxide (quartz, tridymite, and cristobalite) were referred to in Section 2.6.5, and another example in the ceramics field is dicalcium silicate (melting at ~2130 °C), which shows behaviour similar to that of zirconium dioxide at a lower temperature, ~650 °C (Kriven, 1988; Nettleship *et al.*, 1992). These examples illustrate the fact that aspects of phase stability other than simple melting-point can influence the usability, or otherwise, of a ceramic material (Kelly and Rose, 2002). Although many alternative ceramic systems have now been examined, most of these materials are at present of scientific interest only, and the commercially viable products are based on zirconium dioxide.

6.4 Phase equilibrium diagrams

The effective use of di- and tri-valent metal oxides in stabilising the tetragonal (and cubic) phase of zirconium dioxide at room temperature depends on having a detailed knowledge of the phase equilibrium diagrams. This information is needed for determining optimum sintering temperatures and

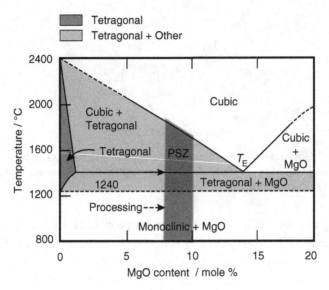

Figure 6.5 The MgO–ZrO$_2$ phase equilibrium diagram, showing the normal processing zone for partially stabilised zirconia (PSZ) type materials. (After Grain, 1967. Reproduced with kind permission of Wiley-Blackwell Publishing.)

possible subsequent annealing treatments, to allow the development of the required phases and microstructures.

After some initial uncertainty, there are now generally accepted versions of the phase diagrams for the systems most commonly used, MgO–ZrO$_2$ (Grain, 1967), CaO–ZrO$_2$ (Hannink *et al.*, 1981), Y$_2$O$_3$–ZrO$_2$ (Scott, 1975), and CeO$_2$–ZrO$_2$ (Duran, 1990; Hugo and Muddle, 1990). All have been studied in detail. The ZrO$_2$-rich sections of the diagrams are shown in Figs. 6.5–6.8. Figures 6.5 and 6.7 also illustrate the temperature ranges used for sintering, and annealing, in order to obtain the required compositions and microstructures in the sintered products. Strictly, the zirconium dioxide phases should be shown as *solid solution* phases. Magnesium oxide and calcium oxide have low solubility in the zirconium dioxide crystal lattice. On cooling after equilibration at high temperature, excess additive tends to precipitate, as magnesium oxide, magnesium zirconate (also known as δ-*phase*, Mg$_2$Zr$_5$O$_{12}$), or calcium zirconate (CaZr$_4$O$_9$) (Hannink, 1983; Green *et al.*, 1989). Yttrium oxide and cerium oxide have much higher solubilities, reflected in the patterns of the phase diagrams, showing that much larger quantities can be incorporated without the formation of binary phases, or the rejection of excess oxide on cooling. Even so, in these systems other binary phases have been identified, the φ-*phases*, with the fluorite structure. These have the generic forms M$_9$O$_{16}$, M$_5$O$_9$, and M$_{25}$O$_{44}$ where M represents any combination of cations. All these binary phases can be formed with high levels of oxide addition, and may

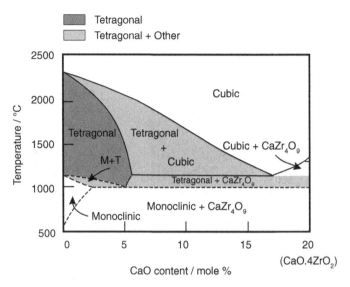

Figure 6.6 The CaO–ZrO$_2$ phase equilibrium diagram. (After Hannink *et al.*, 1981. Reproduced with kind permission of Wiley-Blackwell Publishing.)

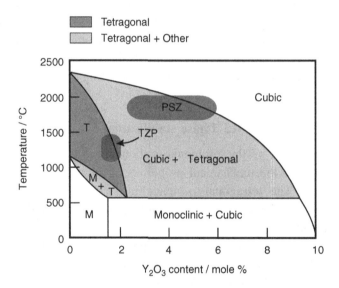

Figure 6.7 Y$_2$O$_3$–ZrO$_2$ phase equilibrium diagram, showing the processing zones for the two types of transformation toughened materials, partially stabilised zirconia, PSZ, and tetragonal zirconia polycrystal, TZP. (After Scott, 1975. Reproduced by kind permission of Springer Science and Business Media.)

play a small part in influencing stability of the zirconium dioxide solid-solution phases. For small amounts (~1 to ~2 mole%) of CaO, Y$_2$O$_3$, or CeO$_2$ at temperatures of ~1000 °C and above °C, or ~1400 °C for MgO, the equilibrated

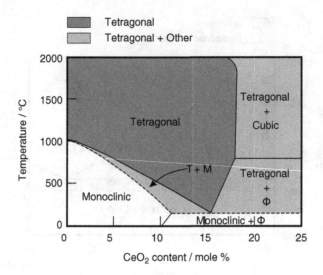

Figure 6.8 CeO$_2$–ZrO$_2$ phase equilibrium diagram. (After Duran, 1990. Reproduced by kind permission of Springer Science and Business Media.)

system should consist either of a single-phase tetragonal solid-solution, or a mixture of two solid-solution phases (tetragonal and monoclinic, or tetragonal and cubic). With larger quantities of oxide, cubic ZrO$_2$ becomes more stable, and this phase can easily be obtained even at room temperature either as a *metastable* phase (but with long-term practical stability) or, by using the more soluble yttrium (and lanthanide) oxide additions, a *stable* phase.

6.5 Phase stabilisation

This discussion focuses on the tetragonal-to-monoclinic transformation, because of its importance for the mechanical properties of zirconia ceramics: the cubic phase is ignored. The tetragonal-to-monoclinic transformation involves only minor adjustments to the positions of the atoms, no diffusion is involved, and it would be expected to happen spontaneously on cooling. Why it does not always do so requires an explanation. This is provided by considering the energy changes involved when the transformation does occur. The ability of an applied mechanical stress subsequently to trigger the transformation also has to be examined, together with how the associated changes in unit cell dimension increase fracture toughness, and thereby strength.

6.5.1 Influence of temperature

The phase diagrams show that for a specific overall composition, there is a critical temperature (T_o), below which the tetragonal phase should not be stable. In

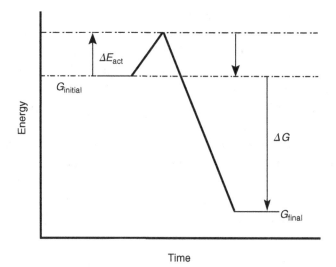

Figure 6.9 The reaction energy hill between $G_{initial}$ and G_{final}: species require the activation energy (ΔE_{act}) for the reaction to proceed. The overall energy change is ($G_{final} - G_{initial}$), or ΔG.

toughened zirconia ceramics, however, the tetragonal phase can exist at room temperature, which means that (when transformation does finally occur) the transformation temperature (T_{tr}) must have decreased below T_o. The starting point for the explanation of this behaviour is to consider the chemical energies of the tetragonal and monoclinic phases, and more importantly, the difference between these energies.

Any species (or phase) has its chemical energy (Gibbs function) G, usually expressed in units of J mole^{-1}. Here, the energy is for unit volume of material, and the unit used J m^{-3}. There is a problem with using this unit, in that unit volume of tetragonal phase will on transformation generate slightly more than unit volume of monoclinic phase, but each monoclinic crystal can be considered to be initially constrained to occupy the original volume of the tetragonal crystal. The primary driving force for the tetragonal-to-monoclinic transformation is the difference in chemical energy (ΔG_{tr}) between the two crystal structures, where ΔG_m and ΔG_t are the Gibbs functions of formation of the two phases from their elements:

$$\Delta G_{tr} = (\Delta G_m - \Delta G_t). \tag{6.10}$$

The material is regarded at this point as mechanically stress-free (though under standard, 1 atm, pressure). For a reaction to be energetically (that is, *thermodynamically*) favourable ΔG_{tr} must be negative, and the final energy level must be lower than the initial level: that is, the reaction is going "down-hill" (Fig. 6.9). Note that we are considering only the final and initial states, and not the pathway between them. There may well be an intermediate energy hill, as shown in the figure, to be surmounted during the transformation (an *activation energy* ΔE_{act}), and there will certainly be the question of *initiating* the transformation process (the monoclinic phase *nucleation* stage), when there is likely to be an energy barrier to the formation of the initial nucleus. However, because these factors influence only the time for the process to take place (or its probability of actually occurring), they are ignored for the moment. They do not affect the final and initial energy states: this approach can therefore be considered to be an "end-point" treatment.

The Gibbs function consists of two components, enthalpy (H) and entropy (S). For a specified amount of material, the chemical energy (G) changes with temperature according to the general expression (Atkins and de Paula, 2006):

$$G = H - TS. \tag{6.11}$$

At a given temperature, the change in Gibbs function (ΔG) during the formation of the monoclinic phase from its elements in their standard states is

$$\Delta G_{\text{m}} = \Delta H_{\text{m}} - T.\Delta S_{\text{m}} \tag{6.12}$$

and for the tetragonal phase

$$\Delta G_{\text{t}} = \Delta H_{\text{t}} - T.\Delta S_{\text{t}}. \tag{6.13}$$

For both phases at the temperatures of interest here (~1200 °C) ΔH is negative and very large (~1090 kJ mol^{-1}): zirconium metal, providing a reference point, burns in oxygen with the release of a large quantity of heat, but at this temperature ΔH_{m} has a slightly larger *negative* value than ΔH_{t} (Barin, 1995). ΔH_{tr} therefore has small negative values, and a small amount – about 6 kJ mol^{-1} – of heat is released during the tetragonal-to-monoclinic transformation. More importantly perhaps, the entropies of formation of both phases also have negative values. Entropy is a measure of disorder; the oxide crystals are more ordered than, notionally, the mixture of zirconium metal and oxygen gas from which they were formed. At ~1200 °C the entropy of formation (ΔS_{f}) of the monoclinic phase (~−182 J mol^{-1} K^{-1}) is slightly larger than that of the tetragonal (~−178 J mol^{-1} K^{-1}). Because ΔS_{f} is negative, plots of ΔG_{f} (=ΔH_{f} − $T.\Delta S_{\text{f}}$) as a function of T for the two phases (shown in Fig. 6.10) both have positive slopes, but that of the monoclinic phase is slightly steeper than that of the tetragonal (Fig. 6.10 greatly exaggerates the differences). For any temperature (in the temperature range of practical interest – room temperature to ~1200 °C) the difference in ΔG values, ΔG_{tr}, is the energy change for the transformation of unit volume of unconstrained tetragonal phase to the monoclinic phase:

$$\Delta G_{\text{tr}} = (\Delta G_{\text{m}} - \Delta G_{\text{t}}) = (\Delta H_{\text{m}} - \Delta H_{\text{t}}) - T.(\Delta S_{\text{m}} - \Delta S_{\text{t}}) = \Delta H_{\text{tr}} - T.\Delta S_{\text{tr}}. \tag{6.14}$$

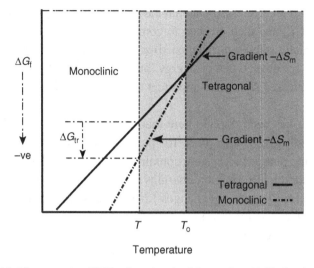

Figure 6.10 The energies (Gibbs functions) of formation (ΔG_f) for the tetragonal and monoclinic phases of ZrO_2, and for the tetragonal-to-monoclinic transformation (ΔG_{tr}). At high temperatures the tetragonal form is stable (shaded zone). The difference between the gradients is considerably exaggerated.

Because ΔS_f is negative for both phases, and because the slope of the line for the monoclinic phase is steeper, the two lines cross at a temperature T_o for which $\Delta G_{tr} = 0$. For this special case (6.14) becomes

$$0 = \Delta H_{tr} - T_o.\Delta S_{tr} \tag{6.15}$$

and

$$\Delta S_{tr} = \frac{\Delta H_{tr}}{T_o}. \tag{6.16}$$

Therefore for any other temperature of interest (T_{tr}):

$$\Delta H_{tr} = \Delta H_{tr}\left(\frac{1 - T_{tr}}{T_o}\right). \tag{6.17}$$

For $T_{tr} < T_o$, ΔG_{tr} will be negative (because ΔH_{tr} carries a negative sign), and for $T_{tr} > T_o$, positive. As the temperature drops further below T_o, the driving force (ΔG_{tr}) for the tetragonal-to-monoclinic transformation steadily increases and, left to the chemical energies, the tetragonal phase would be increasingly unstable with respect to the transformation (at room temperature the transformation driving force is equivalent to about 10 kJ mol^{-1}). This is simply another way of saying that above a critical temperature (T_o), the tetragonal phase is thermodynamically stable, and below this temperature, the monoclinic phase is stable – the phase equilibrium diagram picture. In an unstressed material the martensitic transformation should start spontaneously at T_o and then proceed to completion as the temperature is lowered further. In a partially stabilised (toughened) zirconia ceramic the chemical energy is counterbalanced by other types of energy, and the tetragonal-to-monoclinic transformation is suppressed. The most important of these are surface energy, and lattice strain energy.

As an aside, it has been implied here that the zirconium dioxide is chemically pure (and the data given were those for pure zirconium dioxide); however, the zirconium dioxide phases in a system containing additive oxides are actually solid solutions. The chemical formation energies of the transforming phases (ΔG_t and ΔG_m), and therefore the energy changes for the tetragonal-to-monoclinic transition, will be influenced by the phase compositions. In part this is because the incorporation of new cations, and the creation of defect sites, will alter the lattice energy of the crystal, and its energy of formation. The changes in phase stability with composition, as shown by the phase fields of the phase diagrams, of course reflect this. Any marked differences in solubility of the stabilising oxide in the zirconium dioxide phases mean there will also be requirements during the transformation for the lattice defects themselves (aliovalent cations on zirconium sites, M_{Zr}, and oxygen vacancies, V_o) to rearrange, possibly by involving diffusion processes. New energy terms associated with these processes also contribute to the total transformation energy, and ΔG_{tr} will not be exactly that of the pure material. However, the overall broad picture is unchanged, and serves the present purpose.

6.5.2 Influence of other energy terms

To understand why the (high-temperature) tetragonal phase may be able to exist in real materials at lower temperatures than it should (that is, below T_o), a wider range of factors affecting the internal energy of a crystal, other than the simple chemical (or thermodynamic) energy which forms the basis for the phase equilibrium diagram, must be considered. It was known for a long time that grain, or particle, size could affect the stability of the tetragonal phase: pure tetragonal crystals can exist at room temperature if they are fine enough, which is <30 nm (Garvie, 1970). Even larger particles of tetragonal phase (~1 μm) can exist at room temperature, if they can be constrained in a suitable matrix (Green, 1982). One theoretical treatment focuses on the surface (interfacial) energies, and the strain energies induced by unit cell volume and shape changes, the result of the tetragonal–monoclinic transformation (Lange, 1982a; Budinsky *et al.*, 1983; Garvie and Swain, 1985). This therefore takes the picture away from the simple one, which assumed unstressed, and (in principle) infinitely large, samples of material, to a polycrystalline, and possibly multi-phase, ceramic.

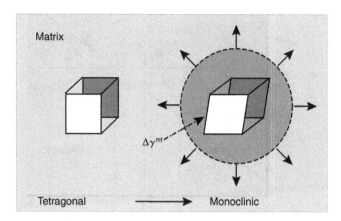

Figure 6.11 Illustration of the stress field developed around a transformed tetragonal grain, held in a rigid matrix, as a result of the expansion and shear strains.

Two new energy terms have to be taken into account in assessing whether an untransformed tetragonal crystal is able to exist at temperatures below T_o. The first is the interfacial energy. Because a new phase (the monoclinic) is nucleated in a rigid matrix of still untransformed material a new interface is created between the monoclinic crystal and the tetragonal crystal in which it is forming. There is therefore an increase in interfacial energy ($\Delta U_{interfacial}$) which, if it is a positive term (and it is in zirconium dioxide), will oppose the tendency for the transformation to occur. The second is the strain energy. Because the transformation involves volume and shape changes, corresponding strains, with associated energies (ΔU_{strain}), will be developed in and around the transforming tetragonal grain (shown schematically in Fig. 6.11). This is also a positive term. In fact, the only energy factor definitely *favouring* the low-temperature transformation is the lower Gibbs function of the monoclinic phase relative to the tetragonal, the *chemical* energy difference, ΔG_{tr}. The overall energy change (*chemical* and *mechanical*) during the transformation (ΔU_{total}) can then be represented by the equation

$$\Delta U_{total} = \Delta U_{strain} + \Delta U_{interfacial} + \Delta G_{tr}. \tag{6.18}$$

For the transformation of a tetragonal grain within a matrix of untransformed material to occur spontaneously ΔU_{total} must be $\leqslant 0$. One way by which this condition can be achieved is through increasing (negative) ΔG_{tr} by lowering the temperature, to a new critical temperature T_o', although for very small crystals this may mean to temperatures well below room temperature. The alternative way is to relax the restraining forces – by the application of suitable counterbalancing (assisting, therefore also negative) tensile or shear stresses, $\Delta U_{applied}$. With a

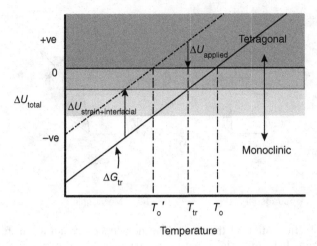

Figure 6.12 The influence of temperature and applied energy on the tetragonal-to-monoclinic transformation. The total transformation energy, ΔU_{total}, shown as a function of temperature: the (negative) transformation energy is decreased by natural strain and interfacial energies, and increased by the strain energy of an appropriately oriented applied mechanical load ($\Delta U_{applied}$).

sufficiently large and correctly oriented applied stress, ΔU_{total} will again become zero, and the transformation occurs, now at T_{tr}, which is lower than T_o, but higher than T_o'. Such a stress can appear at the tip of a crack, and the associated transformation can have a profound effect on the ability of the crack to propagate. These features are illustrated by Fig. 6.12. Here the (solid) line indicating the basic value of ΔG_{tr} (as a function of temperature) can be forced upwards under the influence of the strain and interfacial energies, and forced down again by the applied stress, until the total energy value (ΔU_{total}) is brought back to the zero line at a suitable temperature (T_{tr}), and the tetragonal-to-monoclinic transformation becomes possible again. This process is the basis for the transformation toughening of zirconia ceramics, and will be considered now.

The morphologies of the tetragonal grains (and the monoclinic crystals into which they transform) are complex and varied (Kriven *et al.*, 1981), but the basic principles can be illustrated if the grains are regarded as cubes. For a cube of tetragonal zirconium dioxide of side d the change in interfacial energy during the transformation is given by:

$$\Delta U_{interfacial} = 6d^2 \, \Delta\gamma_{tr} \tag{6.19}$$

where $\Delta\gamma_{tr}$ is the difference in interfacial energy per unit area of tetragonal phase in its matrix, and the monoclinic crystal formed from it (in its matrix). For the case where the matrix is a different phase, such as aluminium oxide or cubic zirconium dioxide, it may be difficult to predict the sign of the $\Delta\gamma_{tr}$ term, but when the matrix is the tetragonal phase, it is positive by ~0.36 J m^{-2} (Holmes *et al.*, 1972). The larger and deformed monoclinic product crystal will tend to generate volume and shear strains within the adjacent matrix. The volume strain energy (ΔU_{vs}) associated with the phase volume change is (Eshelby, 1961; Davidge, 1980)

$$\Delta U_{vs} = E^* \frac{(\varepsilon_{11} + \varepsilon_{22} + \varepsilon_{33})^2}{6} \tag{6.20}$$

where E^* is a reduced modulus, given by

$$E^* = \frac{2E_m E_t}{(1 + \nu_t)E_m + 2(1 - 2\nu_m)E_t}. \tag{6.21}$$

The E and ν terms are the Young moduli and Poisson ratios for the monoclinic and tetragonal phases, as shown by the subscripts, and the ε_{nn} terms are the resolved strains. The corresponding shear strain energy generated per unit volume of transformed tetragonal phase (ΔU_{ss}) is

$$\Delta U_{ss} = \frac{1}{2} \psi\varphi^2 \tag{6.22}$$

where ψ is the shear modulus of the matrix, and φ the shear strain. Combining ΔU_{vs} and ΔU_{ss} gives the total strain energy change per unit volume (ΔU_{strain})

$$\Delta U_{strain} = \Delta U_{vs} + \Delta U_{ss}. \tag{6.23}$$

The total energy change (ΔU_{total}) of the transformation, per cube (of side d) of tetragonal phase is therefore given by

$$\Delta U_{total} = d^3(\Delta G_{tr} + \Delta U_{strain}) + 6d^2\Delta\gamma_{tr}. \tag{6.24}$$

The condition for transformation, $\Delta U_{tr} \leqslant 0$, then corresponds to the relationship

$$6d^2\Delta\gamma_{tr} = -d^3(\Delta G_{tr} + \Delta U_{strain}). \tag{6.25}$$

Using (6.17) to replace ΔG_{tr} by $\Delta H_{tr}(1 - T_{tr}/T_o)$ gives

$$\frac{6\Delta\gamma_{tr}}{\Delta H_{tr}d} = -\left(1 - \frac{T_{tr}}{T_o} + \frac{\Delta U_{strain}}{\Delta H_{tr}}\right) \tag{6.26}$$

and rearranging

$$T_u = T_o\left(1 + \frac{6\Delta\gamma_{tr}}{d\Delta H_{tr}} + \frac{\Delta U_{strain}}{\Delta H_{tr}}\right). \tag{6.27}$$

Because $\Delta\gamma_{tr}$ for a tetragonal zirconium dioxide matrix is a positive term, and ΔH_{tr} is a negative term, the smaller the grain size d, the lower is T_{tr}, and the greater the degree of undercooling required for spontaneous transformation. This is the reason why small grains of tetragonal zirconium dioxide (less than a few hundred nm or so) can be stable at room temperature (and below). Rearranging (6.27) gives

$$\frac{1}{d} = \frac{\Delta H_{tr}}{6\Delta\gamma_{tr}} \left(\frac{T_{tr}}{T_o} - \left(1 + \frac{\Delta U_{strain}}{\Delta H_{tr}} \right) \right). \tag{6.28}$$

Linear relationships between $1/d$ and T_{tr}/T_o are observed (the graph has a negative slope), and the smaller the $\Delta\gamma_{tr}$ and ΔU_{strain} terms, the larger the degree of undercooling which is possible.

Another way of looking at the effect of crystal size is to note that, while ΔU_{strain} and ΔG_{tr} are both determined by crystal (or grain) *volume*, that is, they *scale* with (dimension)^3, the interfacial energy is proportional to crystal *surface area*, and scales with (dimension)^2. A reduction in dimension by a factor of 10 means that strain energy and chemical energy both decrease by a factor of 1000, while the surface energy decreases by a factor of only 100. This means that as the crystal size is decreased, although the strain energy opposing the transformation decreases rapidly (which is helpful), the chemical energy which is driving the transformation also becomes rapidly smaller, while the interfacial energy remains (by comparison) large. At some critical (small) value of crystal dimension, ΔU_{total} will become positive, and the tetragonal phase will be stable, as is the case.

For a more complete picture of the transformation process, it is necessary to consider other microstructural aspects of the transformation. Putting in reasonable values for shear strain (~0.16) and shear modulus (~85 GPa) leads to the conclusion that the shear energy term ΔU_{ss} is extremely large (>1 GJ m^{-3}), so large in fact, that under normal conditions transformation would never be expected to occur. The reason that it does is that we have overestimated the restraining elastic energy terms (ΔU_{vs} and ΔU_{ss}). In practice, two independent mechanisms may operate to reduce strain energy: *twinning* of the product monoclinic crystals, as is shown schematically in Fig. 6.13, and microcracking of the matrix (Lange and Green, 1984). In fact at one time twinning had been considered to eliminate the shear strain energy completely, through the accommodation of long-range shear strain associated with the transformation. However, it is now known that shear strain does have an important part to play in increasing fracture toughness, both in the monoclinic phase nucleation stage, and through the transformation at a crack surface to make crack advance more difficult, that is, to increase *crack shielding* (Chen and Reyes-Morel, 1987).

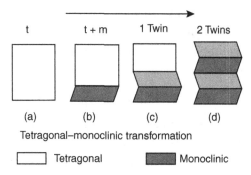

(a)   (b)   (c)   (d)

Tetragonal–monoclinic transformation

☐ Tetragonal   ■ Monoclinic

Figure 6.13 Twinning during the formation of the monoclinic phase; this has the effect of reducing the strain energy. (After Hannink *et al.*, 2002. Reproduced with kind permission of Wiley-Blackwell Publishing.)

The simple discussion above has also treated the transformation purely in terms of the initial and final states, to highlight the influence of grain size and transformation strains. It has not considered phase nucleation, intermediate states, and the mechanisms of processes. The formation of any new phase (monoclinic zirconium dioxide in this case) requires a nucleus, which can then grow. However, the formation of a nucleus, because it is by definition extremely small, tends to be energetically very unfavourable (it has a very high surface energy per unit mass, and the chemical energy available is small). Nucleation may not occur unless the interfacial energy can be reduced in some way. Assistance with nucleation can be made available by microdefects within, or at the surface of, the tetragonal crystal. Because the shear component of the shape strain is large (three to four times that of the dilatational component), shear is also a dominant factor in the initiation of the transformation. An appropriately oriented shear stress can assist the required shear distortion of the tetragonal unit cell into the monoclinic phase.

Even when the overall energy balance for transformation is favourable, the *probability* that a monoclinic crystal will be nucleated within a tetragonal crystal still has to be considered. The larger the tetragonal crystal, the higher is the probability of nucleation occurring, so that there is an additional influence of crystal or grain size on the transformation.

It now remains to apply this information to the question of how the high-temperature transformation, with its volume change and shear components, is able to influence fracture toughness and strength of a zirconia material at room temperature. Part of the answer is that the tetragonal phase may be sufficiently stable to be retained at room temperature, but that transformation can be triggered in response to the presence of a developing crack, and in particular to the tensile and shear stresses at the crack tip. The transformation volume and shape changes then operate so as to increase the energy required for a crack to propagate.

Another possibility is that transformation does in fact occur during cooling, because of inadequate restraining forces. The resulting microcracking can then interact with a propagating crack, and increase toughness.

6.6 Mechanical properties

This section concentrates on the improvements in fracture toughness available in zirconia ceramics, and the related increases in strength. Underlying any toughening process is the existence of a mechanism for the absorption (and dissipation as heat) of the mechanical energy arising from applied forces at a crack tip, so that extra energy must be supplied to allow crack propagation to continue. The energy absorbed is seen as an increase in fracture toughness. However, the natures of the factors involved in the tetragonal-to-monoclinic transformation are complex, and many features of the toughening mechanisms are still not fully understood (Hannink *et al.*, 2000). This treatment presents a very much simplified picture.

6.6.1 Transformation toughening

In order to obtain a high strength and toughness zirconia ceramic, it is necessary to develop a microstructure that at room temperature contains a significant proportion of transformable tetragonal phase. Transformation may then occur spontaneously if the restraining energy is relaxed, or is overcome mechanically, by the imposition of an appropriately oriented and sufficiently large stress. This can take place in the vicinity of an advancing crack. The transformation of each tetragonal grain to a monoclinic grain then generates a volume expansion of ~0.045 and a shear strain of ~0.16, equivalent to local dilatational and shear stresses. It is easy to see that because the transformed tetragonal grains have expanded, they can place a constraining compressive stress on the untransformed material at the crack tip and surface. Propagation of the crack is now inhibited as a result of the extra work that must be done to drive it through the material; fracture toughness, and strength are then increased. The shape of the crack zone as conventionally modelled is illustrated in Fig. 6.14. For maximum effect, transformation of the tetragonal grains must take place not only at the crack tip, but in the wake of the propagating crack. The size of the tetragonal grains, and their degree of stabilisation, are therefore critical. It is important that the grains are neither too small, in which case the required transformation under service conditions will never occur because in effect T_{tr} is below room temperature, nor too large, when the transformation will always occur spontaneously on cooling and the tetragonal phase will not be retained at room temperature (though in this case the resulting localised

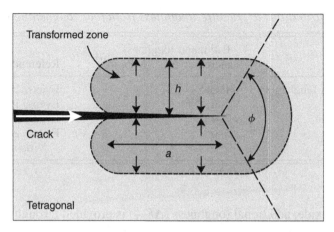

Figure 6.14 The stress field caused by the tetragonal transformation at a crack tip; the compressive and shear stresses then inhibit further crack propagation, or crack shielding. The transformed zone of half-width h and length a behind the crack tip is most influential in increasing toughness. φ is ~120°. (After Hannink *et al.*, 2002. Reproduced with kind permission of Wiley-Blackwell Publishing.)

microcracking may still be able to influence crack propagation, through the microcrack toughening mechanism).

Two groups of models have been applied to the transformation toughening behaviour of zirconium dioxide, based on those which were developed for the transformation induced plasticity (TRIP) steels in the 1950s and 1960s. One group considers the energies associated with an advancing crack, and then dissipated by the transformation. The other group of models is based on linear elastic fracture mechanics, and is used to calculate the stress shielding effects of the transformation on a crack tip. The two approaches give equivalent results, at least during steady state cracking. Before steady state conditions are reached, they are not equivalent (Rose, 1987). The outline discussion below is based on the widely used fracture mechanics approach.

As was described in Section 1.4, strength (σ) and critical fracture toughness (K_{Ic}) are related by the Griffith equation:

$$\sigma = \frac{1}{Y}\left(\frac{K_{Ic}}{c}\right)^{\frac{1}{2}} \tag{6.29}$$

where Y is a numerical factor related to crack shape and loading geometry. For a multi-phase material K_{Ic} can be described by the relationship:

$$K_{Ic} = K_o + \Delta K_c \tag{6.30}$$

where K_o is the matrix toughness, and ΔK_c is the contribution from processes influencing crack propagation, that is, crack shielding (Evans and Cannon, 1986; Evans, 1990). In materials containing metastable, but transformable, tetragonal

Table 6.2 *Contributions to fracture toughness from four toughening mechanisms.*

Mechanism	Estimated toughness increase: ΔK_c / MPa m$^{1/2}$	Reference
Transformation toughening	10–15	Evans, 1985
Microcracking	2–6	Evans, 1985
Crack deflection	2–4	Evans, 1985
Ferroelasticity	3–6	Foitzic *et al.*, 1993; Virkar, 1998

zirconium dioxide, additional toughness (ΔK_{cT}) is provided through the necessity for accommodating dimensional changes taking place in the unit cell during transformation. The overall volume change creates a compressive strain field in the wake of the crack tip, and the net shear component of the transformation strain in the transformation zone also increases the fracture energy. There may be contributions from additional mechanisms, such as microcracking (ΔK_{cM}) and crack deflection (ΔK_{cD}), but these have been considered to be of lesser significance for zirconia materials, with ΔK_c values in the range 2–6 MPa m$^{1/2}$, compared with 10–15 MPa m$^{1/2}$ for the transformation phase change term (Evans, 1985). Estimated values for the increased toughening arising from these mechanisms are listed in Table 6.2. Considerable efforts have been made to model the effects of the unit cell dimension changes, and to predict the volume and shape of the transformation zone at the crack tip, and backwards in the wake of the crack tip, along the crack surface. These models permit the calculation of ΔK_{cT}, allowing comparisons to be made with measured improvements in fracture toughness, which in turn can be used to validate, or refine, the models. The contribution to toughness from the stress-activated tetragonal-to-monoclinic transformation is commonly given by the expression (Hannink *et al.*, 2000)

$$\Delta K_{cT} = XE^* \varepsilon_T V_f h^{\frac{1}{2}} \tag{6.31}$$

in which the numerical term X can also be expressed as ($\eta/(1- v)$). E^* is the effective Young modulus of the material, ε_T is the dilatational strain, V_f the volume fraction of transformed particles, and h the width of the transformed zone measured from the crack surface. η is a variable that depends on the nature of the stress field and the shape of the transformed zone at the crack tip, and v is the Poisson ratio. These terms can be determined by independent measurements, and thus ΔK_{cT} calculated. Values calculated for η range from ~0.2 to ~0.5, depending on the assumptions made about the contributions from transformation volume change, and shear. Values of h are of the order of a few tens of μm.

The volume fraction of transformation (V_f) of the tetragonal phase taking place within a defined transformation zone is an important factor, with major implications for the influence of the transformation on the material. In many common systems, the extent of transformation decreases smoothly from the crack surface to the outer edge of the transformed zone; that is, there is a gradient in the volume fraction of monoclinic phase within the zone. This is described as *subcritical* transformation, in contrast to *supercritical* transformation, where 100% conversion takes place within a sharply defined transformation zone (Budiansky *et al.*, 1983). The earlier (and mathematically simpler) assumption of supercritical transformation led to overestimates in the additional toughening obtained; a more realistic assumption is that the extent of transformation varies within the zone. This has been confirmed experimentally for a number of zirconium dioxide-toughened materials, and it is clear that the stress-activated transformation of zirconium dioxide is always subcritical (Marshall *et al.*, 1990; Yu *et al.*, 1992).

Earlier treatments played down the effects of the transformation shear strain. However, it now seems clear that shear is an essential component, and theoretical analyses show that shear strain plays a very significant part in increasing fracture toughness. While some of the shear strain is indeed accommodated in the formation of monoclinic twins (illustrated by Fig. 6.13 above), a considerable proportion (~40%) of the strain remains to improve toughness by changes in the profile of the crack frontal transformation zone, and the width of the transformation zone (Chen and Reyes-Morel, 1986). Comparisons of measured improvements in toughness with calculated values of η showed that purely dilatational strains could not account for the high values of fracture toughness obtained; in some cases the predicted increase in toughness was only 25% of that actually found (McMeeking and Evans, 1982). It is therefore clear that shear strains must be involved, both in assisting nucleation of the transformation, and after completion of the transformation during subsequent crack movement (Lankford, 1983). Calculations of the shape of the transformation zone at the crack tip show that if only dilatational strains occur, there will be no increase in toughness until the crack tip moves forward, leaving a transformed zone in its wake. Fracture toughness should then be a function of crack length (Fig. 6.15), the property known as *R-curve behaviour*, and there is abundant experimental evidence that this is the case (Swain and Hannink, 1984; Swain, 1986). Calculated values for η assuming contributions from shear and dilatational strains, could be as high as 0.48, and in good agreement with measured values, validating the assumptions made about the contribution of the shear component to toughening.

It is also clear that shear plays an important part in the nucleation of the transformation. It is therefore necessary to treat separately (*decouple*) the influence of shear in the nucleation stage, and in the actual transformation stage.

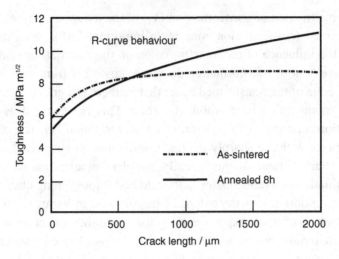

Figure 6.15 Increasing fracture toughness with increasing crack length: the *R*-curve behaviour. In this case it is more effective with the annealed material because of optimisation of the size of the tetragonal crystals. (After Hannink *et al.*, 2002. Reproduced with kind permission of Wiley-Blackwell Publishing.)

Equation (6.31) also shows that, other things being equal, a large value of E^* is desirable. For this reason aluminium oxide ($E \sim 410$ GPa) would be expected to be a better matrix material than zirconium dioxide itself ($E \sim 205$ GPa). Zirconia-toughened alumina, ZTA, materials are now important commercially for their toughness, and have wide applications as cutting materials, because of their hardness, and their ability to retain a sharp edge.

The critical stress (σ_T) required to trigger the transformation is important (Budiansky *et al.*, 1983), and this leads to the concept of *transformation intensity* (Amazigo and Budiansky, 1988), measured by a dimensionless parameter ω, which is defined by the expression

$$\omega = \frac{(1+v)E^*V_f\varepsilon_T}{(1-v)\sigma_T}. \tag{6.32}$$

A material that requires large stresses to initiate transformation has small values of ω and is described as a *weak* transformer. Materials that readily transform have large values of ω and are *strong* transformers (Becher *et al.*, 1987).

A very wide range of zirconia materials is produced, with very good toughness values and strengths. However, it is generally not possible to obtain very high strengths and very high toughnesses in the same material, and choices have to be made. Typical relationships are shown in Fig. 6.16. The reason for this feature is based on the relative magnitudes of the stresses required to cause transformation,

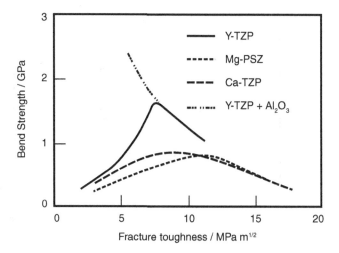

Figure 6.16 Strength as a function of toughness for three types of transformation-toughened zirconia, and a zirconia-toughened alumina. (After Hanninck *et al.*, 2002. Reproduced by kind permission of Wiley-Blackwell Publishing.)

or catastrophic failure, in a tetragonal zirconia material. As would be expected from the Griffith equation, strength at first increases with toughness, with the strength controlled by the (critical) flaw sizes, of between 10 and 100 μm. But as the toughness of the material is made to increase (by changes to the micro-structure to increase its transformability), a point is reached at which the stress required to cause transformation ($\Delta U_{applied}$) is less than that needed for fracture. From here onwards the extent of microcracking generated by the transformation increases, and it is these microcracks which then become the critical flaws. Under loading, transformation occurs preferentially and continues until all the tetragonal phase has been converted to monoclinic, when strength again becomes flaw size limited. In practice, most commercial materials aim to achieve high strength, in preference to very high toughness. Though strengths can be very high, there is a wide scatter of values and corresponding toughness. Figure 6.17 of man-ufacturers' data for four standard commercial zirconia materials shows a wide spread of toughness and strength values. The main weakness of the transform-ation toughening mechanism is that it is fundamentally dependent on the thermodynamic tendency of the tetragonal phase to transform to the monoclinic. For this to exist requires the temperature to be below the critical transformation temperature (T_o). As the temperature is raised, the thermodynamic driving force decreases, until at T_o it is zero, and the potential for a transformation toughening mechanism disappears. Zirconia ceramics are therefore primarily materials for use at (relatively) low temperatures.

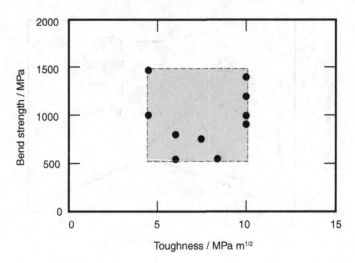

Figure 6.17 Bend strength and toughness values from manufacturers' data sheets for transformation-toughened products.

6.6.2 Microcrack toughening

The spontaneous transformation of tetragonal zirconium dioxide grains in a ceramic matrix is also able to make an independent contribution to transformation toughening. As a result of the very large tensile hoop stresses caused by the volume expansion of the crystal, small cracks will be generated in the immediate vicinity of the transformed grain. These resident microcracks can subsequently extend in the tensile stress field of a propagating large crack or deflect the propagating crack, to increase the energy required for its propagation (Claussen *et al.*, 1977; Rühle *et al.*, 1983). The size of the tetragonal particle is again important, in that it must be large enough to transform on cooling, and yet provide only limited development of the associated microcracks. When the microcracks become so large (around a large tetragonal grain for example) that they are able to interact, their beneficial effect is reduced (as is the case with a siliceous porcelain, containing large quartz grains). Microcracking has long been considered responsible for the high resistance to thermal shock of some zirconium dioxide-based refractories. It also appears to be an important toughening mechanism in composite materials such as zirconia-toughened alumina, although it may not make as large a contribution to toughening as the stress-induced transformation (Evans, 1990).

6.6.3 Surface compression toughening

Because the transformation of tetragonal zirconium dioxide crystals is facilitated when the constraining stresses are reduced, it is easier at the external surface of

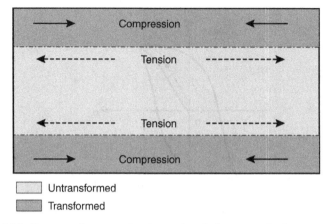

Figure 6.18 Compression toughening: the influence of the tetragonal-to-monoclinic transformation at the surface of the material (see also Fig. 2.21).

the material. Surface transformation may then occur during cutting or grinding when large stresses are applied. As a result of the transformation volume expansion of individual tetragonal grains, the whole surface of the material can be placed into compression over a depth of several μm (Green *et al.*, 1984). This is indicated schematically in Fig. 6.18. Surface crack initiation and propagation is then made more difficult, and the strength of the component is increased, in much the same way in which the compressive surface layer of a "toughened" glass strengthens the glass (Gupta, 1980), and glazing improves the strength of porcelain (Section 2.69) (Xu *et al.*, 1997). The tensile stress required to allow a crack to propagate is increased, and a material whose surfaces have been ground can have twice the strength of the same annealed material (Claussen and Rühle, 1981). The strengthening effect is determined by the severity of the grinding treatment, and the depth of transformation induced below the surface. Ideally, for maximum benefit, the depth in which transformation has occurred should be just larger than the size of the critical flaw, and many attempts have been made to increase the thickness of the surface compressive layer (Green, 1983; Cutler *et al.*, 1987). Residual surface compressive stresses have been generated of up to 400 MPa, giving an enhanced apparent fracture toughness of up to 30 MPa m$^{1/2}$ (Lakshminarayanan *et al.*, 1987).

6.6.4 Ferroelasticity

Ferroelastic switching of tetragonal crystals (see Fig. 6.4) is another possible toughening mechanism. This can occur as a result of applying appropriate stresses to a tetragonal crystal – a compressive stress along the *c*-axis, a tensile

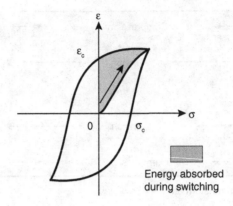

Figure 6.19 The energy absorbing action of the ferroelastic transition: the shaded area reflects the energy of the stress-induced c- to a-axis switch.

stress along one of the a-axes, or a suitable shear stress (Aizu, 1970). During crack propagation, if the stress at the crack tip exceeds the coercive stress required for domain switching in a suitably oriented tetragonal crystal, a- to c-axis conversion can occur, effectively reducing the local stress, and providing additional crack shielding. The energy absorbed in the switching corresponds to the shaded area in Fig. 6.19, and switching just ahead of a crack tip would be expected to lead to an increase in fracture toughness. An approximate indication of the magnitude of this effect on fracture toughness is given by

$$K_c = K_c^o \left(1 + \frac{\sigma_c \varepsilon_c E h}{2 K_c^{o2}} \right) \tag{6.33}$$

where K_c^o is the background toughness, h is the transformed zone half-width σ_c the coercive stress and ε_c the coercive strain. Putting in reasonable values (that is, $h \sim 30$ μm, $\sigma_c \sim 600$ MPa, $\varepsilon_c \sim 0.01$), suggests that fracture toughness could be increased by ~ 6 MPa m$^{1/2}$, which is a significant amount (Virkar, 1998). One potential advantage of the ferroelastic switching mechanism over mechanisms involving the tetragonal transformation is that, because it is less sensitive to temperature, it may be a more effective high-temperature toughening mechanism.

6.7 Other physical and mechanical properties of zirconium dioxide

It is not a simple matter to discuss the intrinsic physical properties of zirconium dioxide (such as density, or thermal expansion) because there are no unique values; the properties of a material are determined by its composition, and the phases present. Other problems arise because of the tetragonal-to-monoclinic phase transformation on cooling from production temperatures, and it is difficult to produce large single crystals of pure monoclinic and tetragonal zirconium dioxide. Most measurements have therefore been made on polycrystalline materials, usually containing varying amounts of stabilising oxides. In contrast, large single crystals of cubic zirconium dioxide can readily be produced by "skull-melting" zirconium dioxide powder to which has been added a sufficient quantity of stabilising oxide (calcium oxide or yttrium oxide), and then allowing the growth of a single crystal by controlled slow cooling. This very high-temperature technique uses induction heating (the zirconium dioxide–oxide solid solution is electrically conducting) in a water-cooled crucible, in which the material in direct contact with the molten oxide is simply a layer of solid oxide (of low thermal conductivity). Single crystals of fully stabilised zirconium dioxide have some attraction as gemstones, because they are reasonably hard and have a similar refractive index to diamond. The refractive index is 2.17 for zirconium dioxide, and 2.42 for diamond. "Brilliance" is induced by the internal reflections, and "fire" is caused by a high dispersion factor that splits white light into its primary colours (Wood and Nassau, 1982; Wood *et al.*, 1990).

6.7.1 Density

Zirconium has the high relative atomic mass of 91.22, and the relative molar mass of zirconium dioxide is 123.22. The density of a sintered polycrystalline zirconium dioxide ceramic will also depend on the amount and type of stabilising oxide incorporated, and the proportions of the higher density tetragonal and cubic phases, as well as, of course, on the volume of residual porosity. Normal densities for partially stabilised zirconia are in the region of 5.8 Mg m^{-3}, while those for a partially stabilised mixed-phase zirconia containing smaller amounts of additive oxide are between 6.0 and 6.2 Mg m^{-3}. Zirconia ceramics are ~50% more dense than aluminium oxide (~4 Mg m^{-3}), but less dense than steel at ~7 Mg m^{-3}. They can contain small amounts of intergranular silicate glass, a consequence of the use of zircon as a source material for zirconium dioxide. The glass is normally distributed at the grain boundaries as amorphous films of thickness 1–10 nm, and along three-grain edges – the "triple points" (Rühle *et al.*, 1984; Hughes and Badwal, 1991), as it is in alumina and silicon nitride ceramics.

6.7.2 Modulus and hardness

The room-temperature Young modulus of pure monoclinic zirconium dioxide is ~244 GPa, and values have the typical exponential dependence on porosity, with the constant *b* in the region of 3.5 (Garvie, 1985). For sintered stabilised

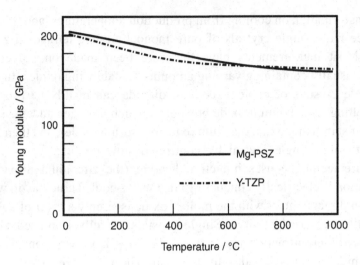

Figure 6.20 Young modulus as a function of temperature for two types of standard PSZ and TZP transformation-toughened zirconia materials. (After Adams *et al.*, 1997. Reproduced by kind permission of Wiley-Blackwell Publishing.)

zirconium dioxide the modulus is ~205 GPa, about half that of aluminium oxide. Modulus varies slightly with the composition of the material, and falls with increasing temperature (Ryshkewitch and Richerson, 1985e; Adams *et al.*, 1997). A typical pattern of behaviour is illustrated for two materials, a magnesia partially stabilised zirconia, and a tetragonal zirconia polycrystal, by Fig. 6.20. Even allowing for the weakening presence of grain boundaries and the several per cent by weight of additive oxide in the polycrystalline materials, this is a relatively low value and the consequence of intrinsically weaker bonding between the larger zirconium atoms, and the oxygen atoms. The room-temperature shear modulus, which is important in the context of the tetragonal-to-monoclinic transition because of the shear distortion involved, is ~96 GPa for the pure monoclinic phase, and ~84 GPa for a partially stabilised material. The Poisson ratio is ~0.3, a value found with many ceramic materials (Selçuk and Atkinson, 1997).

Zirconium dioxide is not a particularly hard material: the mineral baddeleyite has a hardness of 7 on the Mohs scale. Sintered stabilised zirconium dioxide has microhardness values under normal indentation loads in the range of 9–14 GPa, the value depending, again, on the amount of stabilising oxide incorporated (King and Yavorsky, 1968; Ryshkewitch and Richerson, 1985b). These values are high enough to allow zirconia materials to be used as grinding media, and to resist wear generally. For this kind of application, the high density is also beneficial, in giving the moving media high kinetic energy.

6.7.3 Thermal conductivity

Polycrystalline zirconia materials have very low thermal conductivities, in fact considerably lower than those of sintered alumina materials (Garvie, 1976). Typical values for fully stabilised sintered cubic zirconium dioxide are in the region of 1.7 W m^{-1} K^{-1} at 100 °C, rising to 2.1 W m^{-1} K^{-1} at 1700 °C (Fitzsimmons, 1950). The low conductivity is partly a consequence of the high atomic mass of the zirconium atom: it is also reduced by the presence of the additive oxides in solid solution in the zirconium dioxide crystal lattice. These low values mean that zirconium dioxide can be a very effective thermal insulation material, or thermal barrier layer. Many applications have been developed for zirconia thin films of 100 μm or so, to mm thickness (Stiger *et al.*, 1999). They are very widely used as protective films on alloy surfaces in the hot zones of gas-turbine engines. The bulk applications of zirconia as industrial refractories, where thermal conductivity may be a factor influencing the choice of material, though still important, are excluded from this review.

6.7.4 Thermal expansivity

The thermal expansion behaviour of zirconium dioxide, and its interpretation, is also complicated by the transformations between the polymorphs (Fig. 6.21). Each polymorph has its own thermal expansion characteristics, influenced by the symmetry of its unit cell (Garvie, 1970). The thermal expansivity of pure monoclinic zirconium dioxide up to the transformation temperature is low, with a mean value of around 4.5 MK^{-1} (Curtis, 1947). The monoclinic unit cell is highly anisometric; the *b*-axis mean expansion coefficient over the range 25–1100 °C is 2 MK^{-1}, the *a*-axis ~8 MK^{-1}, and the *c*-axis ~13 MK^{-1} (Patil and Subbarao, 1969; Garvie, 1970). In these respects it shows much more extreme characteristics than aluminium oxide. The tetragonal phase is also anisometric (Fig. 6.22) with the *c*-axis expansion higher than that of the *a*-axis, by an amount that depends on the stabiliser content (Schubert, 1986). For a very fine-grain fully tetragonal phase zirconium dioxide, the mean thermal expansivity over the temperature range 25–1000 °C is high at ~10 MK^{-1}. Mixed-phase zirconia materials have thermal expansion values depending on the proportions of each phase (Subbarao, 1981), but a typical mean value for partially stabilised zirconia consisting of tetragonal crystals dispersed in a monoclinic zirconium dioxide matrix is of the order of 8 MK^{-1}. This value is close to that of aluminium oxide, so that resistance to thermal shock might be expected to be similar (and not particularly good). The expansivity of fully stabilised cubic zirconia, over the standard range of room temperature to around 1200 °C, is higher still, in the region of 9–12 MK^{-1}.

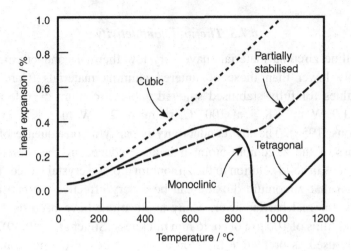

Figure 6.21 The thermal expansion of zirconium dioxide, showing the effect of the tetragonal-to-monoclinic transformation on reducing the overall shrinkage during cooling. (After Garvie, 1970. Reproduced with kind permission, Elsevier Ltd.)

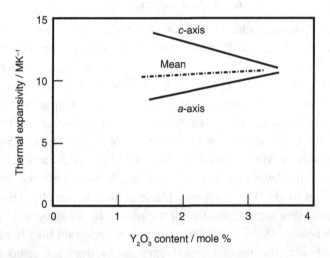

Figure 6.22 The thermal expansion of the tetragonal phase of zirconium dioxide as a function of yttrium oxide content, with extensive anisotropy. (After Schubert, 1986. Reproduced with kind permission of Wiley-Blackwell Publishing.)

6.7.5 Thermal shock

The ability to withstand thermal shock is commonly quantified by the Hasselman R-parameters (Hasselman, 1969). For strong materials for which the R- and R'-parameters can be applied, a key factor is the thermal expansion coefficient.

However, as was realised at a very early stage in the commercial development of zirconia refractories, the thermal shock resistance of materials containing additive (or impurity) oxides can in fact be very good. Values for the critical shock temperature in a standard quench test into cold water are in the region of 250–375 °C (Ryshkewitch and Richerson, 1985e), which is slightly better than with a sintered alumina. Putting in reasonable values for strength (500–1000 MPa) and Young modulus (205 GPa) gives *R*-parameters in the range ~200–400 K, for an assumed mean thermal expansion coefficient of 8 MK^{-1}. The *R'''*-parameter for crack propagation might be applied to microcracked materials, and the values are very low, as would be expected from a (relatively) low-modulus but high-strength material, at 0.3 to 1 M Pa^{-1}.

While the relatively low Young modulus of zirconium dioxide would also mean that thermal stresses would be correspondingly smaller (other things being equal perhaps around 50%), the good resistance to thermal shock was at first attributed to the tetragonal-to-monoclinic transformation, which would have the effect of reducing the overall mean thermal expansion coefficient (Curtis, 1947). An alternative view was that the tetragonal-to-monoclinic transformation was causing the development of annular microcracks around the zirconium dioxide grains (Hasselman, 1969; Claussen *et al.*, 1977). These cracks would further reduce the capacity of the material to develop large stresses, and increase the work required for crack propagation. For this type of material the *R''*-parameter (which does not contain the thermal expansion coefficient) might be more appropriate. In contrast, the phase stability in fully stabilised, and dense, cubic zirconia materials and the high thermal expansivity mean that cubic zirconia components are susceptible to thermal shock, and need to be treated with care.

6.7.6 Thermal stability

The incorporation of stabilising oxides at their usual levels has the effect of lowering the zirconium dioxide melting point. This is still very high – certainly much higher than that of aluminium oxide (~2054 °C) – and the effect is not of significance from the point of view of the use of the materials as structural ceramics. While the very high melting-point of zirconium dioxide has always underpinned the usefulness of zirconia materials as refractories and high-temperature thermal insulators, the successful use of zirconium dioxide as a structural ceramic depends on phase stability. The real importance of temperature for the practical applications of zirconia ceramics is that, as noted above, as the tetragonal–monoclinic transition temperature is approached and the tetragonal phase becomes more stable, the toughening and strengthening action of the

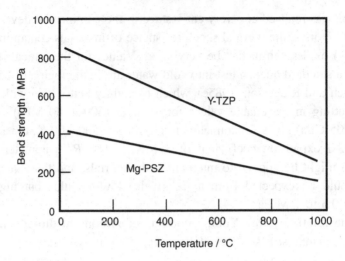

Figure 6.23 Strength as a function of temperature for two standard types of transformation-toughened zirconia. (After Cannon, 1989. Reproduced with kind permission of Elsevier Ltd.)

transformation is lost. Strength decreases approximately linearly with temperature, typically falling to 30–40% of the room-temperature value by 1000 °C (Cannon, 1989). This is illustrated by Fig. 6.23. Above ~1000 °C the mechanical properties of the material revert to those of a normal reasonably fine-grain and low-modulus oxide ceramic. This type of behaviour follows that of siliconised silicon carbide, the strength of which becomes that of a porous material at a specific temperature, in this case the melting-point of silicon (1412 °C). Creep also becomes an important factor determining the usefulness of a zirconia-toughened ceramic. Creep resistance is not as good as that of alumina and silicon carbide at comparable temperatures (Cannon and Langdon, 1988), partly because of the high concentration of oxide lattice vacancies, and the high mobility of the O^{2-} ion in the zirconium dioxide lattice. It is also in part a consequence of the presence at the grain boundaries of intergranular glasses formed from small amounts of impurity silicon dioxide and additive oxides. In this respect, the zirconia ceramics are similar to sintered alumina, and silicon nitride materials. At temperatures less than about 800 °C, the microstructures appear to be fairly stable, and there are no significant property changes with aging. Zirconia materials are therefore best suited for use at relatively low temperatures, near to room temperature, although as with any fine-grain ceramic reasonable strength can still be expected above this point: manufacturers quote values of more than 200 MPa at 800 °C.

However, even at temperatures well below 1100 °C, there are questions of the long-term stability of the (partially stabilised) tetragonal phase, and therefore strength. With low-solubility oxides (MgO and CaO), the tetragonal and cubic phases are both intrinsically unstable over this (relatively low) temperature range: magnesium oxide stabilised cubic zirconium dioxide in particular has a tendency to transform spontaneously to the monoclinic phase when heated above 900 °C for long periods of time. Tetragonal zirconia materials containing the higher solubility Y_2O_3 and CeO_2 also tend to suffer from instability, leading to severe strength degradation on long-term exposure to water vapour or superheated water. This has been attributed to surface chemical reactions of the zirconium dioxide, leading to destabilisation of the surface tetragonal phase, the development of surface stresses, and micro- and macrocracking (Lilley, 1990). Several factors have been considered responsible, but one of the most widely accepted explanations is that the stabilising oxide (such as yttrium oxide) is hydrolysed at the surface of the zirconia, with consequent loss of tetragonal phase stability and the introduction of incipient flaws (Schmauder and Schubert, 1986; Lilley, 1990):

$$ZrO_{2(t)}, Y_2O_3 + 3H_2O = ZrO_{2(m)} + 2Y(OH)_3. \qquad (6.34)$$

A number of ways of reducing the extent of this problem have been examined. The susceptibility to this type of degradation can be minimised by increasing the stability of the tetragonal phase through the use of small grain sizes, and the controlled incorporation of oxides has been the most favoured approach (Watanabe *et al.*, 1984; Lange *et al.*, 1986; Hannink and Swain, 1994).

6.7.7 Wear

Dense zirconia with its high hardness, fracture toughness and strength, is a "wear resistant" material. The fine grain size of many TZP and ZTA materials is also likely to be beneficial (Davidge and Riley, 1995). However, as with most materials, the dominant wear mechanism, and the wear rates, vary with the specific conditions, that is, the load, interface relative velocity, and environment (Buckley and Miyoshi, 1989). Under sliding conditions at low loads, plastic deformation and polishing are important mechanisms for the loss of material. In an aqueous environment, tribochemical wear can give higher rates of removal of material. This process may be related to the tendency for tetragonal zirconium dioxide to transform spontaneously as the result of the loss of its stabilising oxide (Lee *et al.*, 1993; Rainforth, 2004), and it has to be remembered that very high temperatures can be reached locally, particularly under sliding conditions. At higher loads and sliding speeds, fracture mechanisms become more important, leading to detachment of material as a result of crack propagation and interlinking, and the loss of entire grains. Transformation-toughened zirconia materials have important uses where good resistance to wear is required, with many applications involving sliding and abrasive wear under wet conditions.

6.7.8 Electrical conductivity

Very pure zirconium dioxide, like aluminium oxide, has a very large band gap of ~4.99 eV (Bendorati and Soloman, 1965), and is an electrical insulator. Its semiconductivity follows the standard Arrhenius (logarithm as function of reciprocal thermodynamic temperature) relationship. At room temperature, the conductivity is of the order of 1 pS m^{-1}, but it rises to only 1 μS m^{-1} at 1000 °C. The incorporation of di- and tri-valent oxides such as calcium oxide and yttrium oxide as solid solutions in the ZrO_2 lattice does not affect low-temperature conductivity to a significant extent (conductivity does increase slightly, but this is not of practical importance), but above about 900 °C these materials become very good electrical conductors. The conductivity continues to increase markedly with rising temperature, independently of oxygen pressure, with an activation energy of about 100 kJ mol^{-1}. This property of zirconium dioxide has been known since 1897, when it was discovered (Nernst, 1897, 1900) that by preheating stabilised material containing 15% by weight of yttrium oxide to 900 °C, a glowing "self-heating" conductor could be obtained (the "Nernst light" rapidly became obsolete with the development of the incandescent tungsten wire light). It was not until 1943 that the "defect" nature of the stabilised zirconium dioxide crystal structure was established (Wagner, 1943). The lattice defects in this case consist of cations, of valency other than four, occupying zirconium sites, and the creation of counterbalancing charge defects in the oxide ion sublattice. It was later confirmed that electrical conductivity (*ionic* conductivity, in which current is carried by oxide ions) is the result of considerably enhanced oxygen ion mobility, allowed by a high concentration of oxide ion vacancies in the ZrO_2 lattice (Kingery *et al.*, 1959). The formation of (for example) a Y_2O_3–ZrO_2 solid solution can be expressed by a form of the chemical reaction (6.9):

$$Y_2O_3 = 2Y_{Zr}^- + 3O_O + V_O^{2+}. \tag{6.35}$$

For every unit of yttrium oxide incorporated in the zirconium dioxide lattice, the three oxygen ions create, in effect, a small extension to the cubic oxide sublattice. Two yttrium ions take up positions on two zirconium ion sites, and one sublattice oxide vacancy is created. Another way of expressing this is to say that, because a cation of lower formal positive charge (Y^{3+}) replaces one of higher (the Zr^{4+}), less balancing O^{2-} is required and there is a deficiency of oxygen ions in the ZrO_2 lattice. These vacant oxide sites allow, at a sufficiently high temperature (which turns out to be between 800 and 900 °C), an adequate mobility of lattice oxide ions, each carrying a charge of 2–, and permitting the passage of electric current. Electrical conductivity initially increases linearly with the oxide ion vacancy concentration, and values therefore depend on the nature of the added oxide and its concentration (Kiukkola and Wagner, 1957). Conductivity reaches a maximum near the additive level required to just stabilise the cubic phase, and then falls, because of interactions between the inserted cations and the oxide ion vacancies. For cubic zirconium dioxide containing 15% by weight of Y_2O_3, conductivity is of the order of 200 S m^{-1} at 1000 °C (Ryshkewitch and Richerson, 1985b). In this respect cubic zirconium dioxide has some similarity with porcelain, which also becomes an ionic conductor (Na^+ and K^+ in this case) at temperatures above ~300 °C.

The oxygen ion mobility in stabilised cubic zirconium dioxide also provides a second interesting and very useful high-temperature property, in that under an oxygen pressure gradient a surface electrical potential difference is developed, that can be used as a measure of oxygen pressure (Steele *et al.*, 1981), or as the basis of a solid state battery:

$$E - E_0 = \frac{RT}{4F} \ln\left(\frac{p_{O_2}}{p^o_{O_2}}\right) \qquad (6.36)$$

where E and E_0 are the cell wall potentials for oxygen pressures p and p^o, and F is the Faraday (96.485 kC mol^{-1}). With oxygen pressures of 0.21 and 10^{-10} atm the potential difference developed at 1000 °C is ~0.6 volts. These are specialised topics with extensive literature, and because they take the subject into the electroceramics area, they will not be discussed further.

6.8 Ceramic materials

The term *zirconia* covers a wide range of materials of varying composition, which although consisting predominantly of zirconium dioxide have widely varying compositions. As with the silicon carbide and silicon nitride materials, it is therefore more convenient to treat each group separately. Because compositions can be quite complex, a standard system for identifying zirconia materials has evolved. The common notation places the cation symbol of the stabilising oxide before the PSZ or TZP abbreviation, together with the amount used expressed as mole%. A partially stabilised zirconia containing 5 mole% of magnesium oxide would then be represented as 5Mg-PSZ. Symbols indicating any non-stabilising oxides are placed after the abbreviation, with the amounts expressed as weight%. For example, a TZP zirconia, stabilised with 3 mole% yttrium oxide, and incorporating 20 weight% of aluminium oxide, would be represented as 3Y-TZP/20A (Hannink *et al.*, 2000).

Table 6.3 *Zirconia ceramics: terminology and common abbreviations.*

PSZ	**Partially stabilised zirconia**
Mg-PSZ	magnesium oxide doped partially stabilised zirconia
Ca-PSZ	calcium oxide doped partially stabilised zirconia
TZP	**Tetragonal zirconia polycrystal**
Y-TZP	yttrium oxide doped tetragonal zirconia polycrystal
ZTC	**Zirconia-toughened ceramic**
ZTA	zirconia-toughened alumina

A selection of the most common types, and their abbreviated descriptions, are shown in Table 6.3.

Zirconia ceramics are normally produced by sintering compacted high-purity powders, shaped using conventional ceramic processing techniques. Modern zirconium dioxide powder production methods provide very fine powders, often incorporating any required additive oxides on a very homogeneous and fine scale – a requirement for the development of the controlled grain size micro-structures necessary for attainment of the best properties of the zirconia materials (Standard and Sorrell, 1998). The oxides can be incorporated in amounts ranging up to ~10% by weight during the final stages of the zirconium dioxide purification process, using techniques designed to ensure homogeneity of distribution. Fully formulated zirconium dioxide powders, with accurately controlled compositions, are commercially available with mean particle sizes down to 25 nm ($\sim 40 \text{ m}^2 \text{ g}^{-1}$), which are fine enough to be sintered to full density at ~1400 °C, although higher temperatures may be needed to develop the required grain sizes. Otherwise, a range of methods for blending the zirconium dioxide and its stabilising additive, and any host phase, can be used. These include the standard mechanical mixing of precursor powders, or chemical blending and co-precipitation methods, followed by an appropriate calcination treatment. Calcined powders may then be milled further to give smaller particle sizes. An illustrative selection of commonly used powders is shown in Table 6.4. The production process for the PSZ and ZTA types of sintered material generally involves a similar pattern of heating stages. These are designed to produce a uniform dispersion of tetragonal zirconium dioxide crystals of the required size, in a matrix of cubic or monoclinic zirconia, or other ceramic such as alumina or silicon nitride. In principle, practically any ceramic could be toughened by incorporating either stabilised or unstabilised tetragonal zirconium dioxide. Zirconia-toughened alumina, ZTA, is a commercially important material, used because of its very good wear characteristics. It contains an appreciable concentration of the (metastable, and convertible) tetragonal phase at room temperature. TZP materials contain almost

Table 6.4 *Analyses of typical zirconium dioxide powders.*

Analysis / unit	ZrO_2	3Y-TZP I	3Y-TZP II
ZrO_2 (+ HfO_2) / weight%	99.5	94.7	94.1
Y_2O_3 / weight%	–	5.2	5.4
SiO_2 / ppm	1800	40	<1500
Al_2O_3 / ppm	500	70	–
Fe_2O_3 / ppm	300	30	<100
Na_2O / ppm	–	60	<300
TiO_2 / ppm	500	–	<1500
CaO / ppm	300	–	<500
Loss on ignition / %	0.20	0.60	<0.30
Agglomerate size / µm	1.1	0.3	0.4–0.7
Primary crystallite size / nm	–	24	30–150
Specific surface area / $m^2 g^{-1}$	4–8	17	~4–10

100% tetragonal phase. In PSZ materials there are lower concentrations of tetragonal phase crystallites dispersed in a cubic, or monoclinic (or a mixture of the two) phase matrix.

Sintering is carried out in air, at temperatures up to 1800 °C, with schedules designed to give maximum sintered density, and the required tetragonal or cubic phases, and grain sizes. Liquid-forming additives are not normally deliberately added, and the standard stabilising oxides do not form low-melting binary eutectics with zirconium dioxide. Therefore there should be no liquids at sintering temperatures, unlike the silicate additives used for sintering aluminium oxide, and silicon nitride. However, silicon dioxide is a common impurity, generally the consequence of using zircon as the source material, and may be responsible for the formation of small amounts of liquid during sintering. The densification process is therefore notionally that of *solid state sintering*, although it is suspected that small amounts of liquid phase generated by trace amounts of impurity oxide may actually be involved (Standard and Sorrell, 1998). Part of the evidence for this is the presence of the thin, ~1 nm width, amorphous silicate films at most grain boundaries (Claussen *et al.*, 1984). The rate of transport of material within the grains is likely to be controlled by diffusion of the cations, on the basis of the relative magnitudes of their diffusion coefficients. While the Zr and O lattice diffusion coefficients are both extremely small at low temperature (D_{Zr} is up to eight orders of magnitude smaller than D_O, and thus rate-controlling), atomic movement is still possible (Rhodes and Carter, 1966; Simpson and Carter, 1966; Brook, 1981; Park, 1991; Kilo, 2003). Figure 6.24 summarises lattice diffusion data for a number of transformation-toughened and fully stabilised zirconias. Sintering rates are not affected significantly by the type and content of stabilising

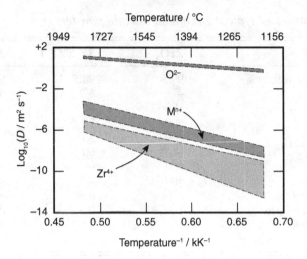

Figure 6.24 Lattice ionic diffusion coefficients (D) as a function of temperature, showing the much greater mobility of the oxygen ion in the lattice. Zirconium diffusion is rate-controlling in processes involving material transport. (After Standard and Sorrell, 1998. Reproduced with kind permission of TransTech Publications.)

oxide. However, grain boundary diffusion is likely to be the predominant process in densification in these fine-grain materials, and the grain boundary mobilities of the zirconium and oxide ions are similar (Badwal and Drennan, 1994). Grain boundary mobility will also be enhanced by the impurity silicate films.

Considerable effort has also been put into creating zirconia microstructures of considerable subtlety, in order to extract maximum value from the tetragonal–monoclinic transformation, and a summary of some of the possibilities has been provided (Claussen, 1984). For example, small polycrystalline units of stabilised zirconia can be created within a matrix of another form of zirconia, or another ceramic. Microstructures are referred to as *duplex* structures when they contain zirconium dioxide in more than one phase. Although quite sophisticated microstructures have been developed (Lutz *et al.*, 1991; Lutz and Swain, 1991), most commercial zirconia microstructures are generally of a simple form.

6.8.1 Partially stabilised zirconia (PSZ)

PSZ ceramics are typically produced by sintering the blended powder at a temperature such that the composition lies just within the cubic phase field (~1800 °C or higher) to give predominantly cubic material. This is followed by rapid cooling (>500 °C / hour) through the tetragonal/cubic region to below 800 °C, when fine

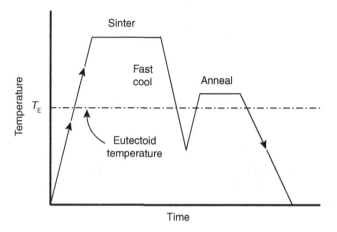

Figure 6.25 The general form of the sintering schedule, designed to ensure development of the optimum size and concentration of transformable tetragonal crystals.

(10–50 nm) tetragonal crystals precipitate homogeneously within the cubic grains. Isothermal aging just above the *eutectoid temperature* at ~1000–1400 °C is then carried out for several hours to allow the development of a high density of tetragonal crystals of the optimum size. Details of the sintering schedule depend on the additive oxide and the amount present, and are critical for the properties of the material: an illustration of this type of heating programme is shown in Fig. 6.25. Because sintering temperatures are high, a PSZ normally consists of a relatively coarse-grained (~30–70 μm) cubic zirconium dioxide matrix, containing an intragranular dispersion of fine tetragonal phase particles. Small amounts of coarse monoclinic grain boundary phase may also be present. The amount of tetragonal phase depends on the stabilising oxide: 30–40% is common in commercial magnesium oxide stabilised materials. Many oxides have been tested as possible stabilising additives for the production of PSZ materials, but magnesium oxide is the most commonly used.

Ca-PSZ

These materials were the first to be developed (Garvie *et al.*, 1975), but are now not so widely used because it can be difficult to control the microstructure. Calcium oxide is typically present at the 6.5–9.5 mole% (3–4.5 weight%) level, with sintering at 1700–1900 °C in the cubic phase field to full density. Rapid cooling (>500 °C min^{-1}) to below ~800 °C precipitates very fine (~10 nm) tetragonal phase crystals, which are then ripened to the critical size by annealing

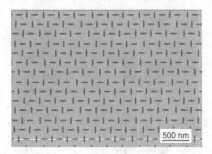

Figure 6.26 An illustration of a typical pattern of a partially stabilised zirconia microstructure, with fine precipitated tetragonal crystals in a matrix of cubic and monoclinic zirconia.

at ~1300 °C. The fully developed tetragonal crystals are cuboid, and must be <90 nm in size for stability at room temperature (Hannink *et al.*, 1981). Controlled growth of the tetragonal crystals in a reasonable time requires careful selection of annealing temperature. At higher temperature, there is a risk of overaging, because the required time is too short to control properly. Room-temperature bend strengths of ~650 MPa can be obtained after several hours' aging (Hannink *et al.*, 1981).

Mg-PSZ

The most common additive used in commercial PSZ materials is magnesium oxide, at the 8–10 mole% (2.8–3.5 weight%) level. These compositions, and the processing temperatures, were shown in Fig. 6.5. Materials are sintered to full density just inside the cubic phase field at 1700–1800 °C, and rapid cooling to ~1000 °C gives small lenticular intragranular tetragonal crystals with aspect ratio ~5, and diameter ~50 nm. Coarsening is carried out above the eutectoid temperature at ~1400 °C. The final crystals are usually lenticular with an aspect ratio of ~5, and size of the order of 200 nm width, and ~1 μm length (Hannink, 1978; Porter and Heuer, 1979; Lanteri *et al.*, 1986). A schematic illustration of the microstructure characteristic of this type of material, containing small tetragonal crystals in a matrix of cubic and monoclinic zirconium dioxide, is shown in Fig. 6.26. A compound micrograph showing large (~50 μm) cubic grains, with ~250 nm tetragonal precipitates, is shown in Fig. 6.27.

Overall microstructures can, however, be quite complex, and there can be several morphologies of tetragonal crystal, three of which have been termed *primary*, *large random*, and *secondary* (Hughan and Hannink, 1986). Figure 6.28 shows the influence of aging time on strength in a PSZ type of material. Over-

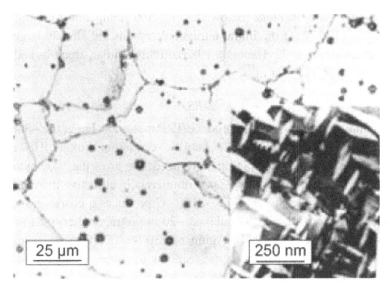

Figure 6.27 The microstructure of a PSZ, showing large cubic grains, and the fine lenticular precipitates of tetragonal zirconium dioxide. (Reproduced by kind permission of the Max-Planck-Institut.)

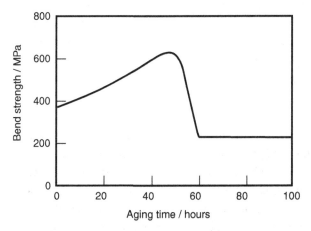

Figure 6.28 The influence of aging on strength, for a PSZ material. (After Hannink *et al.*, 2000. Reproduced with kind permission of Wiley-Blackwell Publishing.)

aging causes excessive tetragonal grain growth, and on cooling to room temperature, spontaneous transformation to the monoclinic phase takes place (Porter and Heuer, 1979). 10Mg-PSZ is commercially available with strengths of 700–1500 MPa, combined with a fracture toughness of up to ~15 MPa m$^{1/2}$.

Alternatively, *sub-eutectoid* (below the eutectoid temperature) aging may be carried out at 1100 °C, if the initial tetragonal crystals are already large enough (Hannink and Garvie, 1982; Hannink, 1983), giving similar strengths and fracture toughness of up to 11 MPa m$^{1/2}$.

Y-PSZ

Yttrium oxide is normally used to produce TZP materials, but at the ~3–6 mole% level it can also be used to partially stabilise zirconium dioxide. However, the higher temperatures and longer aging times needed to grow the 300 nm tetragonal crystals required make this additive less commercially attractive than magnesium oxide. Preliminary sintering at ~1700–2100 °C produces a coarse-grained (>50 μm) cubic matrix, within which equiaxed ~20 nm tetragonal crystals precipitate on cooling (Matsui *et al.*, 1984). Aging at 1300–1400 °C gives a mixture of tetragonal and t'-phases, and properties similar to those of other PSZ systems (Green *et al.*, 1989; Foitzik *et al.*, 1993).

6.8.2 Tetragonal zirconia polycrystals (TZP)

TZP materials consist of a fine-grain size matrix of tetragonal phase. Blended powder is sintered under conditions approximating to the boundary of the cubic and cubic/tetragonal phase fields, and sintering conditions must be chosen to achieve a high density, while retaining a fine grain size. Temperatures are normally in the region 1400–1600 °C. Stabilisation by yttrium oxide and cerium oxide has been widely investigated.

Y-TZP

In commercial materials the most widely used additive is yttrium oxide. With <1 mole% Y_2O_3 the tetragonal phase is unstable at room temperature. With between 1 and ~2 mole% Y_2O_3, although 100% tetragonal phase can be obtained at sintering temperatures, some monoclinic is formed during cooling. With 2–4% additions the compositions lie in the tetragonal–cubic phase field, and the tetragonal phase is readily obtained at room temperature. 3 mole% yttrium oxide is therefore a standard commercial composition (3Y-TZP). The microstructure consists predominantly (60–100%) of equiaxed tetragonal crystals with grain sizes in the region of 300 nm to 2 μm; the remainder of the material is cubic (Lange, 1982b). There is usually also an yttrium silicate rich grain-boundary film. A typical microstructure of 100% tetragonal material is illustrated by Fig. 6.29. Figure 6.30 is a fracture face of this type of material, with the intergranular fracture showing the grain structure clearly. With rapid cooling, the cubic phase

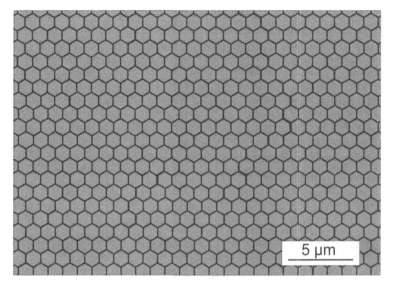

Figure 6.29 A schematic illustration of the typical pattern of a 100% tetragonal zirconia polycrystal microstructure, with small equiaxed tetragonal zirconium dioxide crystals.

can give the non-transformable, high-yttrium oxide, t'-phase, with redistribution of some of the additive oxide. TZP materials can have very high strength at room temperature, with bend strength values of up to 1.5 GPa (Tsukuma *et al.*, 1985; Swain, 1985), and fracture toughness of ~8 MPa m$^{1/2}$. Marked losses of strength tend to occur with aging under wet conditions in the temperature range 200–300 °C, which is an aspect of importance for some applications (Matsui *et al.*, 1984), and referred to in Section 6.5.

Ce-TZP

The use of cerium oxide (CeO_2) as a stabiliser has been widely studied, but the commercial applications of this form of material are limited. Fully tetragonal materials can be obtained with 8–20 mole% cerium oxide: 12% is the optimum level (Tsukuma and Shimada, 1985). Sintering is carried out at 1400–1600 °C. The microstructures of these materials are similar to those of Y-TZP, with equiaxed predominantly tetragonal grains, except that the critical grain size is slightly larger, at 2–3 µm. 12Ce-TZP materials with fracture toughness of 35 MPa m$^{1/2}$ have been produced, but the strengths are relatively low at ~125 MPa. The low strength in this type of very high toughness material is attributed to the surface transformation of the tetragonal crystals to the monoclinic phase on application of stress, and the creation and extensive linking of surface cracks (to

Table 6.5 *Typical property values for four standard zirconia materials.*

Property / unit	8–10Mg-PSZ	6.5–9.5Ca-PSZ	2–3Y-TZP	12Ce-TZP
Density / Mg m^{-3}	5.70	5.80	6.05	6.20
Mean grain size / μm	~ 50	~ 50	~ 0.5	~ 2–3
Bend strength / MPa	800	560	1200	500
Toughness / MPa m$^{1/2}$	8–12	10	8	20
Young modulus / GPa	205	207	205	200
Vickers hardness / GPa	11	13	13	7

Figure 6.30 The microstructure of the fracture surface of a fully dense yttria stabilised (TZP) zirconia. (Reproduced by kind permission of Dynamic Ceramic Ltd.)

generate, in effect, large surface flaws). On the other hand, the sensitivity to moisture is less than that of yttrium oxide stabilised TZP (Tsukuma, 1986).

The addition of small amounts (10–20% by weight) of aluminium oxide as a fine dispersion to a TZP system increases strength even further. Bend strengths of 2.4 GPa at room temperature have been obtained, with the retention of strengths of 1 GPa at 1000 °C (Swain, 1985).

Typical, representative, properties of four types of zirconia ceramics are listed in Table 6.5.

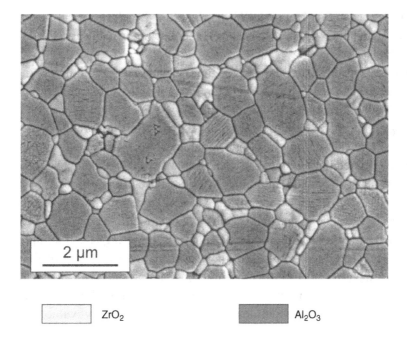

Figure 6.31 The microstructure of a 15% zirconium dioxide toughened alumina (ZTA). The lighter phase is the dispersed zirconium dioxide. (Courtesy Tony Bromley.)

6.8.3 *Zirconia-toughened ceramics (ZTC)*

Zirconium dioxide, in principle, can be used to toughen and strengthen practically any ceramic system. The essential practical requirement is that the matrix phase must not be chemically too reactive towards the high surface area zirconium dioxide particles at sintering temperature, and that mutual solubility is low. Fine zirconium dioxide particles without (Claussen *et al.*, 1977) or with (Becher, 1983) oxide stabilisers, can be incorporated in a wide range of different matrices to produce a toughened version of the host ceramic (a *dispersed oxide ceramic*). Figure 6.31 is a scanning electron micrograph of the polished surface of a zirconia-toughened alumina, containing 15 volume% of zirconium dioxide. The main difference between these two forms of material is that the first is primarily microcrack-toughened, the second primarily transformation-toughened. The most widely used host material is aluminium oxide, and yttrium oxide has been the most widely used stabiliser. For transformation-toughening, aluminium oxide has the advantage of a high Young modulus (in the region of 410 GPa), which generates high restraining stresses, and maintains a low transformation temperature. Aluminium oxide has been used to develop a range of zirconia-toughened alumina

(ZTA) materials (McMeeking, 1986), which are widely used commercially for their strength and wear resistance, for example as grinding media and metal cutting tools.

To achieve the best strength and toughness values the zirconium dioxide grains must be less than 1 μm. Mixed powder production routes tend to give intergranular zirconium dioxide grains; chemical production routes, such as co-precipitation, tend to give intragranular zirconium dioxide grains. It is important for mechanical properties that the zirconium dioxide grains are well dispersed, and that grain growth is avoided during sintering, factors that have to be balanced with the need to obtain full density. The optimum content of unstabilised zirconium dioxide is around 10–12% volume fraction; above 15% the proportion of transformed monoclinic phase increases, and toughness and strength decrease. The reason for the reduced stability of the tetragonal phase at these higher zirconium dioxide loadings is believed to be a result of the decreasing overall Young modulus of the composite material and the decreasing constraining stresses on the tetragonal grains. The incorporation of stabilisers into the zirconium dioxide allows higher loadings to be used, without loss in mechanical property value. Very high strengths and toughness values (1.5 GPa and 6.5 MPa m$^{1/2}$) have been obtained with a 70% aluminium oxide material containing yttria-stabilised zirconia. With CeO_2-stabilised materials (for example, 12Ce-TZP/70A), a fracture toughness of 8.5 MPa m$^{1/2}$, and strengths of 850 MPa can be obtained (Tsukuma and Takahata, 1987). Typical commercial ZTA has lower room-temperature bend strengths of ~500 MPa, and fracture toughness of 4–5 MPa m$^{1/2}$.

6.9 Applications

Zirconia ceramics have two important microstructural features, which in combination allow the production of oxide materials with very good mechanical properties. The first is the potential for very fine grain size. This is usually a desirable feature because, other things being equal, it will tend to produce high-strength materials. Though in fact things are not quite equal because the lower Young modulus of zirconium dioxide would be expected to reduce the strength for a given critical flaw size, by a factor of about 1.4 ($2^{1/2}$), compared with polycrystalline pure aluminium oxide or silicon carbide. However, this is outweighed by the effects of the strains caused by the tetragonal-to-monoclinic phase transformation, producing the large increase in fracture toughness. As the Griffith equation shows, for a given defect size, increasing K_{Ic} has the effect of increasing strength. Alternatively, for a specified strength, a larger K_{Ic} allows larger flaw, and processing defect, sizes to be tolerated. However, it is also the case that accurate control of powder quality, and careful powder processing, is

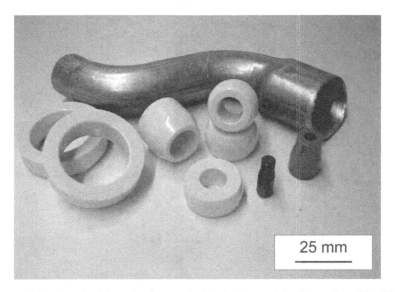

25 mm

Figure 6.32 Metal tubing shaping tools in ZTA material. (Reproduced by kind permission of St. Gobain Advanced Ceramics s.r.o.)

required if optimisation of the materials' properties is to be realised (these features are of course in themselves highly desirable, and likely to yield improved mechanical properties in any ceramic material).

6.9.1 Metal working

One of the first applications for Mg-PSZ was for metal extrusion die nozzles. This is still one application for a hard and wear-resistant material, and now extended into the area of metal-forming more generally (Garvie, 1984). ZTA materials containing ~10% of zirconia, with strengths of up to 800 MPa, and toughnesses of ~5 MPa m$^{1/2}$, are routinely used as cutting tool tips in the machining of cast iron and steel. Tool tip life can be better than with pure sintered, or hot-pressed alumina tips. Figure 6.32 shows an old application where good resistance to wear is required, in the die materials used in the extrusion of metal tubing.

PSZ materials are routinely used for the filtration of molten metal. The filter is in the form of a foam with ~80% of void space, which is prepared by impregnating plastic foam with a slip of the ceramic, and after drying, burning off the plastic, to leave a ceramic skeleton for sintering. The large irregular voids create turbulent metal flow, allowing removal of potential inclusions, and the cell wall thickness of ~250 μm is thick enough to give the whole structure adequate strength. Figure 6.33 shows two examples of small Mg-PSZ filters.

Figure 6.33 Foam filters in Mg-PSZ for metal filtration.

6.9.2 Wear resistance

Commercial markets have been gradually developed for the use of zirconia in areas, where the main requirements are for materials with better wear and corrosion resistance than metals, and ceramics already in use (Cannon, 1989; Birkby and Stevens, 1996; Briggs, 2007). These are for the most part "room temperature" applications, or at least notionally so. Because of the resistance to damage, and the intrinsic hardness of zirconia, a highly polished surface can be retained.

Typical applications for Mg-PSZ and Y-TZP components are in pumps in the petrochemical and process industries, where there is exposure to abrasive and corrosive conditions. Small centrifugal pumps are produced in which TZP materials are the major components. Figure 6.34 shows the components of an external gear pump used for pumping abrasive fluids, made from 3Y-TZP. Figure 6.35 shows sealing rings, the mating faces of which have been ground and polished to a mirror finish. A second area of application is for simple cutting edges able to retain their sharpness better than a steel blade might. The production of a range of scissor and knife blades using zirconia provided an effective illustration of the possibility of having a sharp edge (on a tapered or right-angle base), which resisted the chipping that would have been expected from a brittle ceramic. TZP blades are now routinely used for the industrial cutting of paper and fibres, and are particularly useful when the materials being cut contain hard particles or mineral fillers. Although the blade edge is initially no sharper than with a conventional steel blade, the sharpness can be retained for much longer,

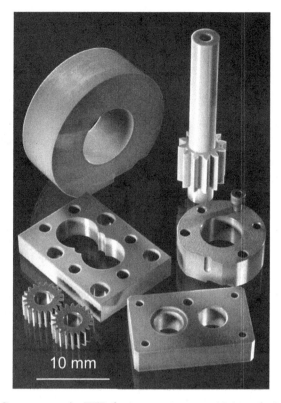

Figure 6.34 Components in TZP for a gear pump, which uses the meshing of gears to pump fluid. (Reproduced by kind permission of Dynamic Ceramic Ltd.)

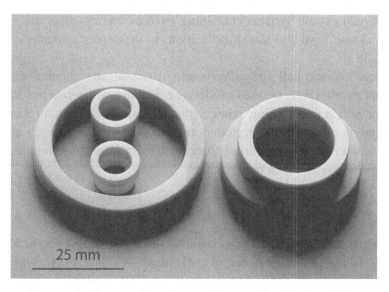

Figure 6.35 Sealing rings for use in pumps for abrasive fluids.

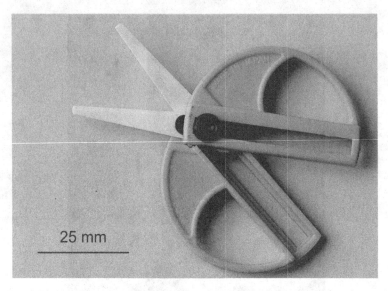

25 mm

Figure 6.36 Novelty zirconia scissors, making the point that transformation-toughened zirconia can be used where ordinary ceramics might not!

without the need for resharpening. They also have the advantage that they can be used under corrosive conditions. The disadvantage is in a higher price, compared with conventional cutting materials. Figure 6.36 shows an early example of a small pair of scissors with TZP blades, more of a curiosity, but marketed to attract attention to these unexpected properties of a ceramic material. Now a wide range of domestic and technical knife and other cutting blades is produced. Examples of commercial cutting systems containing zirconia cutting edges are shown in Fig. 6.37. These have also found applications in the domestic market, as shown by Fig. 6.38.

The main ceramic ball and roller bearing materials are alumina and silicon nitride, but zirconia balls are used to a lesser extent in bearing systems. They have been successful in fluid medium lubricated pumps, when the medium is corrosive. While there is evidence that a ground zirconia surface may show a slow continuing transformation to monoclinic form with loss of strength, a highly polished surface appears to be much more stable.

6.9.3 Bioceramics

The essential requirement for any material used for prostheses is long-term chemical stability in the presence of body fluids (Garvie, 1984), and the biocompatibility of zirconium dioxide is good. 3Y-TZP zirconia was introduced in the 1980s as an alternative to high-purity alumina as a femoral head replacement.

Figure 6.37 A selection of commercial cutting edges, for a wide range of applications. (Reproduced by kind permission of Dynamic Ceramic Ltd.)

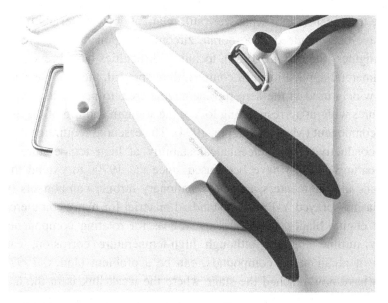

Figure 6.38 Kitchen knife blades in TZP. (Reproduced by kind permission of Kyocera Corporation.)

To date >500 000 zirconia implants have been used. Although zirconia is less hard than alumina, the TZP used had bend strengths of 900–1200 MPa, and fracture toughness of 7–10 MPa m$^{1/2}$ (two to three times those of alumina), with very good wear resistance. Mg-PSZ has also been tested, although its mechanical properties generally do not match those of TZP materials. A different issue giving concern is that of the stability of transformation-toughened zirconia over several decades in the presence of water, and in water vapour at temperatures even below 150 °C (steam sterilisation temperature). It has been suggested that this may not be a problem at body temperatures (Piconi and Maccauro, 1999), but *in vitro* stability is a key issue, and the use of TZP in hip-joint prostheses was discontinued in 2001 for this reason. The ability of materials containing tetragonal phase zirconium dioxide to resist degradation, and their long-term strength, is being kept under review (Clarke *et al.*, 2003; Chevalier *et al.*, 2007).

6.9.4 Transport

Initial interest in high strength and toughness zirconia was encouraged by possible applications in internal combustion engines, where the materials' wear resistance and low thermal conductivity might have seen improvements in engine performance (Dworak *et al.*, 1984; Larsen and Adams, 1985). As with the silicon nitride and silicon carbide groups, these early ambitious expectations were not realised, because of the extremely demanding nature of the applications, and the immature technical state of what were still relatively new materials. While many efforts have been made to incorporate zirconia components into internal combustion engines (Suhr *et al.*, 1984), technical difficulties have prevented significant commercial applications. In contrast, thin, sprayed, thermal barrier coatings are very widely used in the aero-turbine engine area, to permit higher operating temperatures without corresponding loss of the strength and life of the underlying metallic component (Meir and Gupta, 1994). The essential requirement is for low thermal conductivity, and mechanical stability at high temperature. Thermal barrier coatings (TBCs) have been used since the 1970s to extend the life of combustors and augmenters, and now stationary turbine components (Goward, 1998). Plasma-sprayed Y-PSZ is a standard material for many commercial high-thrust jet engine blades and vanes, as well as for rotating components and in stationary turbine engines, although high-temperature corrosion, caused by ingestion of alkali metal compounds, can be a problem (Jones, 1997). These materials have now reached the stage where the weak link is in the underlying bond coat, designed to accommodate differences in thermal expansivity between the zirconia barrier layer and the metal (Clarke *et al.*, 2006).

6.10 Summary

This study of the zirconia structural ceramics has shown another material which is predominantly solid state sintered, with very little intergranular phase.

Zirconium dioxide, like silicon carbide, has considerable versatility, although the fully stabilised, electrically conducting cubic form used mainly for its electrical properties has been excluded from this study. Zirconium dioxide is not exceptionally hard, has only ~50% of the modulus of aluminium oxide or silicon carbide, and does not easily develop the large high-aspect-ratio grains of silicon nitride. Zirconia ceramics might therefore not have been expected to be significantly better than solid state alumina or sintered silicon carbide of similar grain size. As it is, they are able to develop a remarkable combination of strength and toughness, a feature with origins, uniquely, in the zirconium dioxide tetragonal-to-monoclinic transformation. To date zirconia has provided the ceramic systems of the highest fracture toughness values (composites excluded). It has been suggested that zirconia is the ideal transformation-toughened ceramic, and so far, attempts to find a better system have not succeeded; the characteristics of an ideal transformation-toughened system might then be said to be simply those shown by zirconia (Kelly and Rose, 2002). These microstructural features account for the ability of zirconia materials to resist wear, and the high toughness provides damage tolerance; transformation-toughened zirconia cutting edges retain their sharpness better than other materials and are widely used for this reason.

The importance of the high-temperature phase equilibria, and the crystal chemistry of the solid binary and ternary phases appearing in these systems, explains the considerable attention these aspects have received. This change in emphasis contrasts with the efforts made to understand the behaviour of the silicate liquid-forming systems of importance for the liquid phase sintering of aluminium oxide, silicon carbide and silicon nitride, and the grain boundary silicates in these materials.

Because it is a very stable oxide, toughened zirconia has no problems in purely oxidising environments, and its corrosion and degradation at quite low temperatures in the presence of water might not have been expected. However, this feature, like those of strength and toughness, is linked to the reactivity of the stabilising oxide, and thus the stability of the tetragonal phase. Questions have therefore been raised about long-term stability in the human body, and for this reason solid state sintered alumina has for the moment been preferred for prosthetic devices. Transformation-toughened zirconia also suffers from loss of strength as the transformation temperature of 1000 °C is approached. However, the low thermal conductivity of zirconium dioxide provides extensive applications as thin-film, high-temperature thermal barrier layers; these are particularly important for blade protection in the gas-turbine engine, and will become even more important with the need to improve engine efficiency. With this exception, like silicon carbide and silicon nitride, the zirconias are at present used more for structural applications in the lower temperature range, and with their main bulk applications where resistance to wear is needed.

Questions

6.1. In what important respects does zirconium dioxide differ from aluminium oxide, as a potential high-strength ceramic?

6.2. Why should transformation-toughened zirconia be considered more for applications at lower temperatures ($<1000\ °C$)?

6.3. In what way does cerium oxide differ from magnesium oxide, as a solid solution stabilising additive for zirconium dioxide?

6.4. Explain why aluminium oxide has been used as a matrix in commercial transformation-toughened zirconia ceramics.

6.5. Fully stabilised zirconia is the basis of a good sensor for low pressures of oxygen. A potential difference of 1 volt is developed at $1000\ °C$ with an electrochemical cell, when one electrode is exposed to 0.2 atm oxygen pressure. What is the oxygen pressure at the other electrode?

6.6. 12Ce-TZP represents a commonly used composition of a high-toughness zirconia. What is the amount of cerium oxide in this material, expressed in weight per cent? (A_r: O = 16.0, Zr = 91.2, Ce = 140.1.)

6.7. Ceramics have been assessed on the basis of *strength per unit volume*, on the grounds that a component can always be made stronger by increasing its dimensions – though at the expense of weight and cost. How would a sintered alumina, and a tetragonal zirconia polycrystal, perform on this basis?

6.8. To what extent is it true to say that the high toughness values obtained in silicon nitride and zirconia ceramics both depend on phase changes?

6.9. Silicon nitride and zirconia ceramics can both be manufactured with fracture toughness values of $10\ \mathrm{MPa\ m^{1/2}}$. Outline bases on which a choice might be made between them, for a particular application requiring good toughness.

6.10. Would it be sensible to try to develop a (zirconia) transformation-toughened porcelain?

Selected reading

Birkby, I. and Stevens, R. (1996). Applications of zirconia ceramics. In *Advanced Ceramic Materials, Key Engineering Materials*, Vols. **122–124**, ed. H. Mostaghaci. Switzerland: TransTech Publications, pp. 527–52.

Cannon, W. R. (1989). Transformation toughened ceramics for structural applications. In *Structural Ceramics*, ed. J. B. Wachtman, Jr., *Treatise on Materials Science and Technology*, Vol. **29**. Boston, MA: Academic Press, pp. 195–228.

Claussen, N. and Heuer, A. H. (1991). Transformation toughening. In *Concise Encyclopedia of Advanced Ceramic Materials*, ed. R. J. Brook. Oxford: Pergamon, pp. 494–7.

Hannink, R. H. J., Kelly, P. M. and Muddle, B. C. (2000). Transformation toughening in zirconia-containing ceramics. *J. Am. Ceram. Soc., Centennial Feature*, **83**, 461–87.

References

Adams, J. W., Ruh, R. and Mazdiyasni, K. S. (1997). Young's modulus, flexural strength, and fracture of yttria-stabilized zirconia. *J. Am. Ceram. Soc.*, **80**, 903–8.

Aizu, K. (1970). Possible species of ferromagnetic, ferroelectric and ferroelastic crystals. *Phys. Rev. B*, **2**, 754–72.

Amazigo, J. C. and Budiansky, B. (1988). Steady-state crack growth in supercritically transforming materials. *Int. J. Solids Struct.*, **24**, 751–5.

Antolovich, S. D. and Fahr, D. (1972). An experimental investigation of the fracture characteristics of TRIP alloys. *Eng. Fract. Mech.*, **4**, 133–44.

Antolovich, S. D. and Singh, B. (1971). Toughness increment associated with austenite to martensite phase transformation in TRIP steels. *Metall. Trans.*, **2**:8, 2135–41.

Atkins, P. W. and de Paula, J. (2006). *Physical Chemistry*, 8th edition. Oxford: Oxford University Press.

Badwal, S. P. S. and Drennan, J. (1994). Interfaces in zirconia based electrochemical systems and their influence on electrical properties. In *Science of Ceramic Interfaces II, Materials Science Monographs*, **81**, ed. J. Nowotny. Amsterdam: Elsevier, pp. 71–111.

Bansal, G. K. and Heuer, A. H. (1974). Martensitic phase transformation in zirconia (ZrO_2): II, crystallographic aspects. *Acta Metall.*, **22**, 409–17.

Barin, I. (1995). *Thermochemical Data of Pure Substances*, 3rd edition. Vol. II, *La-Zr*. New York: VCH Publishers.

Becher, P. F. (1983). Slow crack growth behaviour in transformation-toughened Al_2O_3–$ZrO_2(Y_2O_3)$ ceramics. *J. Am. Ceram. Soc.*, **66**, 485–8.

Becher, P. F., Swain, M. V. and Ferber, M. K. (1987). Relation of transformation temperature to the fracture toughness of transformation-toughened ceramics. *J. Mater. Sci.*, **22**, 76–84.

Birkby, I. and Stevens, R. (1996). Applications of zirconia ceramics. In *Advanced Ceramic Materials, Key Engineering Materials*, Vols. **122–124**, ed. H. Mostaghaci. Switzerland: TransTech Publications, pp. 527–52.

Briggs, J. (2007). *Engineering Ceramics in Europe and the USA*. Worcester, UK: Enceram.

Brook, R. J. (1981). Preparation and electrical behaviour of zirconia ceramics. In *Advances in Ceramics*, Vol. 3, *Science and Technology of Zirconia*, eds. A. H. Heuer and L. W. Hobbs. Columbus, OH: The American Ceramic Society, pp. 272–85.

Buckley, D. H. and Miyoshi, K. (1989). Tribological properties of structural ceramics. In *Structural Ceramics*, ed. J. B. Wachtman Jr., *Treatise on Materials Science and Technology*, Vol. **29**. Boston, MA: Academic Press Inc., pp. 293–365.

Budinsky, B., Hutchinson, J. and Lambropoulos, J. (1983). Continuum theory of dilatant transformation toughening in ceramics. *Int. J. Solids Struct.*, **19**, 337–55.

Cannon, W. R. and Langdon, T. G. (1988). Creep of ceramics 2. An examination of flow mechanisms. *J. Mater. Sci.*, **23**, 1–20.

Cannon, W. R. (1989). Transformation toughened ceramics for structural applications. In *Structural Ceramics, Treatise on Materials Science and Technology*, Vol. **29**, ed. J. B. Wachtman Jr. Boston: Academic Press, pp. 195–228.

Chen, I.-W. and Reyes-Morel, P. E. (1986). Implications of transformation plasticity in transformation toughened ceramics. *J. Am. Ceram. Soc.*, **69**, 181–9.

Chen, I.-W. and Reyes-Morel, P. E. (1987). Transformation plasticity and transformation toughening in Mg-PSZ and Ce-PSZ. *Mater. Res. Soc. Symp. Proc.*, **78**, 75.

Chevalier, J., Gremillard, L. and Deville, S. (2007). Low temperature degradation of zirconia and implications for biomedical implants. *Ann. Rev. Mat. Res.*, **37**, 1–32.

Clarke, D. R., Levi, C. G. and Evans, A. G. (2006). Enhanced zirconia thermal barrier coating systems. *Proc. Inst. Mech. Eng. Part A*, **220**:A1, 85–92.

Clarke, I. C., Manaka, M., Green, D. D., Williams, P., Pezzotti, G., Kim, Y. H., Ries, M., Donaldson, N. T. and Gustafson, G. A. (2003). Current status of zirconia used in total hip implants. *J. Bone Joint Surg. Am.*, **85-A**: 4-suppl., 73–84.

Claussen, N. (1984). Microstructural design of zirconia-toughened ceramics (ZTC). In *Advances in Ceramics*, Vol. **12**, *Science and Technology of Zirconia II*, eds. N. Claussen, M. Rühle and A. H. Heuer. Columbus, OH: The American Ceramic Society, pp. 325–51.

Claussen, N. and Rühle, M. (1981). Design of transformation-toughened ceramics. In *Advances in Ceramics*, Vol. **3**, *Science and Technology of Zirconia*, eds. A. H. Heuer and L. W. Hobbs. Columbus, OH: The American Ceramic Society, pp. 137–63.

Claussen, N., Rühle, M. and Heuer, A. H., Eds. (1984). *Advances in Ceramics*, Vol. **12**, *Science and Technology of Zirconia II*. Columbus, OH: The American Ceramic Society.

Claussen, N., Steeb, J. and Pabst, R. F. (1977). Effect of induced microcracking on fracture toughness of ceramics. *Am. Ceram. Soc. Bull.*, **56**, 559–62.

Cohn, W. M. and Tolksdorf, S. (1930). The forms of zirconium dioxide dependent on pre-treatment. *Z. Phys. Chem. B*, **8**, 331–6.

Curtis, C. E. (1947). Development of zirconia resistant to thermal shock. *J. Am. Ceram. Soc.*, **29**, 180–96.

Cutler, R. A., Bright, J. D., Virkar, A. V. and Shetty, D. K. (1987). Strength improvement in transformation toughened alumina by selective phase-transformation. *J. Am. Ceram. Soc.*, **70**, 714–18.

D'Ans, J. (1928). *Verfahren zur Herstellung von beim Brennen nicht sinternden hochfeuerfesten Gegenständen, insbesondere hochfeuerfesten Steinen, aus Zirconoxyd*. German Patent 469,204. January 24th 1925.

D'Ans, J. (1932). *Verfahren zur Herstellung dichter und fester Gegenstände und Geräte aus Zirconoxyd oder anderen hochfeuerfesten Oxyden*. German Patent 543,772. June 22nd 1927.

Davidge, R. W. (1980). *Mechanical Behaviour of Ceramics*. Cambridge: Cambridge University Press, pp. 18–30.

Davidge, R. W. and Riley, F. L. (1995). The grain size dependence of the wear of alumina. *Wear*, **186**, 45–9.

Dewhurst, J. K. (1998). Relative stability, structure, and elastic properties of several phases of pure zirconia. *Phys. Rev. B*, **57**, 741–7.

Duran, P. (1990). A new tentative phase-equilibrium diagram for the ZrO_2-CeO_2 system in air. *J. Mater. Sci.*, **25**, 5001–6.

Duwez, F., Odell, F. and Brown, F. H. Jr. (1952). Stabilization of zirconia with calcia and magnesia. *J. Am. Ceram. Soc.*, **35**, 107–13.

Dwivedi, A. and Cormack, A. N. (1990). A computer-simulation of the defect structure of calcia-stabilized zirconia. *Phil. Mag. A*, **61**, 1–22.

Dworak, U., Olapinski, H., Fingerle, D. and Krohn, U. (1984). ZrO_2 ceramics for internal combustion engines. In *Advances in Ceramics*, Vol. **12**, *Science and Technology of Zirconia II*, eds. N. Claussen, M. Rühle and A. H. Heuer. Columbus, OH: The American Ceramic Society, pp. 480–7.

Eshelby, J. D. (1961). Elastic inclusions and inhomogeneity. In *Progress in Solid Mechanics*, Vol. **2**, ed. I. N. Sneddon and R. Hill. Amsterdam: North-Holland Publishing Co., pp. 89–140.

Evans, A. G. (1985). Toughening mechanisms in zirconia alloys. In *Advances in Ceramics*, Vol. **12**, *Science and Technology of Zirconia II*, eds. N. Claussen, M. Rühle and A. H. Heuer. Columbus, OH: The American Ceramic Society, pp. 193–212.

Evans, A. G. (1990). Perspective on the development of high-toughness ceramics. *J. Am. Ceram. Soc.*, **73**, 187–206.

Evans, A. G. and Cannon, R. M. (1986). Toughening of brittle solids by martensitic transformations. *Acta Metall.*, **34**, 761–800.

Farnworth, F., Jones, S. L. and McAlpine, I. (1981). The production, properties and uses of zirconium chemicals. In *Speciality Inorganic Chemicals*, ed. R. Thompson. Royal Society of Chemistry Special Publication No. 40. London: Royal Society of Chemistry.

Fehrenbacher, L. L. and Jacobson, L. A. (1965). Metallographic observation of the monoclinic-tetragonal phase transformation in ZrO_2. *J. Am. Ceram. Soc.*, **48**, 157–61.

Fitzsimmons, E. S. (1950). Thermal diffusivity of refractory oxides, *J. Am. Ceram. Soc.*, **33**, 327–33.

Foitzik, A., Stadtwald-Klenze, M. and Rühle, M. (1993). Ferroelasticity of t'-ZrO_2. *Z. Metallkd.*, **84**, 397–404.

French, R. H., Glass, S. J., Ohuchi, F. S., Xu, Y. N. and Ching, W. Y. (1994). Experimental and theoretical determination of the electronic-structure and optical properties of three phases of ZrO_2. *Phys. Rev. B*, **49**, 5133–41.

Garvie, R., Hannink, R. H. and Pascoe, R. J. (1975). Ceramic steel? *Nature*, **258**, 703–4.

Garvie, R. C. (1970). Zirconium dioxide and some of its binary systems. In *High Temperature Oxides*, Vol. 5-II, ed. A. M. Alper. New York: Academic Press, pp. 117–66.

Garvie, R. C. (1976). Thermal-conductivity of stabilized zirconia. *J. Mater. Sci.*, **11**, 1365–7.

Garvie, R. C. (1978). Stabilization of tetragonal structure in zirconia microcrystals. *J. Phys. Chem.*, **82**, 218–24.

Garvie, R. C. (1984). Biocompatibility of magnesia-partially stabilised zirconia (Mg-PSZ) ceramics. *J. Mater. Sci.*, **19**, 3224–8.

Garvie, R. C. (1984). Structural applications of ZrO_2-bearing materials. In *Advances in Ceramics*, Vol. 12, *Science and Technology of Zirconia II*, eds. N. Claussen, M. Rühle and A. H. Heuer. Columbus, OH: The American Ceramic Society, pp. 465–79.

Garvie, R. C. and Swain, M. V. (1985). Thermodynamic analysis of the tetragonal to monoclinic transformation in a constrained zirconia microcrystal. *J. Mater. Sci.*, **20**, 1193–200.

Goward, G. W. (1998). Progress in coatings for gas turbine airfoils. *Surface Coatings Tech.*, **108**, 73–97.

Grain, C. F. (1967). Phase relations in the MgO-ZrO_2 system. *J. Am. Ceram. Soc.*, **50**, 288–90.

Green, D. J. (1982). Critical microstructures for microcracking in Al_2O_3–ZrO_2 composites. *J. Am. Ceram. Soc.*, **65**, 610–14.

Green, D. J. (1983). A technique for introducing surface compression into zirconia ceramics. *J. Am. Ceram. Soc.*, **66**, C178–9.

Green, D. J., Hannink, R. H. J. and Swain, M. V., Eds. (1989). Crystallography and phase transformations in zirconia and its alloys. In *Transformation Toughening of Ceramics*. Boca Raton, FL: CRC Press, pp. 17–55.

Green, D. J., Hannink, R. H. J. and Swain, M. V., Eds. (1989). *Transformation Toughening of Ceramics*. Boca Raton, FL: CRC Press, pp. 124–37.

Green, D. J., Lange, F. F. and James, M. R. (1984). Residual surface stresses in Al_2O_3–ZrO_2. In *Advances in Ceramics*, Vol. **12**, *Science and Technology of Zirconia II*, eds. N. Claussen, M. Rühle and A. H. Heuer. Columbus, OH: The American Ceramic Society, pp. 240–50.

Gupta, T. K. (1980). Strengthening by surface damage in metastable tetragonal zirconia. *J. Am. Ceram. Soc.*, **63**, 117.

Hannink, R. H. J. (1978). The growth morphology of tetragonal phase in partially stabilized zirconia. *J. Mater. Sci.*, **13**, 2487–96.

Hannink, R. H. J. (1983). Microstructure development of sub-eutectoid-aged MgO-ZrO_2 alloys. *J. Mater. Sci.*, **18**, 457–70.

Hannink, R. H. J. and Garvie, R. C. (1982). Sub-eutectoid aged Mg-PSZ alloy with enhanced thermal upshock resistance. *J. Mater. Sci.*, **17**, 2637–43.

Hannink, R. H. J. and Swain, M. V. (1994). Progress in transformation toughening of ceramics. *Ann. Rev. Mater. Sci.*, **24**, 359–408.

Hannink, R. H. J., Johnston, K. A., Pascoe, R. T. and Garvie, R. C. (1981). Microstructural changes during isothermal aging of a calcia-partially-stabilized zirconia alloy. In *Advances in Ceramics*, Vol. **3**, *Science and Technology of Zirconia*, eds. A. H. Heuer and L. W. Hobbs. Columbus, OH: The American Ceramic Society, pp. 116–36.

Hannink, R. H. J., Kelly, P. M. and Muddle, B. C. (2000). Transformation toughening in zirconia-containing ceramics. *J. Am. Ceram. Soc.*, **83**, 461–87.

Hasselman, D. P. H. (1969). Unified theory of thermal shock fracture initiation and crack propagation in brittle ceramics. *J. Am. Ceram. Soc.*, **52**, 600–4.

Heuer, A. H. and Hobbs, L. W., Eds. (1981). *Advances in Ceramics*, Vol. **3**, *Science and Technology of Zirconia*. Columbus, OH: The American Ceramic Society.

Heuer, A. H. and Rühle, M. (1984). Phase transformations in ZrO_2-containing ceramics: I, the instability of c-ZrO_2 and the resulting diffusion controlled reactions. In *Advances in Ceramics*, Vol. **12**, *Science and Technology of Zirconia* II, eds. N. Claussen, M. Rühle and A. H. Heuer. Columbus, OH: The American Ceramic Society, pp. 1–13.

Heuer, A. H. and Rühle, M. (1985). Overview: on the nucleation of the martensitic-transformation in zirconia (ZrO_2). *Acta Metall.*, **33**, 2101–12.

Holmes, H. F., Fuller, E. L. Jr. and Gammage, R. B. (1972). Heats of immersion in the zirconium oxide-water system. *J. Phys. Chem.*, **76**, 1497–502.

Howard, C. J., Kisi, E. H. and Ohtaka, O. (1991). Crystal structures of 2 orthorhombic zirconias. *J. Am. Ceram. Soc.*, **74**, 2321–3.

Hughan, R. R. and Hannink, R. H. J. (1986). Precipitation during controlled cooling of magnesia-partially-stabilized zirconia. *J. Am. Ceram. Soc.*, **69**, 556–63.

Hughes, A. E. and Badwal, S. P. S. (1991). Impurity and yttrium segregation in yttria-tetragonal zirconia. *Solid State Ionics*, **46**, 265–74.

Hugo, G. R. and Muddle, B. C. (1990). The tetragonal to monoclinic transformation in ceria-zirconia. In *Materials Science Forum, Martensitic Transformations, Part I*, Proceedings of the 6th International Conference on Martensitic Transformations, Sydney, Australia (1989), Vols. **56–58**, ed. B. C. Muddle. Switzerland: TransTech Publications, pp. 357–62.

Jones, R. L. (1997). Some aspect of the hot corrosion of thermal barrier coatings. *J. Thermal Spray Tech.*, **6**, 77–84.

Jue, J. F. and Virkar, A. V. (1990). Fabrication, microstructural characterization, and mechanical-properties of polycrystalline t′-zirconia. *J. Am. Ceram. Soc.*, **73**, 3650–7.

Kelly, P. M. and Rose, L. R. F. (2002). The martensitic transformation in ceramics – its rôle in transformation toughening. *Prog. Mater. Sci.*, **47**, 463–557.

Kilo, M. (2003). Cation self-diffusion of Ca-44, Y-88, and Zr-96 in single-crystalline calcia- and yttria-doped zirconia. *J. Appl. Phys.*, **94**, 7547–52.

King, A. G. and Yavorsky, P. J. (1968). Stress relief mechanisms in magnesia- and yttria-stabilized zirconia. *J. Am. Ceram. Soc.*, **51**, 38–42.

Kingery, W. D., Pappis, J., Doty, E. M. and Hill, D. C. (1959). Oxygen ion mobility in cubic $Zr_{0.85}Ca_{0.15}O_{1.85}$. *J. Am. Ceram. Soc.*, **42**, 393–8.

Kisi, E., Ed. (1998). *Key Engineering Materials*, Vols. **153–154**, *Zirconia Engineering Ceramics: Old Challenges – New Ideas*. Switzerland: TransTech Publications.

Kisi, E. H. and Howard, C. J. (1998). Crystal structures of zirconia phases and their inter-relation. In *Zirconia Engineering Ceramics, Old Challenges – New Ideas. Key Engineering Materials*, Vols. 153–154, ed. E. Kisi. Switzerland: TransTech Publications, pp. 1–36.

Kiukulla, K. and Wagner, C. (1957). Measurement on galvanic cells involving solid electrolytes. *J. Electrochem. Soc.*, **104**, 379–87.

Kriven, W. M. (1988). Possible alternative transformation tougheners to zirconia: crystallographic aspects. *J. Am. Ceram. Soc.*, **71**, 1021–30.

Kriven, W. M., Fraser, W. L. and Kennedy, S. W. (1981). The martensitic crystallography of tetragonal zirconia. In *Advances in Ceramics*, Vol. **3**, *Science and Technology of Zirconia*, eds. A. H. Heuer and L. W. Hobbs. Columbus, OH: The American Ceramic Society, pp. 82–97.

Lakshminarayanan, R., Shetty, D. K. and Cutler, R. A. (1987). Toughening of layered ceramic composites with residual surface compression. *J. Am. Ceram. Soc.*, **79**, 79–87.

Lange, F. F. (1982a). Transformation toughening – 2. Contribution to fracture toughness. *J. Mater. Sci.*, **17**, 235–9.

Lange, F. F. (1982b). Transformation toughening – 3. Experimental observations in the ZrO_2–Y_2O_3 system. *J. Mater. Sci.*, **17**, 240–6.

Lange, F. F. (1982c). Transformation toughening – 5. Effect of temperature and alloy on fracture toughness. *J. Mater. Sci.*, **17**, 255–62.

Lange, F. F. and Green, D. J. (1984). Effect of inclusion size on the retention of tetragonal ZrO_2: theory and experiments. In *Advances in Ceramics*, Vol. **3**, *Science and Technology of Zirconia*, eds. A. H. Heuer and L. W. Hobbs. Columbus, OH: The American Ceramic Society, pp. 217–25.

Lange, F. F., Dunlop, G. L. and Davis, B. I. (1986). Degradation during aging of transformation-toughened ZrO_2-Y_2O_3 materials at 250°C. *J. Am. Ceram. Soc.*, **69**, 237–40.

Lankford, J. (1983). Plastic deformation of partially-stabilized zirconia. *J. Am. Ceram. Soc.*, **66**, C212–13.

Lanteri, V., Mitchell, T. E. and Heuer, A. H. (1986). Morphology of tetragonal precipitates in partially-stabilized ZrO_2. *J. Am. Ceram. Soc.*, **69**, 564–9.

Larsen, D. C. and Adams, J. W. (1985). *Long-term Stability and Properties of Zirconia Ceramics for Heavy Duty Diesel Engine Components*. NASA-Lewis Research Center, US DOE Contract DEN-3-305, NASA CR-174943, Sept. 1985.

Lee, S. W., Hsu, S. M. and Shen, M. C. (1993). Ceramic wear maps – zirconia. *J. Am. Ceram. Soc.*, **76**, 1937–47.

Lilley, E. (1990). Review of low-temperature degradation in Y-TZP. In *Proceedings of the Symposium on Corrosion and Corrosive Degradation, First Ceramic Science and Technology Congress of the American Ceramic Society*, eds. R. E. Tressler and M. McNallan. Westerville, OH: The American Ceramic Society, pp. 387–407.

Lutz, E. H. and Swain, M. V. (1991). Mechanical and thermal shock properties of duplex ceramics. *Mater. Forum*, **15**, 307–23.

Lutz, E. H., Claussen, N. and Swain, M. V. (1991). Thermal shock behaviour of duplex ceramics. *J. Am. Ceram. Soc.*, **74**, 19–24.

Ma, Y. X., Kisi, E. H. and Kennedy, S. J. (2001). Neutron diffraction study of ferroelasticity in a 3 mol% Y_2O_3-ZrO_2. *J. Am. Ceram. Soc.*, **84**, 399–405.

Ma, Y. X., Kisi, E. H., Kennedy, S. J. and Studer, A. J. (2004). Tetragonal-to-monoclinic transformation in Mg-PSZ studies by *in situ* neutron diffraction. *J. Am. Ceram. Soc.*, **87**, 465–72.

Marshall, D. B., Shaw, M. C., Dauskardt, R. H., Ritchie, R. O., Readey, M. J. and Heuer, A. H. (1990). Crack-tip transformation zones in toughened zirconia. *J. Am. Ceram. Soc.*, **73**, 2659–66.

Matsui, M., Soma, T. and Oda, I. (1984). Effect of microstructure on the strength of Y-TZP components. In *Advances in Ceramics*, Vol. **12**, *Science and Technology of Zirconia II*, eds. N. Claussen, M. Rühle and A. H. Heuer. Columbus, OH: The American Ceramic Society, pp. 371–81.

McMeeking, R. M. (1986). Effective transformation strain in binary elastic composites. *J. Am. Ceram. Soc.*, **69**, C.301–2.

McMeeking, R. M. and Evans, A. G. (1982). Mechanics of transformation toughening in brittle materials. *J. Am. Ceram. Soc.*, **65**, 242–6.

Meir, S. M. and Gupta, D. K. (1994). The evolution of thermal barrier coatings in gas-turbine engine applications. *J. Eng. Gas Turbines and Power-Trans. ASME*, **116**, 250–7.

Mercer, C., Williams, J. R., Clarke, D. R. and Evans, A. G. (2007). On a ferroelastic mechanism governing the toughness of metastable tetragonal-prime (t′) yttria-stabilised zirconia. *Proc. Royal Soc. A*, **463**, 1393–408.

Miller, R. A., Smialek, J. L. and Garlick, R. G. (1981). Phase stability in plasma-spread, partially stabilised zirronia-yttria. In *Advances in Ceramics*, Vol. **3**, *Science and Technology of Zirconia*, eds. A. H. Heuer and L. W. Hobbs. Columbus, OH: The American Ceramic Society, pp. 241–55.

Moulson, A. J. and Herbert, J. M. (2002). *Electroceramics: Materials, Properties, Applications*, 2nd edition. Chichester: Wiley.

Muddle, B. C. and Hannink, R. H. J. (1986). Crystallography of the tetragonal to monoclinic transformation in MgO-partially-stabilised zirconia. *J. Am. Ceram. Soc.*, **69**, 547–55.

Nernst, W. (1897). *Verfahren zur Erzeugung von elektrischem Glühlicht*. German Patent No. 104872, July 6th 1897.

Nernst, W. (1900). Electrolytic conduction in solid substances at high temperature. *Z. Electrochem.*, **6**, 41.

Nettleship, I., Slavick, K. G., Kim, Y. J. and Kriven, W. M. (1992). Phase-transformations in dicalcium silicate. 1. Fabrication and phase-stability of fine-grained beta-phase. *J. Am. Ceram. Soc.*, **75**, 2400–6.

Orlando, R., Pisani, C., Roetti, C. and Stafanovich, E. (1992). *Ab initio* Hartree–Fock study of tetragonal and cubic phases of zirconium dioxide. *Phys. Rev. B*, **45:2**, 592–601.

Park, K. (1991). Oxygen diffusion in single-crystal tetragonal zirconia. *J. Electrochem. Soc.*, **138**, 1154–9.

Patil, R. N. and Subbarao, E. C. (1969). Axial thermal expansion of ZrO_2 and HfO_2 in the range room temperature to 1400°C. *J. Appl. Crystallogr.*, **2**, 281–8.

Patil, R. N. and Subbarao, E. C. (1970). Monoclinic–tetragonal phase transition in zirconia: mechanism, pretransformation and co-existence. *Acta Crystallogr.*, **A26**, 535–42.

Piconi, C. and Maccauro, G. (1999). Zirconia as a ceramic biomaterial. *Biomaterials*, **20**, 1–25.

Porter, D. L. and Heuer, A. H. (1979). Microstructural development in MgO-stabilized zirconia (Mg-PSZ). *J. Am. Ceram. Soc.*, **60**, 298–305.

Rainforth, W. M. (2004). The wear behaviour of oxide ceramics – a review. *J. Mater. Sci.*, **39**, 6705–21.

Readey, M. J., Heuer, A. H. and Steinbrech, R. W. (1987). Crack propagation in Mg-PSZ. In *Materials Research Society Symposia Proceedings*, Vol. **78**, *Advanced Structural Ceramics*, eds. P. F. Becher, M. V. Swain and S. Sōmiya. Pittsburgh, PA: Materials Research Society, pp. 107–20.

Rhodes, W. H. and Carter, R. E. (1966). Cationic self-diffusion in calcia-stabilised zirconia. *J. Am. Ceram. Soc.*, **49**, 244–9.

Rose, L. R. F. (1987). The mechanism of transformation toughening. *Proc. R. Soc. Lond.*, **A412**, 169–97.

Ruff, O. and Ebert, F. (1929). Contributions on the ceramics of highly fireproof material, I. The forms of zirconium dioxide. *Z. Anorg. Allg. Chem.*, **180**, 19–41.

Rühle, M., Claussen, N. and Heuer, A. H. (1983). Transformation and microcrack toughening as complementary processes in ZrO_2-toughened Al_2O_3. *J. Am. Ceram. Soc.*, **69**, 195–7.

Rühle, M., Claussen, N. and Heuer, A. H. (1984). Toughening mechanisms in zirconia alloys. In *Advances in Ceramics*, Vol. 12, *Science and Technology of Zirconia II*, eds. N. Claussen, M. Rühle and A. H. Heuer. Columbus, OH: The American Ceramic Society, pp. 352–81.

Ryshkewitch, E. (1928a). *Verfahren und Vorrichtung zur Zerteilung einer Flüssigkeit in Tropfen*. German Patent, No. 519,756. June 24th 1928.

Ryshkewitch, E. (1928b). *Verfahren zur Überfuhrüng von Zirconoxyd in den plastischen Zustand*. German Patent, No. 542,320. December 31st 1927.

Ryshkewitch, E. (1953). Compression strength of porous sintered alumina and zirconia. 9. To ceramography. *J. Am. Ceram. Soc.*, **36**, 65–8.

Ryshkewitch, E. and Richerson, D. W. (1985a). *Oxide Ceramics: Physical Chemistry and Technology*. Orlando, FL: Academic Press, p. 353.

Ryshkewitch, E. and Richerson, D. W. (1985b). *Oxide Ceramics: Physical Chemistry and Technology*. Orlando, FL: Academic Press, p. 379.

Ryshkewitch, E. and Richerson, D. W. (1985c). *Oxide Ceramics: Physical Chemistry and Technology*, 2nd edition. Orlando, FL: Academic Press, pp. 384–5.

Ryshkewitch, E. and Richerson, D. W. (1985d). *Oxide Ceramics: Physical Chemistry and Technology*, 2nd edition. Orlando, FL: Academic Press, p. 511.

Ryshkewitch, E. and Richerson, D. W. (1985e). *Oxide Ceramics: Physical Chemistry and Technology*, 2nd edition. Orlando, FL: Academic Press, pp. 524–5.

Schmauder, S. and Schubert, H. (1986). Significance of internal stresses for the martensitic transformation in yttria-stabilised tetragonal-zirconia polycrystals during degradation. *J. Am. Ceram. Soc.*, **69**, 534–40.

Schubert, H. (1986). Anisometric thermal-expansion coefficients of Y_2O_3-stabilised tetragonal zirconia. *J. Am. Ceram. Soc.*, **69**, 270–1.

Scott, H. G. (1975). Phase relationships in the zirconia-yttria system. *J. Mater. Sci.*, **10**, 1527–35.

Selçuk, A. and Atkinson, A. (1997). Elastic properties of ceramic oxides used in solid oxide fuel cells (SOFC). *J. Eur. Ceram. Soc.*, **17**, 1523–32.

Simpson, L. A. and Carter, R. E. (1966). Oxygen exchange and diffusion in calcia-stabilized zirconia. *J. Am. Ceram. Soc.*, **49**, 139–44.

Smith, D. K. and Kline, C. F. (1962). Verification of existence of cubic zirconia at high temperature. *J. Am. Ceram. Soc.*, **45**, 249–50.

Smith, D. K. and Newkirk, H. W. (1965). Crystal structure of baddeleyite (monoclinic ZrO_2) and its relation to the polymorphism of ZrO_2. *Acta Crystallogr.*, **18**, 983–91.

Sōmiya, S., Yamamoto, N. and Yanagida, H., Eds. (1988). *Science and Technology of Zirconia III, Advances in Ceramics*, Vol. 24. Columbus, OH: The American Ceramic Society.

Standard, O. C. and Sorrell, C. C. (1998). Densification of zirconia – conventional methods. In *Zirconia Engineering Ceramics – Old Challenges, New Ideas. Key Engineering Materials*, Vols. **153–154**, ed. E. Kisi. Switzerland: TransTech Publications. pp. 251–300.

Steele, B. C. H., Drennan, J., Slotwinksi, R. K., Bonanos, N. and Butler, E. P. (1981). Factors influencing the performance of zirconia based oxygen monitors. In *Advances in Ceramics*, Vol. **3**, *Science and Technology of Zirconia*, eds. A. H. Heuer and L. W. Hobbs. Columbus, OH: The American Ceramic Society, pp. 286–309.

Stefanovich, E. V., Shluger, A. L. and Catlow, C. R. A. (1994). Theoretical study of the stabilization of cubic-phase ZrO_2 by impurities. *Phys. Rev. B*, **49**, 11560–71.

Stiger, M. J., Yanar, N. M., Topping, M. G., Pettit, F. S. and Meier, G. H. (1999). Thermal barrier coatings for the 21st century. *Z. Metall.*, **90**, 1069–78.

Subbarao, E. C. (1981). Zirconia – an overview. In *Advances in Ceramics*, Vol. 3, *Science and Technology of Zirconia*, eds. A. H. Heuer and L. W. Hobbs. Columbus, OH: The American Ceramic Society, pp. 1–24.

Subbarao, E. C. and Maiti, H. S. (1988). Oxygen sensors and pumps. In *Advances in Ceramics*, Vol. 24, *Science and Technology of Zirconia III*, eds. S. Sōmiya, N. Yomamoto and H. Hanagida. Westerville, OH: The American Ceramic Society, pp. 731–47.

Suhr, D. S., Mitchell, T. E. and Keller, R. J. (1984). Microstructure and durability of zirconia thermal barrier coatings. In *Advances in Ceramics*, Vol. **12**, *Science and Technology of Zirconia II*, eds. N. Claussen, M. Rühle and A. H. Heuer. Columbus, OH: The American Ceramic Society, pp. 503–17.

Swain, M. V. (1985). Inelastic deformation of Mg-PSZ and its significance for strength-toughness relationship of zirconia-toughened ceramics. *Acta Metall.*, **33**, 2083–91.

Swain, M. V. (1986). R-curve behaviour in ceramic materials. In *Advanced Ceramics* II, ed. S. Sōmiya. London: Elsevier Applied Science, pp. 45–67.

Swain, M. V. and Hannink, R. H. J. (1984). R-curve behaviour in zirconia ceramics. In *Science and Technology of Zirconia II, Advances in Ceramics*, Vol. **12**, eds. N. Claussen, M. Rühle and A. H. Heuer. Columbus, OH: The American Ceramic Society, pp. 225–39.

Teufer, G. (1962). Crystal structure of tetragonal ZrO_2. *Acta Crystallogr.*, **15**, 1187.

Tsukuma, K. (1986). Mechanical properties and thermal stability of CeO_2-containing tetragonal zirconia polycrystals. *Am. Ceram. Soc. Bull.*, **65**, 1386–9.

Tsukuma, K. and Shimada, M. (1985). Strength, fracture toughness and Vickers hardness of CeO_2-stabilised tetragonal ZrO_2 polycrystals (Ce-TZP). *J. Mater. Sci.*, **20**, 1178–84.

Tsukuma, K., Ueda, K. and Shimada, M. (1985). Strength and fracture-toughness of isostatically hot-pressed composites of Al_2O_3 and Y_2O_3-partially-stabilized ZrO_2. *J. Am. Ceram. Soc.*, **68**, C4–5.

Tsukuma, K., Ueda, K., Matsushita, K. and Shimada, M. (1985). High-temperature strength and fracture-toughness of Y_2O_3-Al_2O_3 partially-stabilized ZrO_2-Al_2O_3 composites. *J. Am. Ceram. Soc.*, **68**, C56–8.

Tsukuma, T. and Takahata, T. (1987). Mechanical property and microstructure of TZP and TZP/Al_2O_3 composites. In *Advanced Structural Ceramics*, Vol. **78**, eds. P. Becher, M. V. Swain and S. Sōmiya. Pittsburgh, PA: Materials Research Society, pp. 123–35.

Virkar, A. V. (1998). Rôle of ferroelasticity in toughening of zirconia ceramics. In *Key Engineering Materials*, Vols. **153–154**, *Zirconia Engineering Ceramics: Old Challenges – New Ideas*, ed. E. H. Kisi. Switzerland: TransTech, pp. 183–210.

Virkar, A. V. and Matsumoto, R. L. (1988). Toughening mechanism in tetragonal zirconia polycrystal ceramics. In *Advances in Ceramics*, Vol. **24**, *Science and Technology of Zirconia III*, eds. S. Sōmiya, N. Yomamoto and H. Hanagida. Westerville, OH: The American Ceramic Society, pp. 653–62.

Wagner, C. (1943). Mechanism of electric conduction in Nernst glower. *Naturwissenschaften*, **31**, 265–8.

Wang, J., Li, H. P. and Stevens, R. (1992). Hafnia and hafnia-toughened ceramics. *J. Mater. Sci.*, **27**, 5397–430.

Watanabe, M., Iio, S. and Fukuura, I. (1984). Aging behavior of Y-TZP. In *Science and Technology of Zirconia II, Advances in Ceramics*, Vol. **12**, eds. N. Claussen, M. Rühle and A. H. Heuer. Columbus, OH: The American Ceramic Society, pp. 391–8.

Wayman, C. M. (1981). Martensitic transformations. In *Advances in Ceramics*, Vol. **3**, *Science and Technology of Zirconia*, eds. A. H. Heuer and L. W. Hobbs. Columbus, OH: The American Ceramic Society, pp. 64–81.

Wolten, G. M. (1963). Diffusional phase transformation in zirconia and hafnia. *J. Am. Ceram. Soc.*, **46**, 418–22.

Wolten, G. M. (1964). Direct high-temperature single-crystal observation of orientation in zirconia phase transformation. *Acta Crystallogr.*, **17**, 763–5.

Wood, D. L. and Nassau, K. (1982). Refractive-index of cubic zirconia stabilized with yttria. *Appl. Opti.*, **21**, 2978–81.

Wood, D. L., Nassau, K. and Kometani, Y. (1990). Refractive index of Y_2O_3 stabilised cubic zirconia – variation with composition and wavelength. *Appl. Opt.*, **29**, 2485–8.

Xu, H. H. K., Jahanmir, S. and Ives, L. K. (1997). Effect of grinding on strength of tetragonal zirconia and zirconia-toughened alumina. *Machining Sci. Tech.*, **1**, 49–66.

Yokokawa, H., Sakai, N., Horita, T., Yamaji, K. and Brito, M. E. (2005). Solid oxide electrolytes for high temperature fuel cells. *Electrochem.*, **73**, 20–30.

Youngman, P. (1931). Zirconium. *Bull. Bur. Mines*, Part 2.

Yu, C. S., Shetty, D. K., Shaw, M. C. and Marshall, D. B. (1992). Transformation zone shape effects on crack shielding in ceria-partially-stabilized (Ce-TZP)-alumina composites. *J. Am. Ceram. Soc.*, **75**, 2991–4.

7

Conclusions

7.1 The materials

These five materials were selected as case studies because in one way or another they are amongst the most important and well understood of the whole class of structural ceramics. The choice was made partly on the basis of the wide range and values of their properties, partly on the extent of their practical use (assessed either as weight produced or as sales value). Many very successful industries depend on these materials for their existence. They therefore provide a very good guide to the structural ceramics as a whole, and to the questions to be asked when a new material is met for the first time. The five chosen are all now mature, well-established materials: many of the earlier points of contention have been resolved. So much is known about them, in remarkably fine detail in many cases, that it is possible to talk with reasonable confidence about their properties, and the scientific bases for their behaviour. This does not mean that all uncertainties have been resolved – far from it; there is still much which is not fully understood, and the full potential of many materials has probably not yet been realised.

We started with the familiar and simple picture of "ceramics", as rather hard and brittle materials, which easily break when mishandled. However, it should now be apparent that the materials classed as structural ceramics are not all exactly the same. They have many features in common (they are brittle, but not weak), and there are also marked differences (some are electrical insulators, some electrical conductors, depending in some cases on composition and temperature). Indeed, even within one group, different members can cover a wide range of property values, one of the reasons for which is that they have slightly different compositions and microstructures. There is really no such material as "silicon carbide ceramic", any more than there is "aluminium alloy" or "steel"; there are alloys, and steels, and there are silicon carbides. And there are good materials and not so good, depending on the exact formulation, the care taken by the manufacturer, and the price the customer is willing to pay. It is true that the structural ceramics

384

are, like many other things, unforgiving of any kind of mishandling, including mishandling during the processing stages. If they are properly treated, however, this need not be any more of a problem than it would be with a mobile phone, or a high-powered car.

7.2 History

Porcelain developed, slowly, over many centuries. Progress with the other materials forming these studies has been very much faster, but it has still taken many decades to develop and establish markets for the alumina, silicon carbide, silicon nitride, and zirconia structural ceramics. This development has been in response to the needs for improved or new materials, as other areas of technological and social importance have appeared. It has taken a very large amount of research effort, and funding, to learn how to produce and exploit the properties of these materials, and then consistently to manufacture components on a large scale with the same properties as those of test samples made in the laboratory. The human side also must not be overlooked: it has required very considerable effort on the part of a relatively small number of dedicated industrialists and scientists to convince the engineering community (who by nature tend, sensibly, to be rather conservative) that the new brittle ceramics did have a place in the engineering world as structural materials.

It should now be apparent that even the other four main groups (alumina, silicon carbide, silicon nitride and zirconia), which might be considered to be the more modern, advanced, or technical types, are not really new materials. They all have their roots deep in the earlier half of the last century, going back in the case of silicon carbide and zirconia to the very early discovery of sources of silicon carbide and zirconium dioxide themselves at the end of the nineteenth century. But it is true that the properties of the fine-grain materials marketed today have been improved and refined only relatively recently (that is, in the last 30–40 years or so). This has taken place as a result of the development of a much better understanding both of their basic science, and of the processing of very fine ceramic powders. By way of illustration, Fig. 7.1 shows the improvements in strength over three decades of high-purity medical-grade alumina, partly by the attainment of finer grain sizes through advances in processing techniques (Rahaman *et al.*, 2007), and zirconia ceramics have become three times stronger on the same time-scale. There is validity in the point of view that these structural ceramics are simply refined refractories and are not really new materials at all. Even silicon nitride (which if any ceramic can be said to have been intensively developed to fill a specific structural need, then it has), was well-known as a component of furnace refractories very much earlier than the arrival of large

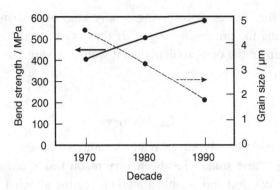

Figure 7.1 Improvements in the bend strength of high-purity medical-grade alumina over three decades, with developments in powder processing to give finer mean grain sizes. (After Rahaman *et al.*, 2007.)

Government-backed high-temperature gas-turbine programmes. However, it is also true that the mechanical properties of these very fine-grain and (usually) high-density structural ceramics are vast improvements on the properties of the refractories from which they grew, so much so that they may be hard to recognise as closely related materials.

7.3 Current application areas

Most of the present structural applications for all these materials are at temperatures not much above room temperature, and if one property is to be singled out as being the most useful, it is probably that of resistance to wear. The broad term "wear" is used to cover those situations involving exposure to wear of all kinds, sliding, erosion, abrasion, under dry and wet, or lubricated, conditions. Examples are the bioceramics (Rahaman *et al.*, 2007; Daculsi and Layrolle, 2008), which include prosthetic components for hips and knees, and teeth – and Europe has an aging population! The demand for cutting tools, ball and roller bearings, and pump seals for use in abrasive and corrosive environments, is also increasing. This class of application might also include impact-resistant components, such as protective armour, which provides one large market for silicon carbide (and boron carbide) in the USA (Franks *et al.*, 2008), but there is no reason why an ability to absorb the energy of a severe impact should not find uses elsewhere. These applications are almost all at or near room temperature, and the important mechanical or physical property requirements are hardness and toughness, coupled with a fine grain size. The materials have to be matched to the particular application: an ability to withstand sliding and abrasive wear is important for pump applications; and complete chemical inertness in the body over many decades is essential for

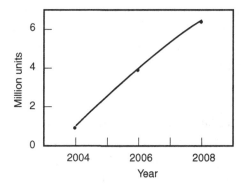

Figure 7.2 Sales of particulate diesel filters in Europe over the period 2004–2008. (Briggs, 2008.)

implants and prostheses. Alumina, silicon carbide, silicon nitride and zirconia have all been widely used, with a trend to prefer the harder solid state or liquid phase sintered silicon carbide to alumina for seals. Even porcelain has important applications in dentistry where wear resistance is needed, though the materials used are the highly glassy feldspathic (*dental*) porcelains.

A second group of applications includes materials used as filters and catalyst supports. These must have porous, or honeycomb structures, and are therefore of high surface area. In the case of liquid and gas filtration these must allow the passage of fluids while retaining small particles of specified sizes (such as smoke particles and bacteria). Intrinsic strength, because of the often low bulk density of the structure, as well as resistance to moderately high temperatures may be required. One application which might not have been foreseen a few years ago is the need to filter particles from diesel-engine exhaust gases (because of concern for air quality). This is particularly important (and of growing importance) in Europe because of the large number of diesel-engine road vehicles, and the number of particulate filters (DPFs) fitted each year, most of which are at present based on silicon carbide, is rising rapidly. Figure 7.2 shows estimates for the numbers of DPFs produced in Europe over the period 2004–2008. Silicon carbide works well in this environment because it is able to withstand the thermal shocks resulting from the high-temperature carbon burn-off cycle. Diesel-engine cars are not so widely used at present in the USA, and the demand there for silicon carbide filters is smaller. However, there is a market for porous supports for catalysts used to remove nitrogen oxides (NOx) from industrial (non-automotive) exhaust gases, and this also is expected to increase. This type of application, interestingly, shows how difficult it can be to predict future needs for materials. The requirement for automotive exhaust gas filtration five years ago was almost unknown, but it is now (2009) becoming a standard requirement for all diesel-engine vehicles and potentially a large market for silicon carbide, and other ceramics.

Applications involving actual use at continuous or near-continuous high temperatures are hot metal filtration, kiln furniture, and special refractory components for metal casting. Silicon carbide and alumina ceramics are used here because of their strength at high temperature, relative inertness, and in the case of silicon carbide, ability to withstand thermal shock. The mixed oxides, cordierite and mullite, are also very useful in these applications, either because of their high-temperature strength (mullite), or their resistance to thermal shock (cordierite), and they are sold in quantities matching that of alumina. However, of the five structural ceramics examined, most applications involve low temperatures, and it is low-temperature aqueous corrosion which is the more serious issue. This may be surprising, given the emphasis generally placed on the high melting-points and high-temperature properties of the structural ceramics. Considerable efforts have been made to develop high-temperature internal combustion engine components in silicon carbide and silicon nitride, but so far without real success. While small silicon nitride components are used in engines (and alumina is the spark plug insulator), these are generally not exposed to really high (>1000 °C) temperatures. These potential applications wait for changes in economic factors, to allow the use of intrinsically much more expensive units than the metallic ones used at present.

This section has examined the materials and their application areas in broad terms. All four of the stronger and harder materials have considerable usefulness as structural materials. It is difficult to allocate one material to one particular type of application; the choice can vary according to the precise details of the operating conditions, the size and complexity of the component and the numbers required. In many cases the materials compete amongst themselves and decisions about which to use are finely balanced.

7.4 The market-place

One reason for choosing these materials for examination is their market size – weight produced or sold, and the return to the producer. But this is not as straightforward a basis as it might seem, because the ceramic component is likely to be hidden in a more complex unit, the price of which is being recorded. The alumina spark plug insulator is only one of four or more separate items making up the spark plug. And a price is not a fixed quantity; it can vary with the size of the batch, the costs of tooling, and the day-to-day price of raw materials and energy. Figure 7.3 (by weight) and Fig. 7.4 (by value) provide indications of the current distributions of sales amongst four members of the group, based on estimates of combined data (2006) for engineering ceramics in Europe and the USA (Briggs,

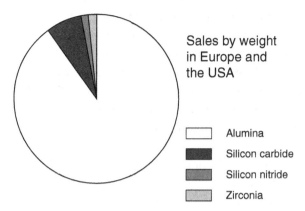

Figure 7.3 Combined figures for the sales in Europe and the USA of alumina, silicon carbide, silicon nitride and zirconia, by weight. (Briggs, 2008.)

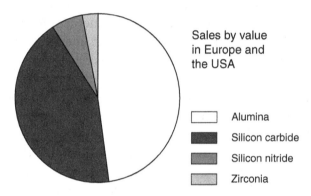

Figure 7.4 Combined figures for the sales in Europe and the USA of alumina, silicon carbide, silicon nitride and zirconia, by market value. (Briggs, 2008.)

2008). "Engineering ceramics" is defined as mechanical and wear parts, filters, catalyst supports, kiln furniture, and parts for the continuous casting of steel. The other three important members of the group of so-defined engineering materials which are not reviewed here are the mixed (that is, binary and ternary) oxides mullite and cordierite, and the non-oxide boron carbide. Data for porcelains are more difficult to obtain and evaluate because of their large domestic markets. Leaving aside porcelain, it is clear that alumina is by far the most widely used of the four binary structural ceramics reviewed here, though the data for sales value are skewed in favour of the intrinsically more expensive silicon carbide, silicon nitride, and zirconia. The second most important of the group seems to be silicon carbide. Silicon nitride and zirconia are a long way behind, though silicon nitride seems to have the slight edge over zirconia (remember that zirconia is almost twice as dense as silicon nitride). It is interesting that the number of research publications over the last 30 years on alumina ceramics is far easily outnumbered by the number of research publications dealing with each of the other three materials, and few international conferences have been devoted recently to alumina. One interpretation of this is that it indicates the difficulties of developing new materials with improved properties for very demanding applications, necessitating detailed examinations of microstructure–property relationships, and the means for developing the required microstructures. Another interpretation might be that alumina, though not perfect, is generally adequate for the current needs for a material with its kind of properties (including reasonably low price), and that the low price does not justify the research expenditure which might be required to improve it. There is certainly considerably higher hardness, strength, and toughness already available in the silicon carbide, silicon nitride, and zirconia groups. It is therefore natural if an alumina does not meet the requirements to consider using one of these materials, in the same way that there was a move from the electrical porcelains to alumina when better electrical insulators were needed. The economic factor has always to be kept in mind; as with everything else, better property values, though always desirable, may not always be affordable because they involve more expensive raw materials or higher production costs. It must always be remembered that a large proportion of the cost of the ceramic component can be in the costs of finishing (grinding and polishing) operations; these are materials which cannot be sintered directly to a precise dimension. To some extent these costs are independent of the costs of the powder processing stages, except that harder and tougher materials which might be more expensive to sinter are also much more difficult to cut or polish. These factors are just as important as mechanical or physical properties. In many cases the choice of material still depends on the costs and performances of other, competing, materials, rather than simply on the merits of the ceramic material itself considered in isolation, but an initially more expensive material will be preferable if it lasts longer, and the overall replacement cost (including that of down-time and the loss of production while carrying out a replacement) is lower.

7.5 Ceramic composites

These studies of the five structural ceramics have referred only briefly (and almost accidentally) to the class of structural ceramics known as ceramic matrix composites (CMCs). In Chapter 1 reasons were given for omitting other classes of ceramic material with structural applications, and the exclusion of composites should be justified too. Composite materials are a large and important subject in their own right and cannot be completely ignored, but because the properties of a composite are developed from those of the matrix phase, modified by the second phase, it is necessary first to understand what these basic phases actually are, which has in part been the function of these five studies. A composite material makes use of the properties of the matrix, which are then modified or enhanced by a second phase, in practice often high aspect ratio particles or fibres. The object is to try to overcome the inherent brittleness and low fracture toughness of the main phase, and ultimate failure made less catastrophic (Evans and Marshall, 1989). One way of achieving this is to incorporate an element of pseudo-plasticity; after the first onset of cracking, a short plateau increases the strain to failure. Alumina–zirconia composites (the ZTA materials) were mentioned in Chapter 6, because the particulate zirconium dioxide, by its transformation, is able to strengthen and improve the fracture toughness of the alumina. Other examples of CMC class currently marketed include silicon carbide fibre or whisker toughened alumina, and silicon carbide strengthened glasses (Lee and Rainforth, 1994). One major market for these materials is for the zirconia toughened aluminas; another is for titanium carbide toughened aluminas, used as cutting tool inserts. As might be expected, there are considerable technical difficulties in forming components from complex mixtures of particles differing markedly in geometry or size, and possibly of widely differing chemical compositions, and therefore sinterabilities and reactivities.

7.6 Summing up

These five structural ceramics have many similarities, and they have many important differences. Each has its own individual features, derived from its compositions, crystal chemistries and microstructures, and each has a range of applications, some shared with the others, some highly individualistic. They have a spread of properties covering a very wide range, from humble porcelain to the thermal shock-resistant silicon nitride, and to the very high toughness, transformation-toughened, zirconias. For this reason they make a very useful set of studies because their features are present in greater or lesser measure in all the members of the class of structural ceramic. The properties of any other structural ceramic being met for the first time can reasonably be assumed to depend on the same sets of principles illustrated by these five. The broad pattern of behaviour is set by the properties of the main crystalline phase, modified by the nature, amounts and locations of individual intergranular and secondary phases. In fact the grain boundary phase in many cases is responsible to a large extent for the mechanical properties of the material. The response to temperature can be influenced strongly by the grain boundary phase, as well as by phase changes and thermal expansion behaviour.

Underpinning these aspects (and all high-quality ceramic materials), is the basic powder production process itself. With all these materials, the reliability and consistency of components (often with large-scale production) is vitally important. These are essentially powder processing issues: how to maintain homogeneity of microstructure in a very fine powder; how to achieve consistency of sintering or other high-temperature treatments; how to control phase changes; how to minimise the size and quantity of strength-controlling flaws. All these aspects form part of efforts to understand and control better the behaviour of very fine, nanodimension, powders, and their processing into ceramic solids. Many of the improvements in material property, and particularly strength, have been the result of improved powder processing, resulting in the elimination, or control, of microstructural defects, and the attainment of the necessary microstructures.

It should be remembered that while there will be theoretical limits set to materials' property values by the fundamental nature of the phases themselves, the practical property values attained so far, and particularly those for strength and wear resistance, may not be fixed at their current levels. A good illustration is provided here by the continuing development of the liquid phase sintered silicon carbides. One reason why the markets for the more expensive structural ceramics might change or expand is the need to reduce energy consumption, and in particular the reduction of greenhouse gas production. Another is the inevitable steep increase in the price of energy, as oil and gas (and uranium) supplies slowly dry up. Both situations will require the more efficient production and conversion of energy, which will require improvements in existing technologies, and the development of completely new ones. Concern for the environment more generally is also rising, with associated implications for the use of materials, which will in turn place more demands on existing materials, and require new, improved, versions. It will also require the ceramics themselves to be produced (and recycled) more efficiently, although the volumes produced are still relatively small on the scale of structural materials as a whole. There is always a need to improve production efficiencies and to reduce costs with existing applications, and new applications emerge with their own sets of requirements: the position is not static. But it also has to be appreciated that there are limits to what can reasonably be expected of any particular material, and that in the past too many expectations have been placed on the "newer" structural ceramics. There is no doubt that in the past they were marketed too enthusiastically, perhaps prematurely, and without sufficient regard to the problems of the consistency of powder processing and mass-production. These mistakes are not likely to be repeated.

The early vision of the 1970s, that structural ceramics would, somehow, become the new high-temperature ("white-hot") materials of the future with widespread engineering applications, has not yet been realised. That it has not been is not because of some basic deficiency in the mechanical or physical properties of the materials, but the challenges of mass-producing components in a brittle material with sufficient consistency and reliability, at competitive prices. This is therefore partly a matter of the quality of powder processing, coupled with the expense of finishing operations on very hard and wear-resistant materials. The structural ceramics already have a very wide range of large-scale applications, as well as many successful smaller niche markets. Such is their current importance in very many ways, it is easy to believe that present-day society in the form that we know it simply could not function without them. Their potential to meet the needs for new materials in a rapidly changing technological world is considerable.

References

Briggs, J. (2007). *Engineering Ceramics in Europe and the USA*. Menith Wood, UK: Enceram.

Briggs, J. (2008). Private communication.

Daculsi, G. and Layrolle, P., Eds. (2008). *Ceramics in Medicine*, Key Engineering Materials, Vols. 361–363. Staf-Zürich: TransTech Publications.

Evans, A. G. and Marshall, D. B. (1989). The mechanical behavior of ceramic matrix composites. *Acta Met.*, **37**, 2567–83.

Franks, L. P., Salem, J. and Zhu, D., Eds. (2008). *Advances in Ceramic Armor III. Ceram. Sci. Eng. Proc.*, Vol. 28. Hoboken, NJ: John Wiley.

Lee, W. E. and Rainforth, W. M. (1994). *Ceramic Microstructures: Property Control by Processing*. London: Chapman and Hall, pp. 533–70.

Rahaman, M. N., Aihua, Y., Bal, B. S., Garino, J. P. and Ries, M. D. (2007). Ceramics for prosthetic hip and knee joint replacement. *J. Am. Ceram. Soc.*, **90**, 1965–88.

Woetting, G., Lindner, H. A. and Gugel, E. (1996). Silicon nitride valves for automotive engines. In *Applications of Advanced Materials in a High-tech Society*, ed. H. Mostaghaci, *Key Engineering Materials*, Vols. **122–124**. Zürich-Ütikon: TransTech Publications, pp. 283–92.

Index

Notes:

1. A distinction is made between pure chemical compounds (in principle the single-crystal form of a material), and ceramics based on that compound. Intrinsic properties are listed under the chemical name of the pure material: entries relating to individual ceramic forms are listed under their ceramic names.
2. Specific physical, mechanical and chemical properties, and applications, are generally listed under the ceramics headings.

Printed in the United States
By Bookmasters